T0227640

DISASTER **SURVIVAL** GUIDE for Business Communications **NETWORKS**

by Richard Grigonis

Strategies for Planning, Response and Recovery in Data and Telecom Systems

CRC Press
Taylor & Francis Group
Boca Raton London New York

CRC Press is an imprint of the
Taylor & Francis Group, an **informa** business

CRC Press
Taylor & Francis Group
6000 Broken Sound Parkway NW, Suite 300
Boca Raton, FL 33487-2742

First issued in hardback 2017

ISBN 13: 978-1-138-41227-9 (hbk)
ISBN 13: 978-1-57820-117-4 (pbk)

Visit the Taylor & Francis Web site at
http://www.taylorandfrancis.com

and the CRC Press Web site at
http://www.crcpress.com

Contents

Acknowledgements

The author is of course grateful to his many colleagues at CMP who contributed information incorporated into this opus, as well as the many engineers, CEOs, CTOs and other industry veterans interviewed whose ideas have had great impact on what the state of the art is regarding fault tolerant telecom and datacom. I especially want to thank my editor, Janice Reynolds, a distinguished author in her own right, who provided tremendous round-the-clock editorial attention to this project (not to mention her adroit prodding of me with phone calls in the wee hours of the morning), and the book's art director, Robbie Alterio, who as always can take a text file and some software and design the resplendent work of art you now hold in your hands.

Preface

FOR MANY YEARS, THE SUBJECTS OF DISASTER recovery, business continuity and "uptime" technologies were something that every senior business manager acknowledged as of great importance – but did little about. Despite giving lip service to such endeavors, actual expenditures to bolster the security and operating resilience of most businesses were minimal or worse.

Since the almost unbelievable events of September 11, 2001, however, the art and science of preparing and recovering from disaster (or averting it entirely) has gained new-found respect. Still, disasters come in many forms, and for every skyscraper that comes crashing to the ground, a thousand new computer viruses, lightning strikes, floods and earthquakes wreak havoc on organizations of all kinds. From the greatest petrochemical multinational to the most diminutive SOHO, the ability to analyze a business' vulnerabilities, set up defensive plans to counteract them, and finally activate them is what separates businesses destined to be in the game for the long haul from the short-term reapers.

When approached by Matt Kelsey, publisher of the Book Division at CMP Media, to do a "telecom / datacom disaster survival guide" I realized there were already a number of security and business continuity books that fell into a certain category, most on how to do tape backups, run anti-viral software, and admonishments not to use wallpaper in your office that gives off toxic fumes in a fire.

That's why this book takes a unique communications and proactive "uptime technology" approach to the subjects at hand. Then there's the "people factor." Establishing the right kind of corporate culture via technology is "human preventive maintenance" and it is just as important has how many disk backups and fire extinguishers will be available in the event of disaster. That's why the reader, to their great surprise, will also discover a massive first chapter on conferencing equipment and know-how. Conferencing is an adjunct to teleworking, and both of these are powered by secure virtual private networks (VPNs) which can keep all of a company's employees from traveling under dangerous circumstances or becoming concentrated in one big death trap of a building. Thus, while I've tried to make this book accessible to a reasonably diverse audience (by for example, including an extensive glossary), IT personnel and telecom managers in particular will feel at home reading it. It is they, after all, who will be on the "front lines" as it were, questioning, analyzing, debating, pondering and finally implementing solutions to protect their fellow workers from a plethora of possible catastrophes.

With any luck at all, and with books such as this and a great deal of resolve, the events of September 11th will prove not to have been a foretaste of the future.

Richard Grigonis
New York City
February 2002

Introduction

SEPTEMBER 11, 2001 – THE WORLD CHANGES

Atop the World Trade Center
Windows on the World
served perfect views
with cocktail dreams
On a clear day it seemed
you could actually see London, Paris
even Rome through those windows on the world
– from the poem, ***Windows on the World***,
by Keenan Pendergrass

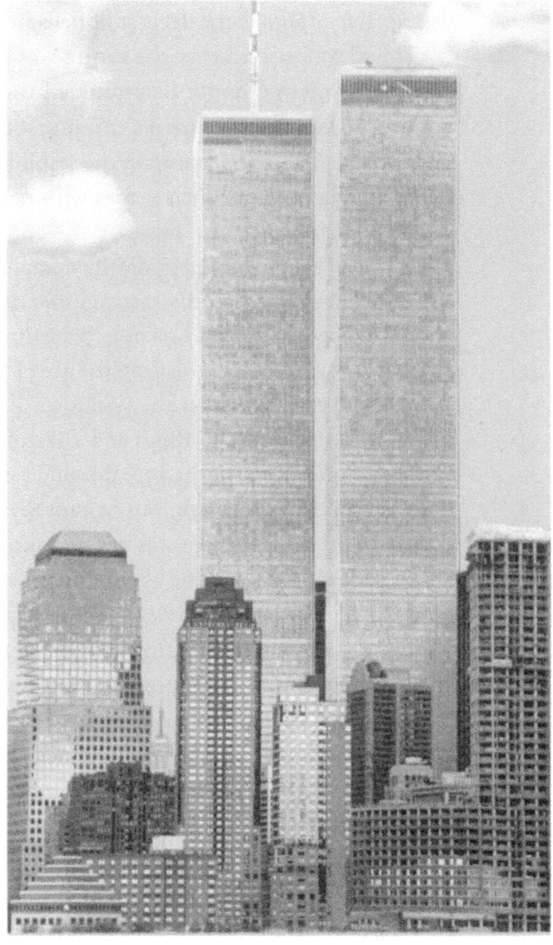

REMEMBERING THINGS PAST

"Welcome to Windows on the World, the highest restaurant on Earth," said the uniformed attendant as the elevator doors opened on the 107th floor of One World Trade Center in New York City.

The 110-story, 1362 foot (417 meter) structure was also known as "Building One", "1 WTC" or simply "the North Tower." Together with its slightly taller (1,368 foot) twin tower a few hundred feet to the south, the World Trade Center buildings became the world's tallest upon the opening of One World Trade center in December 1970 – ironic, when one considers the fact that their architect, Minoru Yamasaki, is afraid of heights. The second tower was finished in April of 1973, but just a month later construction on the slightly taller (1454-foot) Sears Tower in Chicago reached the point where it now could wear the skyscraper celebrity crown. Still, the twin towers had become New York's most recognizable icon. While not an architectural masterpiece, the Twin Towers were the first supertall buildings to forego the use of masonry, and their innovations included elevator "sky lobbies" that freed up additional space for offices. Certainly the towers captured the public's imagination. In the 1976 remake of *King Kong*, the giant ape clambers up one of the Twin Towers. The towers also attracted all sorts of interesting characters. On April 7, 1974, for example, tightrope walker Philippe Petit successfully traversed a rope from one tower to the other seven times in succession. In May 1977 toy maker George Willig spent three arduous hours climbing up the side of one of the towers, only to be arrested by police when he reached the top.

My trip up the One World Trade Center tower that day was less strenuous than Mr. Willig's, involving as it did the mere press of an elevator's button (with a strong hand, mind you). The ride up in the huge express elevator, big enough to hold an automobile, was uneventful, save for a gentle rocking of the building, one consequence of erecting the tallest buildings in New York at the southernmost, windiest spot in the city. As I got out of the elevator, I could feel the usual chilly breeze wafting up and out of the elevator shaft, transforming it into a strange chimney of cold air. For some reason it always brought back memories of my cultural anthropology class in college, and the ancient, unsettling description of *Mitnal*, the deepest of hells in Mayan mythology – eternal cold and darkness.

The Maya believed that the future was predetermined and that Man could not escape Fate, an idea that would no doubt have rankled the pugnacious New Yorkers now swarming past me and heading toward the restaurant's entrance. Still, the Maya believed that they at least had the ability to ward off calamities and maintain the stability of the cosmos, though to achieve these noble ends they were required to honor their deities with ritual human sacrifice.

On this particular day, however, the only thing I was intent on honoring was my growling stomach. Moloch that it is, it periodically demands a plate of delicious offerings, preferably served up by a high priest initiated in the gastronomic rites of the Culinary Institute of America.

It was the year 2001. I had been providing technical advice to a "dot com" company a few blocks away, and whenever I was visiting the area I would stop by for lunch or dinner. But it was always difficult to get reservations at the Windows on the World restaurant. Restaurateur Joe Baum founded the establishment on the 106th and 107th floors of WTC 1, in 1976. Perhaps, even he, was amazed at how it became one of the top-grossing restaurants in the United States – trying to make reservations for dinner at this fabulous restaurant during its heyday in the late 1970s and early 1980s was simply impossible – there was a six month wait for a table! This wasn't too surprising, since it seemed appropriate that the world's highest restaurant would have the highest standards, whether it involved the breathtaking views of New Jersey and New York, the brilliant interior designed by Warren Platner, the purchasing of the finest ingredients, the superlative skills of the chefs, or the attentive service and artful presentation of food on the plate.

One of my best-remembered encounters with Windows on the World was at a private function for telecom experts and magazine editors held one evening in September 1998 by Howard Bubb, the former CEO and president of Dialogic Corporation and now a vice president of Intel. In the days when telecom and the Internet were flying high in the economy, the Dialogic Corporation flew higher than them all. Not surprisingly, Dialogic always held the best parties, called Connection parties, at places such as Gotham Hall in Santa Monica, the Regency Club in Los Angeles and the Rainbow Room in New York. Dialogic would even provide bus transportation from local hotels to the parties, with champagne and *hors d'oeuvres* served by waiters on the bus. When it was Windows on the World's turn to host a Connection party, the food and drink were, as always, fantastic.

Back to my story – That day I was in a hurry (I'm always in a hurry), I don't generally wear a jacket and tie and tend not to make reservations a zillion days in advance, so I did what I usually do under such circumstances. I simply strolled next door into what was another of my favorite restaurants, the lesser-known and more casual sister to Windows on the World, called Wild Blue. Michael Lomonaco was the chef at both Windows on the World and Wild Blue, and both restaurants shared the same 60,000 bottle wine cellar (which didn't mean much to a Pepsi addict like me). A common hallway done in a black and gray style from the 1980s connected the two restaurants as well as a bar/lounge called The Greatest Bar on Earth (this name was sheer nonsense, as New Yorkers all know that the "real" greatest bar on Earth used to be the one at the old Rainbow Room atop the General Electric Building at Rockefeller Center).

Another loss in the World Trade Center disaster was a lesser-known though wonderful restaurant, one of the author's favorites, called Wild Blue.

To enter Wild Blue, one sauntered up a short ramp lit by the icy incandescence of a glowing, deep blue glass wall studded with tiny stars. I soon found myself in the cozy interior, a room paneled in warm burnished wood nestled beneath a gently curved and beamed ceiling. Of the 20 or so generously-spaced tables, I was immediately shown to my favorite at the right rear of the room, next to one of the many floor-to-ceiling windows. Although Wild Blue had windows only on one side of the room, the view, aside from the other tower, was nevertheless pretty good – you could see the Verrazano-Narrows Bridge, the Statute of Liberty, the endless queue of airliners waiting to land at Newark, LaGuardia and JFK airports, and an encyclopedic assortment of ships slowly making their way in and out of the entrance to New York Harbor.

One of my more frivolous purchases accompanied me that day. It was a metal business card case I had picked up at Hammacher Schlemmer, procured solely to impress a visiting Japanese economist and his TV crew who were going to interview me about "why computer telephony and the Internet is good for Japan." (The interview, funded by NASDAQ, would soon change into a "why do you think the NASDAQ stock market model is good for Japan.")

For those readers unfamiliar with Asian formalities regarding business cards or "meishi" ("name cards"), take heed: Whereas in the US. we take business cards for granted, a perfunctory, quickly forgotten activity much like a handshake, in China and Japan the business card is considered a literal extension of the individual's identity – indeed, Japanese government officials will often sign their name and the date on the back of their card to guard against fraud, though it's considered bad etiquette to sign a card in front of another person. As you might expect, the act of exchanging business cards is a solemn ritual of the utmost seriousness. It consists of a verbal introduction, a bow, and the presentation of the card with the front side facing upward toward the recipient (offering the card with both hands demonstrates greater respect). Bilingual cards are offered with the Japanese side up as you simultaneously accept your Japanese counterpart's card. You then study the card with feigned interest (especially anything in Japanese), look at the person, and nod slightly in acknowl-

Happy Times – With no Hint of what was to Come

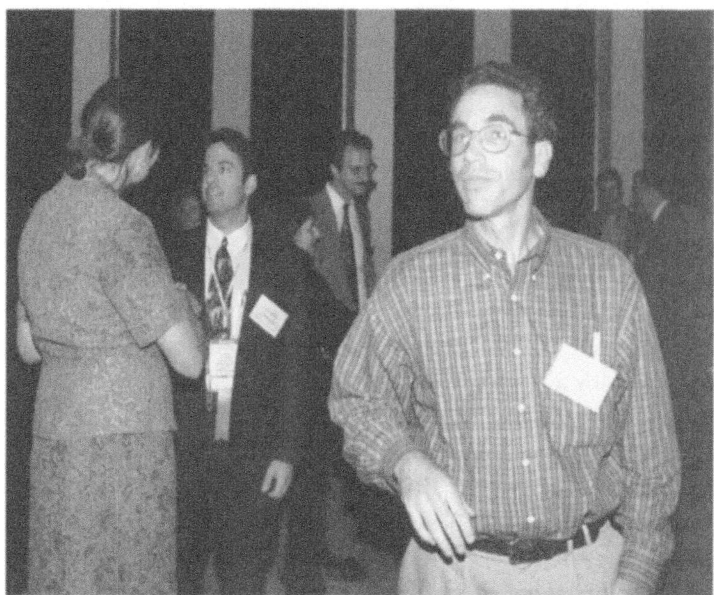

A Dialogic Connection Party held at the Windows on the World Restaurant atop the World Trade Center, September 1998. Note the floor-to-ceiling windows between the perimeter columns. The windows were set in the 22-inch spaces between 18.75-inch wide steel columns that were finished with a silver-colored aluminum alloy which made the towers appear from afar to have no windows. But windows the buildings did indeed have — when they broke on September 11, 2001, they fed large quantities of oxygen to a fire, the heat from which weakened the structural members of both towers and allowed them to collapse. Looking into the camera is John Landau, at the time vice president of strategic marketing for Dialogic (now part of Intel).

In the old days, Windows on the World was a magnet for everyone — movers and shakers, visiting dignitaries, corporate party-givers, tourists, young professionals trying to impress their dates, etc. This unlikely, random collision of somewhat inebriated individuals consists of (from left) Rick Luhmann (Editor-in-Chief of CMP's CommWeb.com, who accidentally wins the Carmen Miranda look-a-like contest by standing in front of some potted plants), the author, Gary Marks (at the time Dialogic's vice president of marketing) and Howard Bubb of Intel.

The party's over, the lights are turned down but the author still has some film in his camera, so he tries taking some time exposures of the premises. The author has used photo editing tools to transform these images into a single panorama. As you can see, unlike the art deco skyscrapers of the early 20th century, the World Trade Center had no setbacks — each floor from bottom to top was the same size, with about 40,000 square feet of space (about an acre, or 0.4 hectare) available per floor with no interior columns, thanks to a structural system pioneered on the earlier IBM Building in Seattle, Washington. The WTC's 208-foot-on-a-side facade was a prefabricated steel lattice, with tubular 18.75-inch wide columns on 39-inch centers acting as wind bracing, leaving each building's central core to handle most of the gravity loads. The prefabricated trussed steel floors that supported the concrete slab on each floor were only 33 inches in depth, yet spanned the 60 foot distance (18.288 meters) from the perimeter columns to the elevator core. The floors also acted as a diaphragm to stiffen the outside wall against lateral buckling forces generated by wind-load pressures.

edgment. When several Japanese are present, the cards are presented according to rank, with, interestingly, the highest-ranking individual presenting his card last. One then presents a gift (I give out copies of my books), which should always be wrapped.

The treasured business cards are of course carried in a distinctive business card case, not in the damp, gauche interior of one's pocket. But my gleaming new card case was stuck – I couldn't get it open. I didn't want to use the butter knife in front of me so I called over a waiter to assist.

Now, New York waiters are the most fascinating sort of people, especially to a seasoned writer and skilled interviewer who over the years has, among other things, made a living cajoling information from people. I usually interject the half-joking question "actor or writer?" when I talk to waiters, and the stories that result could fill a book. Tales of epic journeys, dreams of fame in the face of crushing rejection, privations endured in the name of one's art, tempests in teacups and fatal diseases, despised enemies and lost loves – a million anecdotes at once hilarious and tragic, memorable and forgettable; a fleeting, verbal tapestry of the vicissitudes of urban life, spun and boldly textured by the crazy optimism of youth.

And so it was with my waiter that day, whose face began to light up as he began revealing the real reasons for his day-to-day existence, that date with destiny just around the next corner. But he soon caught himself, and now looked around a bit nervously, doubtless feeling a bit awkward for having revealed his secret aspirations to a curious stranger in the milieu of an upscale restaurant. It was as if I, the harmless, generic customer, had suddenly materialized an intimidating confessional booth or psychiatrists' couch.

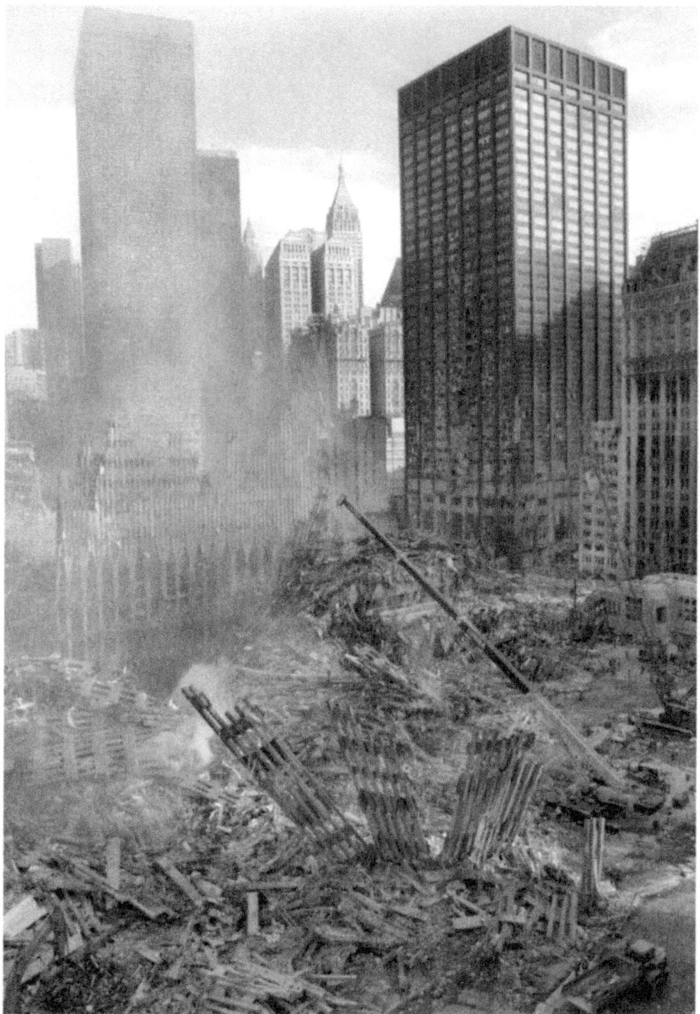

New York City, NY, September 17, 2001 — More than a million tons of rubble had to be dealt with at the World Trade Center. Photo by Andrea Booher / FEMA News Photo.

Before I knew it, my business card case was open and I thanked the fellow, tipping him generously (yes, I'm a big tipper and I love a good story). He smiled, thanked me and took my order.

Often I wonder what happened to that waiter.

As for the entire 85-member staff of Windows on the World (save the owner), their tragic fate would soon be all too well known to every human being on earth.

I sat there high atop 10,000,000 square feet of rentable office space, my feet firmly planted on the restaurant's warm slate floor as I lazily gazed out of the huge plate glass window, enjoying the 55-mile view and smelling with anticipation the always delicious lobster salad, strip steak and seasoned Yukon potatoes about to be served by the cordial, ever-efficient staff.

Little did I realize that this would be my last trip to the World Trade Center. Within a short time all of this architectural and Epicurean grandeur – along with thousands of men and women who worked there and hundreds of America's bravest police and firemen who tried to rescue them – would be gone forever as a result of one of the most dastardly terrorist acts in US. history. Following as it did the collapse of the stock market, it was the final nail in the coffin of America's latest Gilded Age.

Wild Blue was again on my mind on the morning of Tuesday, September 11, 2001. Having just finished my latest article and column for ***Communications Convergence*** magazine (where I am Chief Technical Editor), I had decided to take one of my rare vacations – a working vacation, of course. I was sitting in my home office in New Jersey, just across the river from New York City, hurriedly banging out a book for CMP, one devoted to Voice over DSL technology.

I hadn't eaten breakfast, and my stomach would be alerting me to its needs very soon. A TV dinner would probably take care of lunch, but I hadn't been to the Wild Blue restaurant in a while, so perhaps later on I would treat myself to the congenial atmosphere and delicious food that could always be found there.

Collateral Damage

September 18, 2001 — Ohio Task Force workers anchored this 600,000 pound beam from the World Trade Center lodged in a nearby building. Photo by Michael Rieger / FEMA News Photo.

September 20, 2001 — This building was damaged by the explosions caused by the terrorist attacks on the World Trade Center site. Photo by Mike Rieger / FEMA News Photo.

September 27, 2001 — The World Financial Center, designed by Cesar Pelli and located across West Street from the Twin Towers, opened in 1988. This waterfront building, the center's Winter Garden, was a marvelous, popular public space. Of all the nearby buildings, it sustained the most collateral damage during the collapse of the World Trade Center. Photo by Bri Rodriguez / FEMA News Photo.

Suddenly the phone rang. A friend told me to turn on the TV – pronto. The World Trade Center was gone, the victim of a huge, coordinated terrorist attack that also caused partial destruction of the Pentagon in Washington, D.C.

With my head for trivia, random facts about the World Trade Center began to race through my mind: Although known for its twin towers, the World Trade Center was in fact a seven building, 16-acre complex set around an expansive, five acre stone plaza vaguely modeled after St. Mark's Square in Venice, itself located above an extensive seven-level underground shopping area and subway station. More than 200,000 tons of Japanese steel were used in its construction (half the price of American steel at the time), along with enough concrete to build a five-foot wide sidewalk from New York City to Washington, D.C. The most disturbing fact was that upwards of 50,000 people worked there. Had the attack occurred a few hours later, the number of casualties would have been truly astronomical.

The World Trade Center had been the victim of a bombing previously. On Friday, February 26, 1993, a massive terrorist bomb detonated in the WTC's public parking garage, but the twin towers still stood defiantly (at least until two jumbo jets were flow into its midst). As a result of the explosion, six persons were killed, and more than 1,000 injured. The blast site became one of the largest crime scenes in New York Police Department history. Estimates showed property damage in excess of $500,000,000. At the time, many in law enforcement thought of this investigation as the "case of the century."

The attack of September 11, 2001, however, was far greater in its scope.

Every generation has its special moment in history, a moment evoking such emotional intensity that it leaves reminiscences of the "I was doing X activity at Y when Z happened" variety. Unfortunately, such indelible memories tend to be associated more with horrific disasters than glorious triumphs.

For my generation, most of us still recall where we were and what we were doing the morning of November 22, 1963 when we heard the news that President John F. Kennedy had been assassinated. And no doubt nearly every member of the following generation knows exactly what they were doing shortly after 11:38 a.m. Eastern Standard Time on January 28, 1986, when the Space Shuttle Challenger lifted off from the Kennedy Space Center's Pad 39B on what turned out to be a 73 second flight.

And now, the latest and perhaps most awful spectacle to become permanently lodged in America's psyche are the events of September 11, 2001:

At 8:45 a.m. New York local time, One World Trade Center, the north tower, was hit by a hijacked 767 commercial jet airplane, loaded with fuel for a trans-continental flight. Two World Trade Center, the south tower, was hit by a similar hijacked jet 18 minutes later at 9:03 a.m. In separate but related attacks, the Pentagon building near Washington D.C. was hit by a hijacked 757 at 9:43 a.m., and at 10:10 a.m., a fourth hijacked jetliner probably intended for the US Capitol Building or the White House crashed in Pennsylvania after passengers attempted to retake control of the plane. The south tower, WTC 2, which had been hit second, suffered a complete structural collapse at 10:05 a.m., 62 minutes after being hit itself, 80 minutes after the first impact. The north tower, WTC 1, later collapsed at 10:30 a.m., 105 minutes after first being hit. WTC 7, itself a considerable 47 story office building that had been built in 1987, was damaged, caught fire and also collapsed later in the afternoon.

As I sat glued to my television set, along with millions of other viewers, watching the endless visual montage of chaos and ruins on television, I dimly recalled some of the great words architect Minoru Yamasaki had once uttered about the WTC upon its completion:

". . . the World Trade Center buildings in New York . . . had a bigger purpose than just to provide

September 22, 2001 — The wreckage at the World Trade Center is slowly being cleared away. Photo by Michael Rieger / FEMA News Photo.

room for tenants. The World Trade Center is a living symbol of man's dedication to world peace ... beyond [that], the World Trade Center should, because of its importance, become a representation of man's belief in humanity, his need for individual dignity, his beliefs in the cooperation of men, and through cooperation, his ability to find greatness."

Such magnificent words rang hollow now, as fear, panic and ultimately anger spread throughout the city, America and the world as a whole. The site of the disaster was no longer the World Trade Center but rather a strange place called "Ground Zero," a moniker formerly reserved only for the site of awesome nuclear explosions. When video cameras finally penetrated the virtual moat of security, rubble and smoke to reach Ground Zero, night had fallen on the chaotic scene, revealing how life could imitate art – provided, of course, that the art happened to be Hieronymus Bosch's terrifying Last Judgment triptych from the 16th century. Indeed, Ground Zero's nightmarish scene of tortured rescuers scrambling among burning ruins reminds one of Bosch's phantasmagoric portrayals of tormenting demons, the destructive power of machines and an all-pervading evil. The stately architectural iconography of the New York City skyline had been transformed into the surreal cosmic iconography of Bosch. His grotesque, preposterous visions of hell and terrifying representations of the power of evil had not just given tangible form to the fears haunting the minds of the Middle Ages; they had also now proved to be a distant, though crystal-clear prophesy of new-found quandaries plaguing the modern world.

In the composing rooms of newspapers around the country, headlines were already screaming "Attack on America," but it was really an attack on the whole world, since the World Trade Center

September 28, 2001 — A view of the recovery operation underway from a roof adjacent to the World Trade Center. Photo by Andrea Booher / FEMA News Photo.

served as New York headquarters for companies stretching across six continents, and hundreds of victims were men and women from more than 50 countries.

America would survive, of course, but everything would have to change as a result of the September 11 cataclysm: Changes in air travel, changes in security, changes in the way we do business. That's why I dropped everything I was doing to write this book. Being an expert on fault tolerant communications systems and "uptime" in general, it was time to tell businesses how to plan for the worst, and what technology they should use to stay up and running in the face of disaster.

Just as those ancient Maya thought that proper rituals could change their deities' evil forces into benevolent ones, so you too, by performing today's technological rituals, can protect you and your business from both natural and man-made calamities.

The events of September 11 was an immense wake up call to organizations everywhere, and in this book are detailed steps that can be taken for the protection of communications links, data, electronic equipment and more so that the United State's manna – commerce – can continue.

And if you ever become lax in your duties to your business, just turn on your television. Some political pundit will always be warning of a new threat. New and ominous forces will always be lurking just over the horizon. And above all, the emotion-charged images of September 11 will pop up now and then to remind you, much like the thick, barnacle-encrusted silica tiles from the Space Shuttle Challenger that still occasionally wash up on the sandy shores of Florida's Cocoa Beach, bringing with them a flood of haunting memories.

CHAPTER 1

Disasters Large and Small

ON THE MORNING OF SEPTEMBER 11, Walter Danielsson, senior vice president of technology at Euro Brokers Inc., was driving to work. A breakfast meeting was planned at Two World Trade Center, the south tower where Euro Brokers had its world headquarters on the 84th floor. Although it was "casual day," Danielsson suddenly decided he looked too casual for the meeting. He went back home, changed clothes and arrived downtown late enough to miss being in the building when it collapsed. Sixty Euro Broker colleagues who were there were killed, including eight who reported directly to Danielsson.

Meanwhile, the shift manager for AT&T's global network operations center in Bedminster, New Jersey, was watching not only the center's 141 projection screens showing network status around the world, but

October 4, 2001 — An aerial view of the recovery operation underway in lower Manhattan at the site of the collapsed World Trade Center. Photo by Andrea Booher / FEMA News Photo.

also CNN. She had three close relations within blocks of the World Trade Center (they were unhurt). Like just about anyone else with loved ones in lower Manhattan, she wanted to pick up the phone and hear their reassuring voices. Instead, she stayed on the shift to handle what long-distance carriers euphemistically call a "focused overload."

Says Dave Johnson, a spokesman at the global Network Operations Center (NOC), "When you have a sizable disaster – and it can be what happened to us in New York, a bomb blast in Oklahoma City, a major earthquake in Seattle, or a hurricane in Florida – you get a tremendous leap in calling *into* the impact area." This was particularly true in the case of the World Trade Center attacks, "because NYC has a community of interest that stretches out across the whole country."

It's CNN's coverage, not the event itself, which triggered a focused overload, Johnson notes. In the case of the Twin Tower attacks, media coverage happened so fast that Americans watched the second plane crash live, and the story was picked up nearly simultaneously across all the major news networks. "As a result, people knew instantly. If you saw my call chart for the day, it literally takes off like a spaceship – at about 9 a.m. it just takes off and climbs right on up."

Also, in times of crisis, call center volumes can increase by 20% to 50%, with no business reasons to explain the increase. The average call length can increase from 25% to 100%, as many callers initially ask about business issues and then talk about personal concerns. For example, callers may ask a call center representative about their reaction to the World Trade Center attack, as an opening for them to share their own concerns.

At around 9 a.m. on September 11, AT&T began call gapping, which is to say that they reprogrammed the switches at their POPs, where local carriers feed into the AT&T network, to automatically drop a certain percentage of calls aimed at New York. What percentage? AT&T won't say, but Johnson suggests that a typical starting point for a focused overload is dropping 50% of the calls.

Of course, perceived service-quality suffered. But the most-motivated callers – those willing to keep trying – got through. And as the day wore on, the overload began rolling back. "By about 5 p.m.," Johnson says, "traffic was within a couple percentage points of a normal Tuesday afternoon."

Verizon, by its own account, handled the largest number of customer service calls in the wake of the attacks. The carrier, whose operators normally answer one million calls per day, accepted upwards of three million after the attacks. Most calls were for directory assistance; others dealt with residential and business repair services. Normally, Verizon answers 85% of operator calls in 20 seconds. During the attack, the average wait time was 30 to 40 seconds – extraordinary figures, given the huge call volumes and network congestion resulting from the disaster. (One should note that call centers at other major long-distance carriers helped out during the disaster, and not just for operator services; Sprint's call center helped in delivering PCS wireless phones and free calling services to disaster workers and victims in both New York and Washington.)

Aside from call gapping to keep local switches from becoming swamped, the AT&T long-distance networks were undamaged. Competitive local service offered by AT&T, however, relied on two severely damaged transport nodes in the basement of the World Trade Center.

A few blocks north on Varick Street, AT&T's local switching facility was unharmed but offline: The local Con Edison power substation had just been destroyed. Some lines were rerouted elsewhere; service to other lines was provided by a Lucent 5ESS switch mounted with a portable generator on a semi-trailer. "We've got somewhere between 200 and 300 million dollars invested in disaster recovery systems." Johnson says. "Basically, what we have is a large fleet of specially engineered, specially designed trucks, staged at strategic locations throughout the country. They're designed so they can replace any facility in the AT&T network, data or voice."

The locations taken up by the trucks, the exact equipment on them, and even the location of permanent switching facilities are all details that Johnson says AT&T isn't going to disclose anymore. "Things I would have said a couple years ago or even a couple months ago, I won't be talking about anymore."

Secrecy is now the best policy.

LESSONS LEARNED FROM "GROUND ZERO"

September 11 was a disaster both for the communications infrastructure as well as for individual businesses themselves, most of which were as concerned with figuring out what had happened to their data as getting their phone system back on line. For these companies, "content management" had become "searching for backups" or maybe some scraps of paper scattered in the rubble.

Having survived the 1993 World Trade Center bombing, Euro Brokers had extremely good disaster recovery plans and systems in place. All systems that were running prior to the attack were soon restored and operational in a backup location. Yet for all this preparedness, the devastation reached well beyond "data," leading Senior VP Walter Danielsson and others to rethink their idea of "crucial information."

"In 1993, disaster recovery was, 'get your systems back up,'" Danielsson reflects. "This was different. We were very fortunate to have had our computer systems and data all backed up, but it's hard to imagine how many things are just gone – copies of documents, signatures, things like that."

While many companies displaced by the attack have gone back to business as usual, some have fared better than others thanks to content management technologies. The sight of millions of paper documents clogging the streets of lower Manhattan in the aftermath of the attack testified to the fact that invoices, contracts, letters, statements, memos, checks and other paperwork are still at the heart of business processes and are as important as communications and electronic data storage.

Not surprisingly, Alan Abrina at ActionPoint (San Jose, CA – 408-325-3800, www.actionpoint.com), is a company that offers Input *Accel*, an electronic document capture solution whose customers in the World Trade Center include Lehman Brothers, Nomura Securities and Bank of New York, has seen broader customer interest in electronic document capture and storage since the disaster. "One of our World Trade customers had only digitized certain types of documents," he

➤ Buried Costs

The Twin Towers themselves contained a considerable amount of telecommunications infrastructure. Verizon had a Nortel DMS-100 on the 98th floor. AT&T, which offers local service in lower Manhattan, had two SL-100 switches in the sub-basement of the towers. And building-based local exchange carrier EurekaGGN was midway through the installation of a fiber-optic plant to serve the twin towers. EurekaGGN's mid-June press release had trumpeted "the strategy for ensuring business tenants at the World Trade Center are 'future-proofed' against expensive telecommunications installation costs." The new plant would use a cabling system that uses compressed air or dry nitrogen to carry small, lightweight, bundled multi-fiber optical cables into previously-installed tubes. "Fiber will be air-blown from the 110th floor to the basement in less than ten minutes," the release claimed.

says. "After the disaster, the only documents they could draw upon were those they had scanned in. They now see the value in capturing all their documents."

But the good news is that disaster planning actually works. Companies whose contingency plans were in place could draw on the services of providers such as Comdisco (described later in this chapter) for intermediate carrier service rerouting, to provide temporary contact center services, even to house displaced employees and provide infrastructure. As you'll see below, working with such "preparedness providers" will likely become the norm in coming years, for a wide range of businesses. And the disaster-preparedness field is likely to adapt too, learning to serve smaller customers and businesses with more specialized needs.

Indeed, some organizations recovered far quicker than anyone could have expected, both in terms of communications and data. At Empire Blue Cross/Blue Shield, which had its headquarters and 1,900 employees in the North Tower of the World Trade Center, nine employees were killed, including six from the IT department. But, while the company's headquarters lay in ruin, business was uninterrupted, and the health insurer's sense of purpose was unshaken.

"Even in our memorial services, it wasn't taken lightly that we had the IT infrastructure to immediately appease the concerns of our membership [4.1 million customers] – it was a comfort," says Ann Mottola, assistant vice president of customer service systems. "One of our first public statements expressed our concern about the safety of our employees, but also stated that our customer's records were intact. They didn't need to worry."

The only documents Empire Blue Cross/Blue Shield lost were those in the incoming mail of September 10th and 11th, as well as some regulatory records now being restored by the associated government agencies. All other content, including the 40,000 claims and correspondence normally handled each day at the World Trade Center, had already been sorted, scanned, processed and stored with Captiva's (San Diego, CA – 858-320-1000, www.captiva-software-online.com) claims processing software and Unisys's workflow and imaging systems. Digital document images were archived and backed up on optical discs at a secure remote location.

"Once the servers were restored, most people found their documents and were able to continue where they left off," says Mottola.

Immediately after the disaster, staff at Empire Blue Cross/Blue Shield's Albany office called in vendors to recreate the ruined World Trade Center electronic mailroom. Kodak Document Imaging replaced 13 scanners and Unisys and Captiva provided software, and the imaging system was back up and running within days at a temporary site.

The company's Web site was intact, so it could post press releases right away. The IT team quickly configured hundreds of laptops so that employees could telework. The toughest challenge was working with the phone company to quickly establish dial-up connections. Internal e-mail was up and running within a few days, which helped the now-dispersed staff collaborate and communicate. Most Empire Blue Cross/Blue Shield employees formerly based in the World Trade Center soon were working out of company offices in Melville, Yorktown, Middletown, Albany and Harrisburg, New York. The company plans to secure new offices in New York City for its headquarters.

One key lesson learned from the disaster is that *logical* redundancy in the network is by no means the same thing as *physical* redundancy. Even where businesses had service provided by more than one carrier, the local loops to those carriers tended to be physically routed side-by-side, through the cable egress tunnel into the cable vault at 140 West Street. When that facility was compromised, the redundancy of the carriers upstream was moot.

Carriers generally try to deal with the need for physical redundancy by notations called "diversity flags." These are supposed to alert technicians to the need to avoid routing a circuit and its backup in the same cable sheath. But diversity flags can be overlooked. Furthermore, the geographical reality of a network gets far less day-to-day attention than its logical layout – and is further obscured by the semi-opaque boundaries between service providers and the enterprise. This is especially true during epochs of rapid network build-out, where "getting an application up and running" is the immediate goal, and "planning for disaster" a lower priority.

That this can be obscured by the logical layout of a network means that companies sometimes don't even know where their circuits are located, even if they are trying to be aware of these issues. Jack Norris, head of customer service and networks at financial data and IP services Equant (Reston, VA – 703-689-6300, www.equant.com), for example, would have told you prior to September 11th that his company's circuits to Toronto and Montreal ran through Islip, New York. As it happened, a key connection was routed through 140 West Street in Manhattan.

Equant had built the network with sufficient redundancy that throughput was only temporarily reduced, but they had to be clever to get back up to full steam. Since they'd merged with Global One in July 2001, part of the solution involved dragging cables from an Equant office in Manhattan to a nearby Global One office and using Global One's network as an alternate route for the broken links to Canada.

Still, telecom providers, the channel, consultants, and carriers can coordinate their varied activites, if required; overcoming business process, bureaucratic, legal and financial impediments to get vital work done quickly. The work by Nortel and other infrastructure providers, by Nextira and similar consulting companies (discussed later in this chapter), during the days and weeks after the crisis, is a stellar example of how to "get the job done."

A final bit of good news: The labyrinthine Internet, for all its unruly character, works as advertised. The sudden, and seemingly grandiose destruction of infrastructure at the WTC caused only microscopic changes in the global Internet map: The Net just did its self-healing thing. ISPs saw very little fluctuation in reachability. In the hours immediately following the attacks, hits on DNS service and the Web spiked sharply, but aside from some increased latency and a few news-site server collapses, no damage was reported. You'll notice that there's no Internet component for this chapter; that's because, amazingly, there is no "disaster story" here. The real Internet story – which we'll focus on in later chapters is about security and privacy, and about making the Net more resilient to deliberate attacks, both physical and logical.

Verizon Heroically Rises to the Occasion

The WTC attacks caused catastrophic damage to the telecom infrastructure at the head-end of one of the most prestigious and highest-bandwidth last-miles in the world. Also, at least 14 cellular sites were lost downtown, which added to demand on more distant towers already overloaded with wireless traffic. Private and ISDN dial-up connections used by ATMs and merchants for making fund transfers also failed.

For telemarketers that generate most of their call traffic through outbound campaigns, business collapsed. John Goodman, president of Arlington, VA-based e-Satisfy (formerly the Technical

Assistance Research Program or TARP), notes that public pressure prompted many of the big outsourcers of telemarketing services – Apac, Convergys, Teletek, and Telespectrum, among them – to halt operations for a week following September 11, at a cost of millions of dollars in lost revenue (and those call centers that were running operated often at less than optimal levels).

But incumbent carrier Verizon did what telecom service providers are supposed to do: They dealt – rapidly, efficiently, even heroically. Here's the story, and some lessons learned.

The Twin Towers weren't the only buildings in New York to collapse during the September 11th attack at the World Trade Center. When the fourth building in the complex fell – Building Seven – it slewed against a critical Manhattan telephone Central Office (CO) at 140 West Street. Holes were gashed in the exterior of the CO. Water from burst mains poured into the egress tunnel and cable vault below street level, where some 200,000 local circuits and 3.5 million trunking circuits were served. Steel girders from Seven World Trade Center hit the pavement with such force that they penetrated several feet into the ground, piercing the cables that formed the beginning of one of the most prestigious last miles in the world.

The Pyramid

The New York Telephone Company Building (technically, the Barclay-Vesey building, named after the streets to its north and south) was built back in 1927, when such a thing as the New York Telephone Company actually existed.

It was designed to be an imposing building, a monument to a stable telecommunications system. Verizon Chairman Ivan Seidenberg noted during a press conference that "We joke that [the Barclay-Vesey Building is] built like one of the pyramids." Because the structure was purpose-engineered as a telephone facility and natural light was not a requirement for switching equipment, the 32-story building was designed in an unusually massive form, with a short tower atop its center; and featuring setbacks that do, in some small way, invoke the ancient pyramids of mesoamerica.

It's also a historically significant structure. The building is considered to be the first Art Deco skyscraper, and elaborate Art Deco detailing is found throughout; amusingly, the ornamentation is dominated by bell motifs, the architect's nod and a wink to the Bell Telephone logo. Bronze plates on the lobby floor depict and celebrate the construction of New York's telephone network.

Not only was the building a trendsetter in exterior style, it's also a prototypical CO facility. The basement serves as the cable vault, where both local loops and trunks from other COs and CLECs arrive from sub-street conduits. Before the attacks, these cables were neatly laid across steel racks and hangers, throughout the basement. Within the vault, the cables are three to four inches thick, with multiple thousands of twisted pairs bundled into groups of fifty or one hundred, wrapped in a metallic mesh that provides grounding, then wrapped within stiff PVC sheaths. Also in the basement: redundant emergency power generators.

The cables are split out and routed up through the floor to cross-connect frames, where the 200,000 local loops are punched down. From there, it's up through the building to the switches – there are four in the facility, more than $10 million worth of equipment are in those switches alone.

Early Recovery

If you stood on the fourth floor of the Telephone Company Building after the attacks on the 11th, you'd have been able to look out substantial holes in the building's brick exterior, directly across to the huge pile of debris from the collapsed Seven World Trade Center building. Immediately behind you, you'd see racks and racks of DACS equipment, a whole floor of such equipment, now covered in dust, except at the bottom of the racks, where water had flooded across the floors.

Upstairs, on the seventh floor, a similar hole (here adjacent to the multiple equipment rows of a Lucent 5ESS switch) was in use by fire fighters, who had dragged firehoses up through the building to pour water down on the smoldering pile of rubble below. With windows blown out throughout the building, this meant that water sprayed back in when the wind swirled.

The Immediate Aftermath

Lawrence Babbio, vice chairman and president of Verizon Communications (New York, NY – 212-395-2121, www.verizon.com), was ultimately in charge of the restoration effort. But in the immediate aftermath of the disaster, it was Verizon's emergency procedures playbook – and well-trained local and middle management and trade partners – who responded first. Verizon's operations on September 11th swung into overdrive as soon as the Twin Towers came down. Before it was even declared safe to enter the building at 140 West, workers at nearby COs had created approximately 9,500 temporary local circuits to service rescue and firefighting operations. This first step, Babbio says, was completed within 48 hours.

To let Manhattanites get reassuring words home, Verizon converted some 400 payphones in the area to provide free outbound service within the five boroughs. Like AT&T, Verizon has a significant amount of mobile disaster relief equipment. This included, a Verizon spokesperson said, flatbed trucks loaded with portable cellular payphones. Two hundred and twenty such cellular phones allowed three-minute calls anywhere in the continental US. During a single day of the second week following the disaster, residents made 58,000 calls from these cell phones and another 22,000 local calls from regular payphones.

Once the all-clear had been sounded, getting into 140 West was the next order of business. The truly daunting task, and one that had to be picked up as soon as possible, was restoring trunk switching through the site and local service to the major area businesses – the major trading partners in America's equity markets.

Restocking NYSE Circuits

For starters, there was the New York Stock Exchange, which alone used 15,000 voice and data circuits. Not all of these were cut when 140 West Street was damaged: over the past several years, the Exchange and Verizon had already conspired to move 80% of their circuits to different switching offices, making the market less vulnerable to local equipment failure. Still, 20% of the circuits were handled by 140 West Street, according to Babbio; and several thousand additional circuits were compromised by sub-street damage caused by the collapse, water seepage, and other aftereffects. By Monday the 17th, however, over 14,000 of the 15,000 circuits were operational – a really stunning demonstration of effective emergency planning, flexible logistical thinking, hard work and inspiration.

The 9:30 a.m. ringing of the bell that signified the opening of trading on the NYSE, Verizon Communications' co-CEO Ivan Seidenberg said, "an important signal to the world that America was open for business." And the bell rang on time. "Not only was the network rebuilt, but it was built in a fashion to handle record numbers of transactions." Some 2.3 billion shares changed hands on the trading floor that day, the most ever.

Getting circuits restored to the NYSE was an important symbolic win, but only a start. Babbio says: "If you look at 140 West Street, it takes care of two different things. It takes care of 200,000 customers that are contained within that office geography. We also have a lot of circuits that go through that office destined for other locations. There are about 3.5 million circuits that go through that building." Not all these circuits could be rerouted, given that the switches they routed to were destroyed, but by the 17th, Babbio said, some 90% of the circuits that could be rerouted, had been. Further, "even though traffic today is heavy," Babbio noted, "the blocking is within normal limits of what we would expect under the volumes we have."

Rerouting was a largely a matter of logically reconfiguring switches at other company offices. Beyond that effort, though, lay the enormously difficult task of restoring service on local loops to some 35,000 homes and more than 10,000 businesses. For the most part, the four major switches in the building (two Lucent 5Es and two Nortel DMS units) were operational once dusted off and supplied with electrical power. For several months following the disaster, diesel generator trucks parked outside the building provided the power.

Since power was routed only to the essential switches (about a floor's worth of equipment in each case), the lack of air conditioning quickly posed a problem. "All the windows broke and the building's full of holes," a Verizon spokesperson pointed out, noting that the outside temperature was still fairly warm in mid-September. "So we brought in portable air conditioning units and deployed them on the floors that house the switches."

With switches running, the problem shifted to hooking those switches to the local loops without the benefit of a working cable vault. Part of the solution was simply running cables from the switches and out the windows down to the ground. Once on the ground, shallow trenches were cut into the pavement and the cables were run in the open air out around the ground zero area. In the immediate area, there was no way to patch back into the existing cable plant because the manholes that would normally provide access were buried under 30 to 40 feet of rubble.

Babbio said there were "easily three thousand people working in a very congested area, and on top of that, hundreds and hundreds of contractors for cleanup and digging the street." Part of the work was large-scale hauling to dig down to key access points, but once a replacement cable from the switch had snaked its way to an accessible service point, the work was on a fine scale: punching down thousands of twisted-pair connections onto transfer blocks.

Verizon bore the brunt of this meticulous work and will be at it for a long time to come, because essentially the entire rewiring process will have to be repeated to move from the makeshift trenches in the streets back to repaired underground conduits.

Justifiable Pride

In a press conference on Monday, September 17, 2001, Seidenberg openly admitted that, while plenty of thought had been given to contingency plans for dealing with central office failures, the complete obliteration of thousands of customer premises wasn't something they'd had on the drawing board. "We feel reasonably confident that our restoration activities focused on the right things, but taking planning to the step of an act of war, where buildings are decimated and fires burn all over the place, that's something we'll have to think through."

Still – as most Verizon customers would agree – the company's existing disaster-response mechanisms worked remarkably well during the crisis, even in worse-than-anticipated conditions. Equipment was on hand. People knew what they had to do to get service restored and they did it with remarkable speed. Through a dark, apocalyptic moment, most of us stayed connected, and the "real," full recovery of communication services in Manhattan is now well underway.

Customer Backup

The World Trade Center attack threw a spotlight on companies whose sole purpose is disaster recovery. Their services include the duplication of all incapacitated datacom and telecom infrastructures and end devices. While the scale of loss on September 11 eclipsed anything the two most prominent disaster recovery companies, Comdisco and Sungard, had ever dealt with before, it did show how preplanning had paid off for some of their clients. It also put into action agreements made well in advance between enterprises and long distance carriers. These included rerouting plans, so that as soon as disaster declarations were made, the new emergency locations were already on file as the destination of rerouted 800 and other phone numbers.

Regular systems integrators and interconnects, under standard maintenance contracts, also took up the task of recovering dial tone for New York City's emergency services and for dislocated – or, as often, disconnected – financial and healthcare clients. This required round-the-clock, emergency enlistment of designers, technicians and installers from offices as far as Maine; the rerouting of trunking switches to New Jersey locations, and even the cooperation of Canadian customs, which expedited the importation of Nortel switches.

Nortel Networks (Brampton, Ontario, Canada – 905-863-0000, www.nortel.com), maintains a year-round, 24-by-7 Emergency Recovery Center in Research Triangle Park, North Carolina, that normally employs over a hundred people. When disaster struck, Nortel's ER immediately contacted all customers in the World Trade area to determine stability and to analyze any service or equipment requirements. These calls were then escalated to Nortel's service and support organizations (Nextira, see below, among them).

Tornadoes have taken out central offices before, and Nortel's ER has also responded to floods and fires, but obviously, no previous damage has ever been as extensive as that wrought in the Sept. 11 assault. But having a plan of action, numbers to call, and a recovery center certainly did help dislocated enterprises meet immediate needs.

By Monday Sept. 17, Nortel Networks reports that 432 orders had been loaded and serviced with a 48-hour turn around from order to on-site; 34 customers had been serviced, including carriers Focal, Genuity, and Sprint, as well as Goldman Sachs, Bloomberg, and the Pentagon.

In Verizon's heavily damaged, 32-story, 140 West Street central office, two Nortel DMS 100 switches, servicing 60,000 of the office's 300,000 lines, went down when power went out and basement power supplies were flooded. Nortel's team was allowed to enter the building on Friday, escorted by security offers and wearing respirators. They worked 18-hour shifts and often had to evacuate over the weekend as fires flared up and areas were cleared due to reported bomb threats.

The Nortel switches proved to be fortunate, as they were located near the wall furthest from the damage caused by the toppled neighboring Building Seven. According to Denny Miller, director of emergency recovery, "They had some damage, but nothing we couldn't work around. We had those back up [with backup generators and temporary air conditioning units trucked in] by Sept. 18." Verizon sealed off the part of the floor where the DMS switches were housed.

Nortel ER teams also worked with optical MAN specialists to provide Optera Metro 5200s multi service platforms on Optical MANs and fully restore service by Wednesday. With AT&T, it worked to restore Merrill Lynch's optical ring service between New York City and New Jersey.

Doubling Up in Branch Offices

Employees of Nextira (Houston, TX – 800-324-2222, www.nextira.com), while not in the disaster recovery business *per se*, found themselves in the thick of the New York recovery effort. A major systems integrator for Nortel, Nextira is the convergence solutions company owned by Platinum Equity Holding and founded from the merger of the spun-off Williams Communications Solutions, Timeplex, and Milgo. Nextira employees work in central offices for Verizon and colocated local and long-distance carriers, and also on behalf of enterprise clients.

Interviewed ten days after the attack, Frank Pilon, Nextira executive director, operations, for eastern territories, says, "We're seeing a lot of expansion on existing [CPE] switches where companies are moving dislocated personnel to double up in other, nearby offices." In other places, calls and dial tone were being routed off remote switches connected via T-carrier fiber.

On the day of the World Trade Center attack, Nextira established an emergency command center in Connecticut and a secondary staging area in New Jersey for customer support rerouting voice and data networks. One day following the attack, the company had successfully restored communications networks for the City of New York Command and Control Center, Nasdaq, and Merrill Lynch.

New York City Command and Control Center

The City of New York Command and Control Center had been housed in Building Seven of the World Trade Center complex, which was the next building to collapse after the Twin Towers. "Our New York office got a call from Nortel at 6:30 or 7:00 Thursday night," says Pilon. "They said they had to have 300 phone lines in the relocated New York center, by the next morning. We pulled a team together, scrambled, got design specialists to work around the clock. We took two Option 11 Meridians from a warehouse in Maspeth, Queens. We networked them together and by 6:30 in the morning we had 280 lines wired and ready to go. We just had to wait for Verizon to do their work, which they had up by later that morning. Usually that would have been a three-week job." There's now a third Option 11 and 300 lines; all three cabinets were replaced with one Option 81.

It ordinarily takes 15 to 16 hours to build an Option 81 on order, says Pilon. Then it has to be shipped from Canada and through customs, which typically takes 50 to 60 hours. In the disaster relief effort, the switches took 36 hours after order to arrive.

Nextira also restored critical communications functions for Nasdaq over a weekend. The exchange squeezed workers dislocated from the WTC into pre-existing space at other Manhattan offices. At 125 Broad Street and 4 Times Square, Nextira installed Nortel PBXs. At 33 Whitehall Street, Nextira installed an additional Nortel Option 11 Meridian 1 and several hundred phones (also over a weekend), according to Pilon.

Nasdaq locations were also handling calls for the then still-disabled American Stock Exchange.

For Merrill Lynch, Nextira served in its role as maintenance subcontractor for British Telecom.

They replaced a BT switch and some of the turrets – special trading-desk phonesets – lost in the disaster. The first 200 of the 800 turrets and the switch went in overnight in Merrill Lynch's new office space in Jersey City, NJ. Five hundred were ringing by the following Monday.

Customer requirements are described as constantly changing in the wake of the Twin Towers collapse. "We are doing projects today on the fly that generally take weeks and months to plan, in terms of loads, connectivity. That is all being compressed into extremely short periods of time," says Bob Wentworth, Nextira CEO. Nextira technicians from Maine, New Hampshire, Massachusetts, and Philadelphia have been transferred to New York and New Jersey in the recovery effort. Disaster recovery company Comdisco, itself (see more below), has been a Nextira client, getting an additional 400 station sets at two locations on September 12.

➤ A Look at NYC's Collaborative Command Center

Watching Mayor Giuliani calmly brief reporters in the aftermath of World Trade Center attack, one could only wonder how his key officers and crucial city agencies could be so organized. The task of managing thousands of police officers, firefighters, emergency medical services workers and construction crews was compounded by the challenge of working out the logistics of bringing in equipment and emergency supplies. Besides strong leadership, the city is relying on Web-based collaboration software.

More than 1,700 workers in over 200 federal, state and local agencies in New York are using Eteam (Canoga Park, CA – 818-932-0660, www.eteam.com), an emergency management system based on Lotus Domino, to communicate with each other, keep information up to date and make decisions – first for the rescue effort and now for the cleanup and recovery activities. A hosted version of the Web-based application was in operation within twelve hours of New York City's Office of Emergency Management finding and assembling a downtown operations center (its headquarters in 7 World Trade Center were destroyed).

The system was used to create and access infrastructure reports for hospitals, fire houses, police stations and city agencies. Personnel were assigned to constantly update incident reports for each building affected by the disaster, keeping agencies and officials apprised of casualties, rescue priorities and emergency resources available. The system was also used to submit and process requests for blankets, shoes, dump trucks, cranes and other resources needed for the emergency. Workflow features helped track the costs associated with each request. Representatives from Con Edison, the local power utility, and Verizon, the local telecommunications provider, used Eteam to track outages.

At ground zero, rescue workers accessed the Eteam system from special ruggedized laptop computers rigged with wireless network cards so that updates could be transmitted to the command center and to a staging area and supply warehouse. To keep bandwidth demands to a minimum, the system transmits only the changes made to report rather the full content called up on a remote computer.

In the days after the attack, city managers were able to see which agencies were up and running, which were out of service and which needed help simply by glancing at the reports available through this collaborative application. This came in handy for Mayor Giuliani's twice-daily briefings. Instead of having to read individual one-page reports from 68 different agencies, the OEM director reviewed their status in one organized report gathered by the system. According to John Hughes, a former deputy directory of New York City's Office of Emergency Management and now a vice president at Eteam, "The agencies were able to all be on the same page and utilize the same reports, like a big whiteboard in the sky."

Bridge and Tunnel Telecom

We talk all the time about not having all your eggs in one basket," says Pilon. "A good example is the Port Authority of New York. Look at your airports, La Guardia and Kennedy, Newark. They have services that come from different routes, so they're not shut down if they have a disaster."

The Port Authority users at airports and tunnels did not lose dial tone, but did lose quite a lot of trunks and such services as voice mail, since their switching and services were provisioned via remote switches which were, in turn, "homed" out of the Nortel SL100 switch that Nexitra maintained in the basement of the World Trade Center. Pilon: "While that was wiped out, it didn't put anyone totally out of service. We're now restoring those services out of an SL100 that we have in Middletown, NJ. Again, Verizon and AT&T had to provide the networking to get out to those locations, but that's been done."

With some clients, a standard maintenance contract may have covered the cost, whole or part, of the recovery effort. Where maintenance contracts have lapsed, time-and-materials bills will eventually be applied, presumably on a case-by-case basis. "We are in the mode of act first, and deal with transactions later, says Wentworth.

Disaster Specialists

Comdisco (Rosemont, IL – 847-698-3000, www.comdisco.com), disaster recovery specialists, received a volume of disaster declarations three times larger than anything in its corporate history. Comdisco's Rich Maganini reported that 47 clients had submitted 90 September 11-related declarations as of September 20, and, according to Maganini, all 30 of the customers that Comdisco is presently "recovering," and those still waiting, are holders of preexisting contracts and pay ongoing subscriptions against calamity.

In some cases, the clients' business offices remained intact and functional, but due to the knockout of the local Verizon CO, connections to remote data centers were cut. In these cases, says Maganini, businesses have decamped to one of Comdisco's seven recovery sites in the New York-New Jersey area. Its largest, in Carlstadt, New Jersey, has room for 1,000 work-area positions.

"The bulk [of the World Trade Center disaster recovery work] is desktop, back-office recovery," says Phil Saglimbeni, Comdisco's senior network specialist. "There probably wasn't that much space in the towers used as purely data center, due to the price of real estate. There is some call center recovery going on."

Comdisco has 23 recovery sites around North America, handles data mirroring and restoration from backup, and offers contracts that typically cover work stations or platforms. In a typical example, perhaps one contract (and hence, one declaration) covers the principal mainframe and restoration from backup; another contract may cover voice services.

Some companies prepare for data recovery by maintaining a presence to their network at Comdisco's site at all times. "Perhaps a router and a couple of circuits that tie back to the data center," says Saglimbeni. "Then once they get their people safe, they can just plug their PCs in."

Comdisco also contracts to terminate rerouted 800 and DNIS numbers, a plan that requires pre-event coordination with the long distance provider as well. These services are marketed under several names, like "service assurance." After a disaster, the long distance carrier is simply given the go ahead to reroute 800 numbers and DNIS numbers to the trunks of the recovery site; in this case, Comdisco's in Carlstadt. The same goes for Verizon for DID service.

This prearrangement must be written into the plan, tested and updated; Comdisco recommends twice yearly. "As clients test throughout the year, we keep records on both the voice and data side. As we learned [after September 11] who was declaring and where they were going, we were pulling out these files and configuring the environment for them as they were coming in. As they got settled, we fine-tuned the process," says Saglimbeni. A main number is also assigned to each client for a manual call attendant. Some clients took advantage of contracted voice mail support and fax from stand-alone fax machines after September 11.

Brokerage house Morgan Stanley was profiled in a ***New York Times*** story as one of Comdisco's recovering clients in the wake of the disaster. Some of its employees set up temporary shop in Comdisco's Carlstadt, New Jersey hot site.

➤ Avaya Assesses Customer Needs

Nearly 70% of companies formerly based in the World Trade Center were Avaya (Basking Ridge, NJ – 908-953-6000, www.avaya.com) customers. An army of 400 Avaya service techs worked to help these customers restore their communications infrastructure and get back to business. To date, they've installed or moved more than 40,000 phones and 20,000 voicemail boxes.

Twenty-five technicians dedicated to the New York Stock Exchange worked to provide communications support for about 1,000 people who were relocated to NYSE from offices in the area of the World Trade Center.

Avaya has also teamed with a number of government agencies to meet their short-notice requirements – including FEMA, the FBI, and the IRS in the New York area, and the State Department and the Executive Branch in Washington, D.C.

In the process, Avaya has been interrogating customers about their shifting priorities and worldview. Here are ten emerging trends they're seeing:

1. Avaya customers want to ensure the security of their communications, their corporate information and their people.
2. They want seamless voice, data, and video to support employees at geographically dispersed locations – including those working from home.
3. They want backup systems at a separate location so that they can quickly move and reestablish communications in the event of a disaster.
4. Those who can't replicate an entire system want a satellite facility near their central headquarters so that they can quickly transfer voice, video, and data capability in a "hot swap."
5. They want emergency communication capability, including voice-messaging systems that allow their teams to communicate with each other via voice mail during a disaster.
6. They want video conferencing and audio conferencing as an alternative to travel.
7. They are reluctant to have a single call center supporting their business and want to establish additional sites as a precaution.
8. They see the need to revive and to update their "Y2K" planning infrastructure and to make it an ongoing part of their business.
9. They are more interested than ever in "one number" reachability – a single phone number that can follow an individual whether they're in or out of the office.
10. They've developed a new appreciation for onsite service technicians and for remote diagnostic systems that automatically find and fix network problems.

For some time after the disaster, Comdisco was supporting Morgan Stanley's recovering call centers and offices on a Nortel Option 81C Meridian PBX; taking in "almost 120" T1s for various voice services. They had added switch capacity to support the additional lines and were assessing long-term plans to decide whether to bring in another PBX. Those users not familiar with Nortel station sets and its ACD were given quick training by vendors and Comdisco personnel.

The Carlstadt facility was designed in 2000, in time to incorporate the eggs-in-several-baskets lessons learned from Hurricane Floyd. "All our call center recovery areas in the building are fed from two different COs," says Saglimbeni. "Typically if DIDs for that area are from central office A, then outgoing are from CO B – and all of our trunks are set up for two-way. The facility, in general, is fed from four different central offices."

AT&T's "SAFER," or Split-Access Flexible Egress Routing service, uses a call allocator in the network to drive 800-number traffic from two different COs. They also duplicate tandem connectivity, eliminating sole dependence for long distance on one LEC's tandem office.

One major Comdisco customer is The Bank of New York, which faced many challenges in returning to normal business operation. Banks are ultramodern in that they have advanced telecommunications systems with redundant links, but they're also old-fashioned in that they deal with a lot of paper that's electronically scanned and stored in batches instead of in real time. When disaster hits, you may have to put back together enough confetti to hold a ticker tape parade down Broadway. In the case of The Bank of New York, it had four locations in the World Trade Center area. Its 1 Wall Street headquarters suffered power and telecom outages and was inaccessible for three weeks. The bank's major data and operations center at 101 Barclay Street, across the street from 7 World Trade Center – the building that collapsed on the afternoon of September 11 – took damage to its atrium and the windows on the south-facing side were shattered. The 3,000 employees who worked there have been displaced but will eventually return. Two employees who were in the vicinity were killed.

The bank immediately set up a temporary command post at 1290 Avenue of the Americas. Many of the technical support teams that had worked in the operations center reported to a West Patterson, New Jersey location to start rebuilding the damaged data center from a bank backup center in Teaneck, New Jersey and a Comdisco recovery site. Hierarchical storage management and software disaster recovery tools have helped the business get back up and running. Yet some of the disaster recovery processes were incomplete, and some were untested at the time of the attack.

According to Joe Gerbino, vice president of technology, The Bank of New York had imaging systems in place to support four distinct lines of business-stock transfer, trade services, accounts payable, and benefits disbursement – and these have been indispensable in the recovery process. "We potentially would have lost all those documents otherwise," says Gerbino. "Instead, those records are in electronic storage and can be viewed in multiple locations, so this has helped us quite a bit."

As one of its lines of business, the bank acts as a securities agent for major corporations, handling all recordkeeping and correspondence for their registered shareholders: issuing dividend checks, handling proxies and sending out mailings. The bank uses ActionPoint capture software and an Eastman imaging and workflow system to capture, store and retrieve digital copies of stock certificates, statements and correspondence. These imaging facilities were also knocked out by the Trade Center collapse and are being rebuilt.

While no one could anticipate such a despicable and atrocious attack, the incident has led many executives to rethink planning and preparedness. "We would have more duplicate systems in multiple locations and be more diligent about disaster recovery," says Gerbino, adding that document imaging would be more of a priority. "Being a bank, we deal with a lot of paper, and we want to get the paper out of the office. In the future, we might want to get the online documents backed up in real-time rather than batch mode."

Disaster Drills

Like Comdisco, Sungard Recovery (Wayne, PA – 610-641-8810, www.sungard.com) had disaster declarations from companies around the World Trade Center. As many as 800 clients went on disaster recovery "alert," signaling possible imminent need for some sort of recovery.

Two Sungard customers – one in healthcare and one in finance – had their voice services recovered, in Sungard's Jersey City emergency metro center. Others went there for data recovery, although all customer workstations on the premises are equipped with dial tone.

Both customers had rerouting plans tested and ready. Sungard's voice infrastructure, too, is based on the Nortel 81C Meridian; in this case, more than one, says Jeff Annechini, the company's director of voice technology.

With test files in place, Sungard can reestablish calling queues on site in thirty minutes, he says.

Sungard also maintains adjunct telephony servers (Dialogic equipped) with station set emulation to support remote extensions. The idea here is to let displaced workers dial in, log on as remote agents, and use their home analog phones as PBX and even ACD extensions. However, as of Sept. 21, all switch usage was by displaced workers on-site in Jersey City, a facility of around 220 workstations, according to Annechini.

Sungard's recovery center in Jersey City also offers call recording and remote access to the usual collection of call center performance statistics, and at least one of its voice recovery clients is using auto attendant and voice mail for overflow callback queues. Emergency announcements are quickly recorded and attached to selected mailboxes.

Three weeks post-disaster, Comdisco had 23 out of the original 47 recovering clients working in three NYC-area centers – in Queens and North Bergen in addition to Carlstadt. Some clients, waiting for their buildings to be reoutfitted with windows and interior equipment, would stay for as long as a year, in expanded Comdisco office space. Others found new corporate homes in Manhattan and as far west as Morris County, NJ, soaking up some of an office-building glut that has prevailed for years. The COs of closer-in New Jersey communities, in Jersey City and elsewhere, are seeing floods of orders for new circuits to meet the demand of the new tenants.

Another company that worked to meet their customers' demands in the aftermath of September 11 is Unisys (Blue Bell, PA – 585-742-6865, www.unisys.com). The company's Managed Network Services (MNS) offers desktop and network outsourcing. MSN provides IT support and management services, which include network management, management of IT assets and help desk support to its clients' end users. Unisys has between 25 and 30 call centers located worldwide, eight of which serve MSN's customers.

One of Unisys' offices is located on Wall Street, a short distance from the World Trade Center. "One of our first priorities was locating all of our people from that office," says Marge Trachtenberg, Unisys' director of managed service center operations for North America.

According to Trachtenberg, the initial silence in the call centers after the attacks was deafening.

Once Unisys' clients, some of who were located in the World Trade Center, began to regroup, the calls started coming in. Employees in New York City and in the company's Blue Bell, PA, call center began checking on the status of their clients and finding out what their immediate needs were.

"We had clients that we normally provide a certain level of service for and they called overnight and said they needed us to be there to answer the phones 24x7," she says. "We rolled up our sleeves and figured out how to get things done as we went along. We didn't worry about having every 'i' dotted and every 't' crossed."

Unisys' call centers in Farmington, NY, and Austin, TX, set out to assist customers. "For our key clients we took immediate direction based on what their needs were," says Trachtenberg. "We offered solutions and put any ancillary activities into motion and monitored them every couple of hours."

Trachtenberg reports that overall, call volumes across North America were down. This made it easier for Unisys to concentrate on its affected customers. "We were able to use resources that were experiencing less call volume and move them over to help us," she says.

"We had to do some creative staffing adjustments and move people around," Trachtenberg adds. The company even had about 40 employees who were located in other mid-western offices drive to Austin to help handle the call volumes.

Considering the nature of the disaster, Trachtenberg reports that the help desks were able to respond quickly to customers' needs.

"Now that the absolute critical immediacy is dying back a little, the next step is refining our emergency processes into standard operating processes. We are going back, revisiting clients and building repeatable processes."

The WTC attacks disrupted not only voice communications, but also in many cases access to corporate data and applications, particularly for displaced workers. Though many companies had back-up servers at sites unaffected by the attacks, employees often couldn't tap into the servers because of the disrupted communications.

To help displaced employees access corporate LANs, help desk staff at Getronics (Billerica, MA – 978-625 5000, www.getronics.com) extended a remote computing program to affected clients (chiefly financial houses). The initiative called for help desk workers to send out CDs to clients' employees. The CDs contained the clients' corporate apps, remote connectivity software to link through a dial-up connection with back-up servers, plus user authentication software. The effort went smoothly, with dozens of users up and running within two days after the attacks. The real challenge was to churn out the CDs and user IDs in a timely manner. Getronics actions demonstrate that remote computing programs are not just productivity tools, they're critical to a disaster recovery plan."

THE END OF THE BEGINNING

For telecom, datacom, IT and the business world in general, the Twin Tower attack will force a great number of companies, both scathed and unscathed, to reconsider – or just consider – their survival plans.

The remainder of this book will examine the new technologies that must be employed and new procedures that need to be implemented to get organizations out from under the dark shadow of September 11.

CHAPTER 2

Conferencing Technology — Addressing the Fear Factor

For years Yours Truly has been extolling the wonders of audio, data and video conferencing, mostly in the pages of ***Computer Telephony*** magazine (now called ***Communications Convergence***). After all, why waste time and money by physically traveling anywhere – or not traveling, depending on the vagaries of the weather or, as is the case now, a new fear of flying and delays in airport security spawned by your not-so-friendly local terrorist cell. Audio and video conferencing services and products, that can facilitate pow-wows for as few as three people or as many as 3,000, are now within the reach of any sized company.

Moreover, the teleconferencing paradigm itself has changed. Whereas the traditional view held that conferencing technology could simply replace travel for group-to-group meetings, the contemporary view looks upon this technology as providing a more dynamic, synergistic "visual collaboration" to accelerate information sharing and business decision making.

The electronic conferencing industry got some unexpected help in the early years of the new millennium: First a recession, then The Fear Factor following the horrifying events of September 11, 2001, and now The Hassle Factor, as airline travel becomes ever more difficult. All of these factors have spurred an increasing demand for alternatives to traditional in-person business meetings

President Bush's vow to the American public on the evening of September 11, 2001 that America's economy would be open for business the next day, was almost immediately accompanied by press reports that many companies, including Intel and Sun Microsystems, were suspending air travel indefinitely for the safety of their employees.

For example, a September 18, 2001 *Associated Press* dispatch revealed that "a number of companies, including Salt Lake City-based Kennecott Minerals and German automaker Mercedes-Benz, which has a factory near Tuscaloosa, Alabama, said they would likely rely more heavily on telephone and video conferences."

Geoff Bobroff a prominent consultant to the financial services industry and president of Bobroff Consulting (East Greenwich, RI – 401-886-1194), was quoted in a September 17, 2001 dispatch written by Alex Zavistovich for Business Wire (www.businesswire.com) as saying that with corporate bans on business travel and heightened security measures, "American air travel will become much more complicated. Businesses will need to look for alternatives. For organizations with global communications needs, it makes tremendous sense to use teleconferencing and any other technological tools at their disposal to quickly return to business as usual."

Also quoted was Andrew W. Davis, managing partner at Wainhouse Research & Consulting (Brookline, MA – 617-975-0297, www.wainhouse.com), a market research consultancy specializing in conferencing technologies and applications. Davis said that "American power is derived from economic strength more than anything else, and we can't maintain economic strength if we crawl into a business activity hole with our heads down. With technology, we have credible, viable, effective alternatives to travel, alternatives that can enable us to go about rebuilding America and American business."

Davis, like Bobroff, cited advances in teleconferencing as a practical way to circumvent corporate travel bans.

In a similar vein, a *Newsbytes* dispatch from Chicago, Illinois on September 12, 2001, quoted a study from the outplacement firm Challenger, Gray & Christmas (www.challengergray.com) that "the business world will increasingly turn to technology that facilitates telecommuting and video conferencing in order to keep work going even in times of national crisis."

The Challenger firm's (which has offices throughout the US and Canada) report, "Workplace Response: America in Crisis" said teleconferencing equipment will allow companies to keep communication lines open to partners, suppliers and customers in the eventuality air travel is not possible. The report went on to state that teleworking can "insure seamless customer relationships if each department is evenly split – half working out of company offices, the other half as telecommuters, each with a manager in charge."

Even if a section of staff is forced to vacate a building, a "telecommuter squad" can take over, it added.

John Challenger, the firm's CEO, said in a statement: "US business will never be operated in the same way again. Companies nationwide, under siege by the New York and Washington attacks, will employ the latest technology to assure that continuity of operations and customer service are maintained even under the most unfathomable circumstances."

Through the use of readily available devices such as cell phones and laptop computers, churches and other buildings that are empty for long periods of time can be used as temporary offices if an emergency prompts an evacuation, the study said.

IN A NUTSHELL. . .

"Conferencing" or "multimedia conferencing" are really a catch-all terms that includes everything from electronic bulletin boards and chat rooms to real-time video conferencing, application sharing / data collaboration, audio conferencing and even instant messaging or "IM" (AOL, MSN Messenger, ICQ). Formerly considered the sole provenance of circuit-switched telecom, conferencing technology is now slowly evolving into various forms of IP-based ***groupware***. Groupware itself falls into two main categories: synchronous and asynchronous. Synchronous groupware involves users who are all connected and interacting simultaneously, such as live chat rooms or video conferences. Asynchronous groupware is "non-simultaneous" so people can only share information by leaving messages on a Web site or dial-in Bulletin Board System (BBS) for others to read when they log in (an Internet newsgroup or e-mail list could even qualify).

WHY USE CONFERENCING TECHNOLOGY?

- **Safety**. No harm is going to come to an employee, customer or student if no one has to travel via a train or plane. Executive and customer meetings don't have to be held at conspicuous places. University classes need no longer take place at well-known, stately, facilities that have been easily locatable for the last hundred or so years.
- **Timeliness**. Rather than take a day or more to travel back and forth to distant locations, one can immediately set up a conference thanks to telecommunications technology. This is great for people who often engage in various forms of emergency consultation, such as those involved with computer technical support and life-or-death telemedicine scenarios. Also, companies with multiple branches are becoming more commonplace. That could make emergency "meetings of the minds" difficult to facilitate. Companies with a need to link multiple facilities for audio or video conferencing have the options of either buying the bridging equipment and servers and do it themselves, or they can hire a service provider to do it for them. Each has its merits.
- **Scalability**. During the course of a conference you may realize that another outside expert or other partners should be attending the conference. With current technology, it's simple to page them and bring them into the discussion rather than stop the conference while the person is physically transported to a meeting place.
- **Reduces cost**. It's less expensive to establish an audio-, data- or video conference than to actually transport people to a location, have them rent a car when they arrive, put them up at a hotel, feed them for a day or two, etc. Cost savings related to conferencing technologies are similar to the savings accrued by having employees telework. (See Chapter 9).
- **Less hassle**. Do travelers really enjoy losing their luggage, digesting bad food at an altitude of 35,000 feet, and squeezing into what the writer Tom Wolfe so eloquently calls "haunch-to-paunch-style" seating? As they say, business class travel has its rewards!
- **Levels the playing field**. Although some people think that "there's nothing like being physically at a conference," one executive the author knows, once commented that audio conferencing changes for the better the means of actual discourse by eliminating body language such as the "finger point" which can become a convenient escape or device for posturing, destabilizing actual progress in a meeting.

PRE-DISASTER TRENDS

During 2000 and 2001, the dwindling economy led to tightened travel budgets and thus helped bring about increases in all forms of conferencing, even video conferencing.

A survey by Runzheimer Reports (Rochester, WI – 262-971-2200, www.runzheimer.com) on Travel Management in 2001 revealed that 36% percent of companies cut domestic travel costs by using video conferencing. The research firm TeleSpan (Altadena, CA – 626-797-5482, www.altadena.com) reported that more than 13 million audio conference calls were placed through North American services in the first half of 2001, up 43% over the same period in 2000. Also in 2001 the research firm Frost & Sullivan (New York, NY – 888-690-3329, www.frost.com) reported that US video conferencing service revenues may climb from $1.48 billion in 2000 to $5 billion by 2007. (Towards the end of 2001, Frost & Sullivan analyst Roopam Jain revised these figures to $1.99 billion in 2000 and $7.8 billion in 2006.)

Jain also stated that this expansion is "followed closely by the video application service provider (ASP) marketplace, which is quickly emerging as the new standard for video conferencing applications. While many firms prematurely jumped on the ASP bandwagon, the ASP proposition is a promising area that will eventually succeed in creating a sustained value for end users."

Frost & Sullivan has also indicated that the US Web conferencing services market as a whole may grow from $62 million in 2000 to $800 million by 2007. Major service providers such as WorldCom, AT&T, Ameritech, Sprint, and Global Crossing are beginning to offer Web conferencing services as a complement to their existing audio conferencing services rather than as a "new" service. Approaching the market from the other end, various existing Web-based conferencing services like WebEx.com, PlaceWare.com and Centra.com are contemplating or are in the process of adding more audio and video capabilities to their portfolio, leading us to speculate that "multimedia conferencing" or just plain "conferencing" will become the standard umbrella term for this technology.

POST-DISASTER TRENDS

On top of the economy-driven conferencing uptrend, the World Trade Center disaster in all its horrific enormity terrified the business community, which responded by immediately plunging headlong into conferencing technology. The conferencing industry had previously seen jumps in demand immediately after the crashes of Pan Am Flight 103 and TWA Flight 800, and during the Gulf War. But since the events of September 11, 2001, there's been an upsurge in audio, data and video conferences as never before. In the short-term, the figures were astonishing, and still are quite impressive:

- Brad Levy, VP of Global Crossing Conferencing Products Global Crossing (Westminster, CO – 303-633-3000, www.globalcrossing.com), told the author that "Our video business has had over 80% growth in call volume going through our call centers. Our audio segment, especially our Ready-Access service, has also seen over 50% growth."
- France's Genesys Conferencing (Montpellier, France – +33 (0) 4 99 13 27 67, www.genesys.com), one of the world's major teleconferencing service providers, has seen a surge of around 40% in demand for audio conferencing and even larger increases in video conferencing.
- PlaceWare Inc. (Mountain View, CA – 650-526-6100, www.placeware.com) who offers Web-based conferencing, said its business is up 49% since the disaster.
- On September 25, 2001 Raindance Communications, Inc. (Louisville, CO – 303-928-2400, www.raindance.com), a provider of integrated, reservationless Web and phone conferencing, announced more than a 35% increase in service usage for the previous week.
- V-SPAN Inc. (King of Prussia, PA – 610-382-1000, www.vspan.com), a conferencing company that handles 3,500 corporate clients a day, said traffic rose 35% to 50% for some conferencing services.
- The ***Wall Street Journal*** reported that Ernst & Young for the first time in its history used all of its 100 video conference room systems to full capacity.

Aside from the upwelling of interest in conferencing technology by "fence sitters," even those companies with legacy systems are requesting upgrades and expansions.

Perhaps the ultimate seal of approval for conferencing technology comes from Wall Street. While most of the stock market plummeted, stocks soared for providers of conferencing products and services, such as Act Teleconferencing Inc. (Denver, CO – 303-233-3500, www.acttel.com), Polycom (Milpitas, CA – 800-765-9266, www.polycom.com), PictureTel Corp. (Andover, MA – 978-292-5000, www.picturetel.com) which Polycom has acquired, Ptek Holdings (Atlanta, GA – 404-262-8400, www.ptek.com) the parent of Premiere Conferencing (Lenexa, KS – 913-661-0700, www.premconf.com), and WebEx Communications (San Jose, CA – 408-435-7000, www.Webex.com).

For example, publicly traded shares of the leading video conferencing device maker, Polycom, soared in price from below $20 to $40 a share during the eight weeks following the September 11th disaster, then declined by about a third.

TYPES OF CONFERENCING TECHNOLOGY

The advantages of, growing necessity for, and just sheer convenience of electronic conferencing can't be beat. Now that I've (hopefully) convinced the reader of the necessity of adding electronic conferencing to the corporate communications network ecosystem, let's look at the sometimes bewildering array of hardware and software technologies that can be found in the conferencing infrastructure and htose technologies that will dominate the future of conferencing.

Ostensibly, there are three main categories of conferencing technology: ***audio conferencing, data conferencing*** (also known as ***whiteboarding*** or ***application sharing***), and ***video conferencing***. The characteristics of video conferencing, however, often include the capabilities of the other two technologies.

Circuit-switched versus Packet-switched

One can now further divide each of these three technologies into two "flavors," ***circuit-switched*** and ***packet-switched*** or ***"IP-centric"***. Traditionally, two-way audio, data and video have traveled through the legacy ***Public Service Telephone Network (PSTN)***, which is circuit-switched, meaning that a number has to be dialed that sets up a dedicated full duplex (bidirectional) circuit to a destination via a series of telephone company switches for the exclusive use of two parties. Nearly all voice calls and dial-up modem calls are circuit-switched. The conferencing equipment designed to interface with these conventional telecom networks tends to be pricey (as does the circuit-switched "call" itself), though these prices have come down drastically in recent years.

Packet-switching, on the other hand, gained prominence in the 1980s and 1990s with the rise of Local Area Networks (made possible by the Ethernet protocol) and the Internet (powered by the Internet Protocol, or IP). Instead of "nailing up" an exclusive, dedicated circuit, a packet-switched network breaks information into bundles called packets, gives each packet a destination address where it should go, and sends them off on their way. Each packet may take a different path to get to its destination, whereupon arrival they will be assembled in the order in which they were sent.

Even packet-based voice alone, or Voice over IP (VoIP), has advantages over circuit-switched voice: the bandwidth needed for a voice call can be reduced by using voice activity detection (VAD) techniques (which removes the silence that accounts for about 40% of wasted bandwidth) and low bit rate codecs that can lessen the amount of bandwidth for a voice call from 64 Kbps to as little as 5.3 Kbps.

Capital investment for packet-based platforms is significantly less than the circuit-switched equivalent, while operating costs are significantly reduced through the consolidation of platforms and network management applications.

Packet-based networks and communications systems are less expensive to install and operate than circuit-switched systems, and since they're based on open standards, there's a faster time-to-market for new products and their enhancements. Packet-based communications systems are also great for sending data but not-so-great for real-time communications such as audio- and video conferencing, since packets sometimes arrive too slowly or not at all, which leads to an occasional Quality of Service (QoS) problem. This is because data-centric packet networks rely on routers that use completely different techniques and protocols and algorithms for signaling than does the PSTN. Router-based networks, for example, use routing algorithms, such as Open Shortest Path First (OSPF) and Intermediate System to Intermediate System (IS-IS) to manage data traffic. None of this was designed for real-time voice transmissions, so natural QoS over packet-switched networks suffer somewhat.

There are more expensive packet-based network data services offered by phone-companies that do offer some degree of quality of service, such as ***Frame Relay*** and in particular the ***Asynchronous Transfer Mode (ATM)***. But packet networks are carrying more and more voice and video traffic, and so more suitable signaling protocols are being developed and deployed to manage the unique demands of real-time, interactive conferencing sessions. Manufacturers and carriers are working mightily to adapt multimedia conferencing to inexpensive and pervasive IP networks such as the Internet, since this will allow companies to throw out their expensive external equipment and just use desktop PCs and servers to handle data, audio and video in one system. Towards this end, Microsoft has bundled some simple video conferencing software into its newest Windows XP platform.

Why Now?

Still, for all its efficacy (and just plain sex appeal), exotic conferencing technology such as video conferencing decidedly failed to catch fire in the 1990s. Even towards the end of the 1990s, though video conferencing began to be heavily used in certain niche markets (e.g., telemedicine) and in certain high-end applications (e.g., boardroom-to-boardroom), general market acceptance was elusive.

Now, however, conferencing – and video in particular – has hit, with the crucial watershed of mass business market acceptance having occurred during 2001. And as with certain other "obvious killer apps" that have somehow failed to catch on with the public (e.g., unified messaging), the lever enabling video conferencing's big push will be rendered by the Internet (and its attendant economy, logistics, and culture) which provides a technology platform and a business environment wherein video and other forms of high-level conferencing will thrive. There are actually many reasons for this:

High bandwidth is becoming available everywhere. Most forward-looking companies have already implemented broadband Internet access from discrete locations. Inter-office communications are starting to switch from low-bandwidth Switched 56, expensive leased T-span, frame relay, and similar technologies to more-rational "IP everywhere" VPN technologies. Consumers, too, are beginning to move towards broadband IP, as cable and DSL service-offerings become available. The result: Increasingly, any two points you might wish to connect are likely to be served by cheap, plentiful IP bandwidth.

Quality of Service problems are being addressed. The effort to provide QoS for IP voice is forcing swift implementation of network elements laden with sophisticated technologies such as priority-marked packets, weighted fair queuing, FECN / BECN, RSVP, and similar protocols and strategies. Video – similar to voice in that it demands effective isochronicity for proper transmission – will be a direct beneficiary.

The Web is everywhere. Conferencing is already benefiting hugely from Web standards – both in terms of the Web providing a ubiquitous user interface to conferencing systems, and in terms of its facilitating official and *de facto* standards (e.g., RealVideo) that enable certain types of conferencing.

Mobility is still on the rise. More and more work is done out of the office. The demand for "work at home" privileges is increasing, since urban areas – particularly landmark buildings – may become the scene of terrorist activities. Also, it saves money. Video and other conferencing solutions play directly into these growing trends.

E-commerce is changing the call center. The rise of e-commerce underscores a dramatically increasing need for companies to provide remote customers with efficient means of obtaining service. Enhanced conferencing provides a global solution-set that can exploit the e-commerce infrastructure directly: From customer Web browser to server to call center.

Next-gen "B to B" will use video. Net technologies such as the eXtended Markup Language (XML) are profoundly changing the way businesses communicate with one another. Older, less flexible techniques such as EDI are rapidly falling by the wayside in favor of easier-to-manage, swifter-to-implement solutions. It appears that "B to B video" will soon emerge as a key component of any technologically-enabled business partnership, and will eventually be viewed as ***de rigeur***.

The "telecom culture" is changing. Perhaps the most important benefit the Internet (and the broader concept of "convergence") confers is in changing the nature of telecom management – both in corporations and among service providers. In the past, voice- and PSTN-oriented telecom managers have been confused and put off by the data-centric demands of video and other forms of high-level conferencing. Telcos, by the same token, have only slowly started learning how to package and present data-centric forms of connectivity – mostly in response to the sudden demand for Internet-related services. Today, as the function of telecom management comes more and more under the aegis of IS, the technical challenges of video will be easier to overcome. At the same time, the benefits of video will likely be apparent to IS managers, schooled in thinking of all forms of connectivity as strategic assets, rather than simply as "cost centers."

Standards are now in place. Several signaling protocols have been developed by different groups of standards organizations to satisfy the need for real-time session signaling over packet-based networks. Each of the protocols has differing origins and supporters (each having their own agenda). Still many, such as H.323, T.120, SIP, MGCP, Megaco/H.248 and their related standards, are now solid and widely available. For example, video conferencing began to mature not only as higher bandwidth connections became available, but when call control protocols began to appear as well as special signal processors and/or software capable of compressing the video and speech to manageable bandwidths. Such a compression / decompression algorithm is called a codec (COmpression-DECompression). Codecs can fit on plug-in PC boards or can be run as software on host PC processors if they're fast enough. An additional I/O (input-output) board or NIC (Network Interface Card) connects the video conferencing system with the networks of the outside world: the

PSTN, ISDN, ATM, or Frame Relay. Or the system may just be connected to an Ethernet LAN Intranet network.

The H and G "Standards"

Equally important for the success of packet based audio-, data- and video conferencing is for different manufacturers' codecs to interoperate or "talk to each other." This interoperability problem existed long before the rise of packet-based networks and was a major bone of contention in circuit-switched environments. Getting manufacturers around the world to "march in step" technologically involved much deliberation by the world's two leading communications standards organizations, the International Telecommunications Union – Telecommunications Sector, or ITU-T (www.itu.int/ITU-T) and the Internet Engineering Task Force or IETF (www.ietf.org).

H.320 and H.261

Video conferencing was "about to appear on everyone's desktop" in December 1990 following ITU-T's approval of the first two standards for video interoperability over ISDN lines, H.320 and H.261, but the expensive nature of the equipment and lack of usability slowed adoption.

Whereas H.320 defines a standard for simple image transfer for "narrow-band visual telephone systems and terminal equipment" (ISDN-based), H.261 provides the picture format, error correction, and a standard discrete cosine transform compression algorithm for data compression. It also acts as an intermediary reconciliation entity when different equipment and software must interoperate.

H.261 is also called ***p*64*** (pronounced "p star 64") since it defines video signal compression / decompression in multiples of 64 Kbps (digital transmissions over modern landlines are allocated bandwidth in increments of 64 Kbps, the size of one digitized voice channel) where p = 1 to 30 or bandwidth equals 64 Kbps to 2.048 Mbps.

H.320 and H.261 are but two of what the ITU-T has defined as the ***Series H Recommendations*** for "Line transmission of non-telephone signals." This encompasses what we call video conferencing. Other "H" recommendations include H.221 framing, H.233 encryption, H.323 for real-time packet-based multimedia communications over LANs and the Internet, H.324 for crude multimedia communication over ordinary analog telephone connections, and H.242 for setting up and disconnecting audiovisual calls using digital channels up to 2 Mbps in bandwidth. Although called "recommendations" these technical specifications can for all intents and purposes be treated as standards.

By building H.320 and H.261 compliant equipment, for example, different manufacturers' video conferencing systems could communicate with each other if a point-to-point connection was set up similar to a telephone call, relying on a dedicated circuit-switched ISDN connection. Big, expensive conference room and giant screen-type multipoint solutions immediately adopted and have followed the H.320 spec religiously ever since.

Moreover, much of what was called "video conferencing" in the early and mid-1990s became associated with multiple-channel ISDN circuit-switched networks and relied upon the ITU-T's H.320 / H.261 video encoding standards. Indeed, to this day, among corporations and US government offices with legacy equipment, "business quality" video conferencing is still delivered by H.320-compliant systems using a 384 Kbps bandwidth delivered by a triple Basic Rate Interface (BRI) ISDN line, which is a combination or "aggregation" of six 64 Kbps "Bearer" or "B" channels combined with an inverse multiplexer and is variously called 6B, 3xBRI, or "switched 384."

Before the rise of packet networks, the "business quality" video conferencing made possible by combining six ISDN B channels was (and is) called the ***Common Intermediate Format*** (CIF) or ***Full CIF***. It consists of 30 frames per second (fps) of a 352 x 288 pixel frame or 15 fps of CIF when only two B channels (a "2B bandwidth") are used. In comparison, a commercial NTSC compliant television in the US displays a full video frame as 525 lines, while a PAL compliant television in Europe displays video as 625 lines; thus, video conferencing images are all smaller than current commercial television sizes, and television images themselves are smaller than current workstation screen sizes which are commonly on the order of 1200 x 1000 pixels.

Higher bandwidth systems are capable of supporting ***Super CIF*** (SCIF) format (704 x 576 pixels). The H.261 framing protocol also supports a lower resolution ***Quarter CIF*** (QCIF) format of 176 x 144 pixels. The bandwidth saved by sending a low resolution image means that QCIF can be transmitted at a higher frame rate than CIF. In fact, video conferencing frame rates can be as high as 30 fps and as low as zero fps, depending upon available bandwidth and network congestion. The frame rates available under H.320 are 7.5, 10, 15, and 30 fps. Needless to say, the more frames per second sent, the more fluid and natural the motion of the image (just like commercial television), but this means that more bandwidth is necessary. One way to conserve video bandwidth is by using ***motion compensation***, a way of reducing the amount of video data sent by encoding only the areas of each frame that have changed since the previous frame.

In packet-based conferencing systems, sound is transmitted using audio protocols that are codecs designated under the ITU-T's so-called "G" recommendations. For example, G.711 uses between 48 and 64 Kbps of bandwidth. It's rather wasteful as it doesn't compress voice very much, but it does provide excellent telephone quality sound and is the most essential audio codec required for a video conferencing system to be considered either H.320 or H.323 compliant.

G.722 also needs between 48 and 64 Kbps but also offers stereo sound.

G.728 uses only 16 Kbps, which means that there's more bandwidth available for video.

G.729 is a speech codec that can compress human speech to only 8 Kbps, but the sound quality starts to deteriorate at these levels.

H.323

H.323 is an ITU standard for real-time multimedia audio and video conferencing over IP that specifies the technical requirements enabling a complete "call" or communication sequence. H.323 was originally developed in the Enterprise LAN community as a video conferencing methodology and has much in common with ISDN signaling protocols such as Q.931. Indeed, ISDN-based H.320 served in many ways as the foundation for H.323's more advanced features that relate to packet-switched networks (e.g. the Internet) and improvements in compression and signaling methods, such as the H.263 standard algorithm for video compression. Just as H.323 builds on H.320, H.263 builds on H.261, has a more aggressive compression algorithm and is optimized for low bit rates. At any given bandwidth, a packet-based H.263 compliant system displays better picture quality than does an ISDN-based H.261 compliant system (still, if packet bandwidth is available, H.263 can be used as an option under H.323).

So, around 1996, with the rise of packet-switched networks, the H.323 interoperability standard became recognized as an important emerging standard for video conferencing over packet networks that don't normally guarantee a specifiable level of Quality of Service – networks such as the Internet as well as existing WANs and LANs. Since these packet-switched network technologies (TCP/IP and IPX over Ethernet, Fast Ethernet and Token Ring) have always dominated corporate desktops, it was (and still is) thought that conference room-sized audio and video conferencing systems would eventually be replaced with one-user, desktop PC-based systems.

H.323 products have been developed on various hardware and software platforms. An H.323 protocol stack is software that can be run as part of a computer operating system or the protocol can be run on single hardware devices.

Just as H.323 is part of the "H" family of real-time communication protocols developed under the auspices of the ITU-T (the H.32x family), H.323 itself is not an individual protocol, but an "umbrella" standard – a complete, vertically-integrated suite of protocols associated with each of the conferencing network's components that serve as building blocks for assembling collaborative multimedia communications applications. Amongst others, H.323 uses the following protocols:

Q.931 – call setup and termination.

H.225 – call signaling.

H.235 – security and authentication.

H.245 – exchanges terminal capabilities and negotiates media channel usage.

H.261, H.263 – video codecs that compress and decompress media streams.

RAS – manages Registration, Admission and Status information exchanged between H.323 endpoints.

RTP / RTCP – sequences audio and video packets.

G.711, G.712, G.723.1, G.728, G.729 – audio codecs.

T.120 – data conferencing. Being a data conferencing protocol, T.120 is the most important transmission protocol standard for document conferencing over transmission media ranging from analog phone lines to the Internet. It's network independent, so it will work over the Internet, the PSTN, LANs, ISDN, etc. T.120 can do conference control, application and data sharing. T.120 didn't initially cover application sharing *per se*, which is still sometimes handled differently by various manufacturers.

To set up a call over a packet network, H.323 juggles the use of these protocols. Initially, a supported client queries an H.323 gatekeeper (see below) for a new user's address. The gatekeeper retrieves the address and forwards it to the client, which then establishes a session with the new client using H.225. Once the session has begun, the H.245 protocol negotiates each client's available functions. Since H.323 must first establish a session before it negotiates the features and functions of that session, call setup is not an instant process by any means – the length of the delay depends upon the type of network being used.

H.323 calls use media streams that are transported on RTP / RTCP. RTP. The ***Real-time Transport Protocol***, is what carries the actual media; it "stamps" real-time packet data such as video and audio, giving it a higher delivery priority than ordinary "connectionless" data. It also allows out-of-order packets to arrive in the proper sequence, at the correct time, and synchronized between streams (such as audio and video). As for RTCP (***the RTP Control Protocol***), it is a control channel that carries status and control information via TCP (so that the quality of data delivery can be monitored) involving large multicast networks.

The H.323 architecture also defines four infrastructure elements that have become firmly established and have influenced the design of various types of packet-based audio and video conferencing systems: ***terminals, gateways, gatekeepers,*** and ***multipoint control units*** (MCUs). The gatekeepers, gateways and MCUs are logically separate components of the H.323 architecture but can be implemented as a single physical device. Here's what each component does:

The Terminal. This is the desktop PC or other endpoint at which the communications line ends, or is connected to other endpoint circuits of a network.

The Gateway. Your audio or video conferencing system will be confined to one type of communications network (analog, ISDN, IP, ATM, Frame Relay, etc.) unless you have a special endpoint called a ***gateway***. Gateways enable conferencing to be extended beyond any particular network and can integrate various network resources into a single flexible platform. Essentially, gateways handle "media streams," or what information is actually transmitted and received over a communications line. They translate protocols, convert media formats and transfer information.

For example, while ISDN connections are used at the terminal endpoints in older H.320 systems, some video service providers such as Sonic Telecom (Chantilly, VA – 703-818-0057, www.sonictelecom.com) terminate "last mile" ISDN connections at the central office's switch, handing off the calls to a high-speed Asynchronous Transfer Mode (ATM) gateway, which yields more stable bandwidth (meaning no dropped ISDN channels or other glitches).

In later packet-based systems, gateways link H.323 endpoints on packet-switched networks to endpoints in the PSTN and other networks. Gateways can perform call setup and clear on both sides of an IP-to-switched-circuit connection. Early H.323 systems were not interoperable with the large installed base of ISDN H.320 systems. This required a gateway to interconnect H.320 and H.323 systems. Because many existing video conferencing systems are still using ISDN connections, gateways will continue to be a vital part of any IP-centric conferencing network.

And now, the continued rise of H.323 and packet-switched networks has fostered the phenomenon of ***convergence***, which changes the traditional role of the gateway. For many years some conferencing systems involved the coordinated use of separate networks. The PSTN was used to transfer the voice portion of the conference, a data network transferred the files, and the higher bandwidth ISDN lines handled the video portion of the conference. Real-time IP-centric conferencing allows voice, video and data communications to be handled by one system over one network, such as the Internet or a company's Ethernet LAN. This is possible because H.323 also supports an option for data sharing applications (such as file transfers and application sharing), all of which must comply with the T.120 standard. Indeed, T.120 is the default basis of data interoperability between an H.323 terminal and other H.323 (packet), H.324 (POTS), H.320 (ISDN), or H.310 (ATM) terminals.

The Gatekeeper. Gateways that convert media streams between different kinds of circuit-switched networks (or between legacy circuit-switched networks and packet-switched networks such as the Internet) should not be confused with ***gatekeepers***, the "brains" of an H.323 communications system. Gatekeepers process calls by tracking and managing call progress, provide call transfer and forwarding, maintain call detail records (for billing), as well as handling conversion between addressing schemes so that the confusing numbers of an IP address can be read as a user-friendly LAN directory alias, allowing "e-mail-like" or "phone number-like" names to be used. Endpoints register with the gatekeeper and request permission to place a call to another endpoint. If granted, the gatekeeper returns the transport address for the call signaling channel of the called endpoint. One or more gatekeepers may reside anywhere on the network. They can be standalone devices, client software running in a desktop computer, or fully integrated into another networking device, such as a gateway. A gateway and a gatekeeper could both be housed in a server, for example.

The Multipoint Control Unit (MCU). Whereas simple continuous presence video conferencing handles point-to-point media streams between two sites, a multipoint video conference among three or more H.323 or SIP endpoints generally necessitates use of a "video bridge" or multipoint control unit (MCU), which is a software-based or hardware-based media server that enables group video conferencing similar to the way that a voice bridge enables a large audio conference. The most advanced of these bridges run unattended and allow users to simply "dial" a number, use a Web browser or otherwise connect to the unit and it will automatically set up the conference. Passcodes can be provided so that only the designated participants can join the conference.

The MCU generally consists of a multipoint controller (MC) and a multipoint processor (MP). The MC is the conference controller, handling conference setup including negotiations with all conferencing terminals to determine their common capabilities and the opening and closing of channels for the audio, video and data streams. The MC doesn't directly process any of the media streams, that job is done by the MP. Running under the control of the MC, the MP centrally processes audio, video and/or data. It mixes, switches, transcodes and in whatever other way necessary processes the streams controlled by the MC. "Switching" by the MP means that a certain data flow is sent if several data flows are available (for example with the matching video sequences, if the current speaker in a conference changes via voice activation, or if a change is requested by a participant non-audially via the H.245 protocol). "Mixing" allows several data flows to be brought together so that a single image can be created that's split into several segments, with each segment showing a different speaker. This single composite image is then re-coded to be sent back out to each participant.

One MC is mandatory while one or more MPs are optional. An MP is usually combined with an MC in a single MCU device, but one or more MPs can be placed throughout the network – for example, a standalone MC could be running a multicast conference as each terminal in the conference mixes the audio.

MCU controlled multipoint conferences can be held in lecture mode, where a principal speaker is in control of the conference, or the conference can be voice activated, so each participant can interject comments at will. Conference control can be done out-of-band, where, for example, a Web browser is used to set up and control a conference, or it can be done in-band, or in the same channel used for the conference streams, wherein the MC component would use a protocol such as H.243 for conference control.

One of the best MCUs for video conferencing is the remarkable MGC-100 MCU from Polycom (Milpitas, CA – 800-765-9266, www.polycom.com). When 1,000 people in the universities of Sweden, Finland, Russia, Norway and Lithuania decided to attend the TechNet Baltic 2001 virtual conference in Stockholm, Sweden, the many Polycom ViewStation 512 and VS4000 group video communications systems used in the conference were tied together thanks to an MGC-100.

The MGC-100 can function as a multimedia processor and a H.323 (IP), H.320 (ISDN) or H.321 (ATM) gateway. It allows sites with different frame rates, connection speeds, audio algorithms, screen resolutions, and network protocols to transparently connect with each other. Each user can be configured with port bandwidths ranging from 56 Kbps to 2 Mbps. The unit is designed to incorporate new standards as they emerge and it will interoperate with multiple MGC systems.

This super-flexible MCU is also super fault tolerant. Aside from its remote on-line diagnostics, all modules are front-accessible, hot-swappable and self-configuring. Versions with up to three hot-swappable, load sharing power supplies and NEBS compliant chassis are available.

Some conference room units by Polycom and Tandberg have a built-in MCU / conference bridge. Gateways, gatekeepers and an MCU can be made part of the same server, and future versions of such technology will undoubtedly handle multiple protocols to achieve interoperability among different kinds of conferencing terminals. Additional IETF protocols appearing in packet-switched audio (and soon video) conferencing are the Session Initiation Protocol (SIP), Media Gateway Control Protocol (MGCP) and the Media Gateway Control / H.248 (MEGACO / H.248) protocol.

The Session Initiation Protocol (SIP)

A relatively new and evolving protocol that competes with H.323 is the *Session Initiation Protocol* (SIP). Whereas H.323 and T.120 are standards formulated by the ITU-T, the ascendancy of packet-switched Internet-friendly protocols such as SIP have come under the provenance of the Internet Engineering Task Force (IETF) – specifically, the Multiparty Multimedia Session Control (MMUSIC) Working Group within the Transport Area of the IETF.

SIP is an application-layer control (signaling) protocol that can do "call control" or establish, modify and terminate multimedia sessions or calls. These multimedia sessions include multimedia conferences, Internet telephony, distance learning, and similar applications. Since SIP can be used to control Internet multimedia conferences, Internet telephone calls and multimedia distribution, it is used increasingly in both the core and the periphery of communications networks.

SIP includes the following features.

- SIP invitations are used to create sessions and carry session descriptions that allow participants to agree on a set of compatible media types. In this way SIP is not restricted to any particular media type, and can therefore handle the expanding range of media technologies.
- SIP enables user mobility through a mechanism that allows requests to be proxied or redirected to the user's current location. Users can register their current location with their home server.
- SIP supports end-to-end and hop-by-hop authentication, as well as end-to-end encryption.
- SIP can be used to initiate sessions as well as invite would-be participants to sessions that have been advertised and established by other means. Sessions can be advertised using multicast protocols such as the session announcement protocol (SAP), electronic mail, news groups, Web pages or directory protocols such as the Lightweight Directory Access Protocol (LDAP), among others. SIP can invite people to sessions that are multicast using a multipoint control unit (MCU), or initiate multi-party calls using a fully-meshed interconnection (where every system is literally connected to each other and has bidirectional communication to every other), or a combination of multicast and meshed unicast participants. Internet telephony gateways connecting parties via the Public Switched Telephone Network (PSTN) can also use SIP to set up calls between them. The conference initiator doesn't have to be a member of the session to which it is inviting participants. Media as well as participants can be added to an existing session.
- SIP is independent of the lower-layer transport protocol, which allows it to take advantage of new

transport protocols, such as the Stream Control Transmission Protocol (SCTP), a reliable transport protocol that offers acknowledged error-free non-duplicated transfer of datagrams (messages) and operates on top of a potentially unreliable connectionless packet service such as IP.

- SIP can be extended with additional functionality.

SIP reuses many familiar Internet elements and is, more or less, equivalent to the Q.931 and H.225 components of H.323, the protocols that take care of call setup and call signaling. Consequently, both SIP and H.323 can be used as signaling protocols in IP networks for initiating interactive communication sessions between users, but SIP is a more lightweight and general-purpose, text-based protocol based on HTTP (H.323 is more complex and slower to set up a call than SIP, since it contains more code and it requires more messages to set up a call).

H.323 was initially designed for interactive multimedia communication, while SIP started out as something to handle Voice-over-IP (VoIP) applications. SIP doesn't offer conference control services such as floor control or voting and doesn't prescribe how a conference is to be managed, but SIP can be used to introduce conference control protocols and can be used in conjunction with other call setup and signaling protocols. In such a context, an endpoint system uses SIP exchanges to determine the appropriate end system address and protocol from a given address that is protocol-independent. For example, SIP can determine that a party can be reached via H.323, obtain the H.245 gateway and user address and then use H.225 to establish the call. Or, video and Internet telephony gateways can also connect parties on the PSTN using SIP – SIP can determine that the called party is reachable via the PSTN and indicate the phone number to be called, perhaps along with the Internet-to-PSTN gateway that could be used.

A SIP entity may operate in one of the following modes:

- A **User Agent** is the end-point of a SIP call. It initiates SIP requests as instructed by the user and, on receipt of a SIP request, contacts the user and responds to the request on their behalf.
- **Proxy** is used to route requests and enforce policy or firewalls. It accepts requests on behalf of a user and passes them on, modified as necessary, to the user.
- **Redirector** may be used to provide user mobility. A Redirector accepts SIP requests and returns zero or more new addresses that should be contacted to fulfill the request. A redirector does not initiate SIP requests or accept SIP calls.
- **Registrar** accepts registration requests. These enable users to update their location and policy information as may be used to provide user mobility.

SIP is now starting to overshadow its main competitor, H.323. As Level 3 Communications' Senior Architect, Jon Peterson, once quipped at an industry conference in 2001: "Regardless of what technology carriers are using to do VoIP today, no one has H.323 on their long-term roadmap." Pundits who, not long ago, declared that H.323 and SIP would share the spotlight for years now view H.323 as a legacy technology – good enough to keep using while we work the bugs out of the business models, but slated for replacement by SIP as soon as things pick up.

At the same time, development of and investment in SIP blossomed during 2001, despite market reverses. During the 2000-2001 period alone, two major US carriers – Level 3 Communications and WorldCom – announced commercial availability of SIP-based VoIP services on their networks, and they have aggressive plans to expand their reach in months to come. To varying degrees, other providers, including telecom giant Qwest, are allocating near-term resources to the deployment of

SIP infrastructure and SIP-based applications. In the vendor community, the protocol has seen a groundswell of support with Microsoft's inclusion of a SIP User Agent in Windows XP, and Microsoft's next generation of instant messaging clients will also be based on SIP. Microsoft's move will put the protocol right on the desktop of almost every Internet user (and potential audio and video conference participant) in the world.

Although H.323 appeared first, it started out as a sort of "subscriber protocol" used for office-to-office connectivity within enterprises and for IP telephony hobbyists setting up individual IP telephony calls or conferences using such applications as Microsoft's H.323-based NetMeeting. SIP not only scales up better, but it's so "lean" that it can be used in lightweight clients for mobile devices such as Personal Digital Assistants (PDAs).

MGCP and Megaco / H.248 enable Decomposed Gateways

The IETF's ***Media Gateway Control Protocol*** (MGCP) and ***Megaco / H.248*** (developed by both the IETF and the ITU) are complementary protocols to both H.323 and SIP. MGCP and Megaco / H.248 were developed under heavy influence by the telco community in order to address the issue of integrating VoIP with the conventional PSTN's switch-to-switch trunk signaling network, called Common Channel Signaling System 7 (SS7 or C7). The SS7 standard defines the procedures and protocol by which network elements in the PSTN exchange information over a digital signaling network (the SS7 network) to perform wireless (cellular) and wireline call setup, routing and control. In short, SS7 makes possible the modern Intelligent Network (IN). The ITU definition of SS7 allows for national variants such as the American National Standards Institute (ANSI) and Bell Communications Research (Telcordia Technologies) standards used in North America and the European Telecommunications Standards Institute (ETSI) standard used in Europe.

The "convergence" phenomenon has led to a certain popular way in which calls move back and forth between the PSTN and packet networks. Called ***packet tandem bypass***, or ***transparent trunking***, if one or both communicating parties are using a standard telephone attached to the PSTN but the call must ultimately be transferred over a long-distance IP network, then the call must "hop on and hop off" from the PSTN to the IP network and back again. Thus, two IP telephony media gateways must interoperate with two PSTN switches. For large installations, this transition is done using SS7 in a signaling gateway that provides features such as call setup / teardown and database lookups. Since the H.323 initiative had evolved out of the world of LAN technology, it was incompatible with the PSTN and SS7. Having to contend with multiple gateways and the SS7, the H.323 architecture simply couldn't scale up to public network dimensions.

To address this problem, the ETSI's TIPHON (Telecommunications and Internet Protocol Harmonization Over Networks) committee came up with a new concept, which was to "explode" or "physically decompose" the gateway / gatekeeper model, separating the signaling control from the gateway and distributing switching intelligence in the network, thus effectively decomposing the system to its SS7 equivalents. Such a ***physically decomposed gateway*** is therefore a protocol converter whose component functions can take the form of several distinct devices at different locations in the network. Decomposed gateways "converge" legacy telephony networks with an IP-based, next generation network infrastructure. The decomposed gateway architecture was originally proposed by ETSI's TIPHON committee and was then adopted by the IETF.

Unlike H.323, which is founded on a peer-to-peer paradigm where the switching "intelligence" of the network is found in the endpoints, both MGCP and Megaco / H.248 assume that the bulk of the switching intelligence resides in the network itself, similar to the way that the conventional PSTN's intelligence is located in the big switches and in the SS7 signaling network. Carriers should find this attractive since audio and video conferencing endpoint terminals don't need so much software and circuitry, which means that they can be made smaller and less expensive than if they had to function in an H.323 environment. This also permits the possibility of hand-held packet-based devices.

In most decomposed telephony gateway scenarios, there will still be a ***Signaling Gateway*** (SG) which now terminates the SS7 network call control data governing calls that come from the circuit-switched telephone network; the signaling gateway also manages the signaling and control interfaces between packet networks and the more conventional circuit-switched networks. The signaling gateway thus has the greatest concentration of switching intelligence, and so this is where manufacturers spend time and effort ensuring fault tolerance because it's going to control a lot of other equipment.

The signaling gateway works with one or more ***Media Gateway Controllers*** (MGCs), also known as ***call agents, session agents*** or ***softswitches*** (indeed, the signaling logic is now often combined with the MGC into a single unit), which are the "call control entities" that provide the signaling and service capability, handling call processing by interfacing with the circuit-switch network via the signaling gateway and the IP network via communications with an H.323 gatekeeper. The MGC issues commands to send and receive media from addresses, to generate tones, and to modify network configurations. MGCs have lots of ports and are where the codecs reside. Standalone MGCs are less expensive than signaling gateways since they don't have to be fault tolerant – data can be re-routed to another box in case of hardware failure.

The MGCs use MGCP or Megaco / H.248 to control the media logic which is now housed in ***Media Gateways*** (MGs). The media gateways are the "media processing entities" that are part of the physical transport layer of the ***Open Systems Interconnection Model***, or ***OSI Model***, the internationally accepted framework of standards for communication between different systems made by different vendors. The media gateways are used for the call control and processing of media streams (such as voice traffic) coming from circuit-switched networks. MGs packetize the data into IP packets, then deliver these packets to the IP-based packet network. MGs also adapt the packetized traffic (using compression and echo cancellation), create and attach IP headers, set up media paths and send the packets on their way through the network according to the softswitch's instructions. Media gateway technology has been at the center of the "convergence" phenomenon that is allowing voice traffic onto packetized networks. MGCP and Megaco / H.248 derive their technological philosophy from the telco engineering world and are closely associated with the interface between (or the "intra-domain control" of) these "dumb" media gateways and the new generation of "intelligent" MGCs or softswitches.

In essence, then, signaling has been separated into a media gateway controller (MGC), also referred to as a call agent or "softswitch," which in turn controls multiple "media gateways" (MGs). MGCP and Megaco / H.248 are the protocols used to communicate between the softswitch and the media gateways. MGCP and Megaco / H248 are used by the MGC / softswitch to instruct the media gateways to connect streams coming from outside a packet network onto a packet stream. By separating the call control layer from the service and transport layers using a common API across different vendor products, carriers, service providers and developers now have the freedom to mix and

MGCP, Megaco / H.248 Architecture

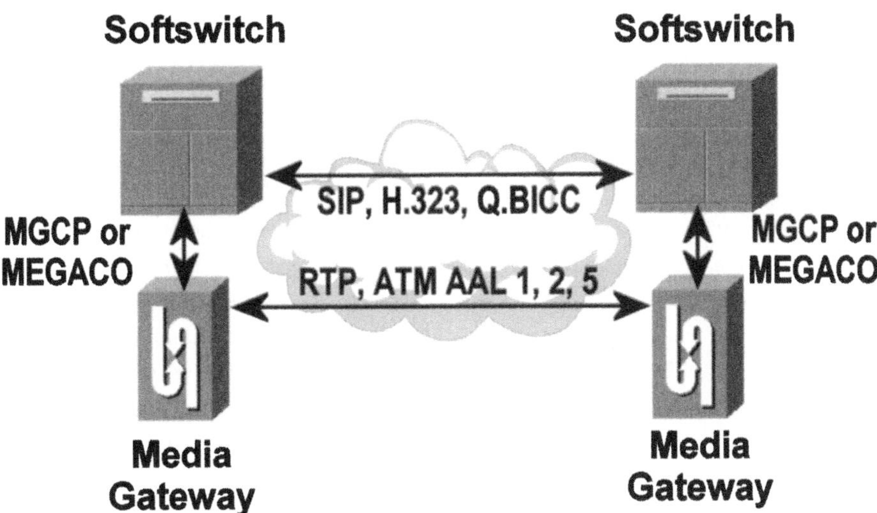

MGCP and Megaco / H.248 communicates from the softswitches (media gateway controllers) to the media gateways but cannot perform all duties in the network. For networks having more than one softswitch, the architecture still needs something like the Session Initiation Protocol (SIP) for communication between the softswitches.

match best-of-breed products and applications.

By mirroring the SS7 architecture and making possible "PSTN over IP" scenarios, MGCP and Megaco / H.248 take on the functionality of central offices circuit-switched networks. This enables an "IP Central Office" that relies on these protocols as control protocols to deliver services across the network via the media gateways. This differs from SIP's (and the Internet's) distributed service model. MGCP and Megaco / H.248 pre-supposes the existence of switching hardware and "dumb" terminal endpoints, while SIP abstracts the signaling layer from the network and pre-supposes "smart" terminals.

Internet-based audio and video conferencing systems will, in the near future, most likely have to deal with these protocols.

MGCP was the first of the two protocols to appear, in November, 1998, as a merger of Telcordia's Simple Gateway Control Protocol (SGCP) and the IETF's Internet Protocol Device Control (IPDC), which was first proposed by Level 3 Communications, 3Com, Cisco, Alcatel, and others. MGCP can be found in devices such as ATM routers, cable modems, and set-top boxes.

MGCP's messages are simple, transmitted in ASCII text, and are designed to allow an MGC / softswitch to fully control a media gateway.

Although MGCP has won wide acceptance in the IP telephony world and was adopted by the CableLabs' PacketCable initiative for establishing telephony services over cable TV networks as well as by the Softswitch Consortium, MGCP didn't take certain multimedia conferencing functions into account. For example, MGCP commands apply only to one connection at a time, not groups of connections. These and other shortcomings led to the MEGACO / H.248 protocol. The specification was first published as RFC3015, MEGACO Protocol Version 1.0 by the IETF in November 2000 and was then tinkered with by the ITU, which calls the protocol Recommendation H.248. Both the IETF and the ITU is now collaborating on what will probably end up being called H.248.

Megaco / H.248 is basically a "superset" or enhancement of MGCP, with many more features. It permits greater scalability than is allowed by H.323, and is capable of supporting thousands of ports on multiple gateways. Megaco / H.248 is also gaining favor because it can handle a broader range of networks such as ATM and technologies that rely on time division multiplexing (TDM) found in conventional telephony networks (another related protocol, Q.BICC, specifies SS7 function mapping for ATM). Even more important, it addresses multimedia conferencing functions left out of MGCP, so Megaco / H.248 can more closely couple with media applications than MGCP and may be a better base for non-media-centric applications, such as MPLS-based session control.

Megaco / H.248 has the same architecture as MGCP and its core commands resemble those of MGCP, but Megaco / H.248 introduces several new concepts (such as Contexts, Ephemeral Terminations, Transactions etc.) that not only provide multiparty multimedia conferencing but also enable sophisticated supplementary services normally enjoyed only in the circuit-switched world, such as Call Waiting. Whereas in MGCP commands apply to individual connections, and in Megaco / H.248 commands apply to ***Terminations*** (streams entering or leaving the media gateway such as analog telephone lines, RTP streams, or MP3 streams) relative to a ***Context*** (a mixing bridge that supports multiple media streams for enhanced multimedia services and helps support multimedia and conferencing calls). Connections are achieved by placing two or more Terminations into a common Context.

A Termination may have more than one stream, and therefore a Context may be a Multistream Context. While a simple call may have two Terminations per Context a conference call might have 10, 20 or more. For business use, it's expected that most audio, video, and data (for example, T-120 shared whiteboard) streams will exist in a Context among several Terminations.

Besides making basic connections by placing Terminations in Contexts, media gateways can generate tones, announcements, ringing, and other signals. Megaco / H.248 can apply signals to Terminations and control them, enabling gateways having Interactive Voice Response (IVR) functionality to be controlled with Megaco / H.248.

Even though MGCP was deployed first, Megaco / H.248 is expected to become the official standard for decomposed gateway architectures endorsed by both the IETF and ITU. As for MGCP, it's doubtful whether the specification will be further enhanced by any of the international standards organizations.

Does MGCP and Megaco / H.248 get along with SIP?

Since SIP is starting to overshadow H.323, one might wonder how SIP relates to MGCP and Megaco / H.248. This is still unclear and is being worked out, since SIP is complementary in certain ways to MGCP and Megaco / H.248 and mutually exclusive in others. As we've seen, MGCP and Megaco / H.248 are device control protocols, where a "master" directs a "slave." The master is a media gateway controller or softswitch, while the slave is a media gateway – this can be connection-aware or session-aware devices such as a VoIP gateway, but could just as easily be an IP phone, a Digital Subscriber Line Access Multiplexer (DSLAM), an optical cross-connect, a PPP session aggregation box or a

MultiProtocol Label Switching (MPLS) router used to speed up data communication over combined IP/ATM networks. This "vertical," centralized, and tightly controlled paradigm is far different than the more "horizontal" peer-to-peer nature of SIP and other Internet protocols where a client can establish a session with another client. A protocol based on this paradigm is needed because while MGCP and Megaco / H.248 are part of a model that mirrors the signaling and control architecture of the circuit-switched Intelligent Network (IN), the protocols by themselves can provide only minimal IN services since they can't be used to communicate between softswitches – they're "internal protocols." For more advanced services, a peer-to-peer protocol such as SIP must be present both in the endpoints and above the signaling network, acting as the service intelligence. To SIP, the media gateway controller looks like another "native" SIP device, just a node with a large number of connections. Similarly, the media gateway is not cognizant that the call between media gateway controllers is established via SIP. Only the media gateway controllers themselves need to understand both protocols.

Of course, with sufficient bandwidth and QoS, one might question the merit of relegating the call control (signaling) on a dedicated general-purpose computer and media processing to a specialized device, along with the protocols that communicate between them. In theory, end devices could be made so intelligent that all signaling could terminate in the media gateway itself. In a SIP phone, for example, the call control signaling runs directly on the terminal endpoint device.

This brings us to the looming problem over where the services themselves should originate in the network. Softswitch manufacturers favor, in the short-term, to place the service intelligence in the softswitch architecture, since during the present "convergence" era all of the action takes place at interconnect points between the circuit-switched and IP networks. In such an interim scheme, SIP application servers reside with the softswitches in the "IP Central Office" with MGCP or Megaco / H.248 controlling multiple media gateways across the network, delivering services to the endpoints.

However, as the legacy circuit-switched network is replaced more and more by IP networks, the interconnect points between the two will lose their importance. A final, pure IP environment will allow for service creation to be distributed throughout the network. Provided that sufficient bandwidth is available and QoS issues are worked out, Application Service Providers (ASPs), ISPs and even end-users could create their own voice-, data- or video-type services. MGCP or Megaco / H.248 would be used only for internally controlling an IP multimedia gateway, while SIP application servers would distribute services throughout the network via SIP proxy servers.

Any way you look at it, conferencing platforms will have to deal with a plethora of hardware and protocols. Media gateways, gatekeepers and softswitches will remain the main components of the "converged" hybrid PSTN-IP network for the near future, which means that they must consistently support the current wide range of protocols in order to operate seamlessly end-to-end with consistent services and reliability.

In the future, not just purchasing, but testing a combination of next-generation conferencing terminals and networks will be a daunting task. That's why, in 2001, RADVision (Mahwah, NJ – 201-529-4300, www.radvision.com) and Spirent Communications (Crawley, West Sussex, UK – +44-1293-767-676, www.spirentcom.com) came to an agreement where Spirent would use RADVision's protocol toolkits to develop solutions for testing VoIP network equipment. Spirent's product is the first of its kind to support SIP, MEGACO, MGCP, and H.323 protocols in a single test platform and to also be the highest capacity solution for testing VoIP equipment on the market.

Radvision's SIP, MEGACO, MGCP, and H.323 protocol toolkits have been incorporated into Spirent's Abacus product line to test functionality and performance of media gateways, softswitches, signaling gateways, and IP clients, such as SIP phones. Already widely used for PSTN stress and feature testing, Abacus is now able to generate VoIP protocol signaling and RTP traffic to determine load handling capabilities and voice quality characteristics of packet-based communications equipment.

Web Collaboration

Another aspect of the rising dominance of IP and the Web in conferencing is its use in what's called ***Web collaboration*** or ***Web conferencing***. Broadly defined, Web collaboration is the modern incarnation of what was once called "audiographics," data or applications sharing. It can mean sharing documents or other visual materials with some degree of further interaction among the parties involved. On the more basic end, you have the ability to share a PowerPoint presentation in sync between two Web-connected PCs. Slightly more complex are applications sharing and collaborative whiteboarding. In many instances, streaming media technologies also play a role, whether they involve audio only or audio and video – investor relations calls, press conferences, and corporate training are all using streaming to reach large audiences.

Increasingly, carriers are offering collaboration apps on an ***ad hoc*** basis to enterprise users. One company that built a whole business around it is Raindance Communications, Inc. (Louisville, CO – 800-878-7326, www.raindance.com). The company was founded in April 1997 as Vstream Inc., changed its name to Evoke Inc. in February 2000, and to Evoke Communications a few months later, only to yet again change their name to Raindance in May 2001. Using a kind of next-gen conference bridge supplied by Voyant Technologies (New York, NY – 212-727-2020, www.voyant.com), Raindance users can set up on-the-fly conferences from a Web site, and combine multipoint audio over the PSTN with visual content shared via the Web. Further, any conference can be recorded and streamed, either in real-time or archived format, to a potentially vast number of individuals. While it may not be loaded with frills and real-time interactive features, such a service is extremely pragmatic: It is low cost, convenient, uses absolutely ubiquitous technology (phone and Web browser), and it integrates processes that are very much a part of the way most any business operates.

As this book went to press, Raindance now offers two services:

- **Web and Phone Conferencing** – Instant, reservationless phone conferencing with optional Web controls allows users to dial out to a new participant, with one click, view participant lists and engage participants with Q&A.
- **Web Conferencing Pro** – With completely integrated phone conferencing, a single-window interface, enhanced application sharing and per-minute pricing, Raindance Web Conferencing Pro has more features and offers more flexible Web conferencing.

Polycom (Milpitas, CA – 800-765-9266, www.polycom.com) offers a somewhat more specialized type of product, though similar in concept, aimed mainly at customers with an installed base of conferencing equipment. The StreamStation, for instance, hooks up to existing audio or video conferencing gear and captures the content of a call for unicast or multicast streaming to any users equipped with a Real Player G2 client. For presentation of documents and visual materials, Polycom makes a system that IP-enables LCD projectors, as well as an IP-based projector that can read Microsoft Office documents.

What these products and services, and several others like them, have in common is that they use the Web and IP as a complement to existing technologies that people use and trust. By making a broader range of content available to a larger group of people, they broaden the channels of communication without demanding fundamental changes in them.

An IP Video Renaissance?

While Web collaboration can be an extremely useful supplement to standard conference calls, informal workgroups, and a myriad of other environments, things are still somewhat problematic when attempting to fearlessly use the Web for effective real-time, interactive video conferencing. At the same time, there has been a good deal of progress in freeing up traditional video conferencing systems from the barriers that have in the past slowed their deployment.

Aside from streaming media, most of the attention has centered on H.323 as a replacement for H.320 and SIP as a replacement for H.323. From the current standpoint, there seems to be no question that IP will eventually replace ISDN as the method for transporting video traffic, and that video will become "just another app" on the converged broadband network. The protocol shift, however, is not in itself adequate to solve many of the concrete problems associated with video conferencing. Rather, much of the progress being made in this direction is happening to some degree independently from the protocol debate, or at least on a course parallel to it.

A good deal of video conferencing endpoint vendors have released "protocol agnostic" products that can support both IP and ISDN. As importantly, however, many of them have introduced products which, even if they are mainly still used for ISDN conferencing, have come significantly down in cost and are much easier to use and integrate. The set-top box, for example, is a viable, inexpensive product that lets even small companies who need video conferencing actually start to use it. Similarly, more intuitive user interfaces and especially browser-based tools have made the process of scheduling and initiating a conference a rather less arcane task.

On the backend, and further along in the network, a number of positive changes are happening. We have already mentioned Polycom's MCU with built-in IP / ATM / ISDN gateway functionality, as well as multi-way transcoding that tailors available bandwidth, frame rate, and compression to individual users' endpoints. But there is also, First Virtual Communications, also known as FVC.com (Santa Clara, CA – 408-567-7200, www.fvc.com), which is building out managed network operations centers to let carriers integrate video as a service offering directly into their broadband networks.

To get to a point where we can have video conferencing "dial tone," with instant connectivity, there is still a good deal of network integration and enhancement that needs to take place. With players like Cisco Systems (San Jose, CA – 408-526-4000, www.cisco.com) getting involved, however, by building the necessary QoS, IP multicasting, and directory services capabilities into their routers and switches, the industry appears to be edging closer to that point of convergence.

Cisco's recent activity in conferencing, in fact, can serve as an interesting roadmap, or at least a rough outline, of the path conferencing is taking in general. On one level, the company is using its IP/TV streaming product to address the particular needs of large IP broadcasting. Thanks to an OEM deal with RADVision, Cisco gained an H.323 MCU and 323-to-320 gateway. Such OEM products are a necessary migratory step – they are the type of product that lets businesses get into video conferencing immediately, and they have the forward looking slant of being IP-based. At the same time, these products are not positioned in isolation, but rather as one part of one layer of an enterprise architecture – what Cisco is calling AVVID (Architecture for Voice, Video, and Data). And, ironically, some of the most interesting aspects of AVVID for IP-based conferencing wouldn't necessarily be recognized as conferencing products at all.

Presently, AVVID centers around an IP-based PBX, which Cisco has named the Call Manager. As it evolves, however, the architecture is becoming much less a phone system and much more a distributed collection of network elements that inter-relate in a modular, rather than a strictly hierarchical, fashion. The whole point of this design is to remove the applications layer of the network from the transport and access layers. And if we accept that in the future, "the network will be the phone system," would it not, therefore, be possible to imagine that the network will also be the conference bridge? Given the potential of IP multicasting, there does not seem any theoretical reason why this should not be.

As IP telephony begins to take root at the enterprise level, we'll probably start to see this hypothesis crystallize first as a tool for simplified, more efficient audio conferencing, and perhaps later as a way of blurring the line between audio and video.

VENDORS AND SERVICE PROVIDERS

Now that you understand the basics of conferencing technology, let's take a look at some conferencing enablers – vendors and service providers.

Audio Conferencing is Still the Workhorse

Video conferencing, be it circuit- or packet-switched, always gets wonderful publicity because its development was foretold decades ago by science fiction writers and so it tends to easily capture the public's imagination. But old-fashioned circuit-switched audio is actually the largest segment of electronic conferencing in terms of number of users. Indeed, although all forms of electronic conferencing increased in the wake of September 11th, the biggest increase was in audio conferencing.

Besides, while you can have a meeting without video, you obviously can't have one without audio. In any case, more and more, people are realizing that they can indeed get their business done with an audio conference.

Audio Conferencing Services

Audio conferencing on the PSTN started becoming popular many years ago, arranged at first by human agents working for conferencing services and now increasingly by automated, "reservationless" Web browser or DTMF-controlled technology, which allows you and your cohorts to set up a conference with little more than a credit card. Usage of these services has been particularly heavy since the September 11th disaster.

Whereas many audio conferencing service providers, such as AT&T cover the whole US or even the world, some specialize in "local" metropolitan areas. For example, BT Conferencing (Boston MA – 866-266-8777, www.btconferencing.com), which is a part of British Telecommunications, has introduced a local area conference call service called 212 Conferencing (www.212conferencing.com). It lets businesses within the New York City 212/646 area codes conference call with other businesses (or their branch offices) within those area codes for 10 cents per minute. Based on technology from Octave Communications (Nashua, NH – 603-459-5200, www.octavecomm.com), the service gives users access to conferencing features available through 800 number conference services that often cost closer to 20 cents per minute.

212 Conferencing users each get a conference calling card with phone number and passcode. The system makes a virtual conference room available to each user 24-hours a day and can connect up to 40 people within the 212/646 area code. The system is reservationless and calls can be set up via touchtone phone. A 24-hour operator is available.

Eventually BT plans to roll out the service throughout other US major metropolitan areas.

One of the most unusual and interesting audio conferencing services is called Mr. Conference. You set up and use Mr. Conference (Miami, FL – 305-503-1841, www.mrconference.com) in a fairly predictable manner: you go to the Web site and register, where you'll be given your own conference number. You can fit as many as 30 people in a conference room.

But there's a twist: conference attendees simply pick up the telephone when they're ready to attend, and dial a long distance number to the bureau. The conferencing element doesn't cost a penny. The only charge a user ever sees is his normal long distance charge from his own carrier.

So where are the strings, you ask? This service bureau is actually a Competitive Local Exchange Carrier (CLEC) that created Mr. Conference as a means of boosting traffic on its network. The CLEC was founded back in 1992, but the bureau has been around since 1999. You have to be 18 or older to use this service, and it's designed for business-to-business communications. If you want to give it a go; it's quite simple to do. There's a sample number to use, which at the time of this publication was 305-503-6666 Ext. 681.

Upon entering the conference "room," you are asked to announce yourself, and (if you're the first one in) you'll hear some pleasing "soft jazz" until another caller joins you there in that virtual room. Admittedly, one might be a little apprehensive about talking in too much detail about important issues since one never knows who is listening.

Indeed, some businesses are uncomfortable using ***any*** conferencing service regularly, which is why audio conferencing is generally under utilized by some members of the business community. Even amongst those who use it regularly, the cost and inconvenience of high-quality options or low-quality alternatives limit use. Most frequent users employ a service bureau to administer the teleconference by initiating and facilitating the conference. Major telcos also provide the service through "conference centers." Both can increase the cost of doing business. The average cost of three-party conferences, for example, can be over $100 an hour, far above the normal call charges for long distance and operator-assisted calling.

Relying on a conference service means that you're renting space on the provider's own conference equipment. To initiate a conference, you contact the service, schedule the time of your conference, and receive a PIN and/or toll-free number, which then must be distributed to the conference participants. This means more calls, more e-mails, more faxes, and more potential for technical trouble and information loss and misinterpretation. That may not sound like too much trouble for a well-run business, but if participants are late, absent, or they cancel, you're spending your money for nothing. Security may also be a concern because you're using the conference service provider's equipment or Web site to share your business information. If you're talking about sensitive information, corporate or personal, who else may be listening?

Still, audio conference service providers keep making improvements to their technology, adding features and lowering prices, and they're more popular than ever before. Recognizing that there are businesses out there who don't want to make the time or dollar investment in a desktop system or a conference bridge, or who want to be able to conference from anywhere, any time, many conferencing service providers have made it as easy as possible for you to conference with their services at your convenience by offering "reservationless" services, a kind of "self-outsourcing," if you will. In most cases you need to become a registered user and provide credit card or billing information, but then you're on your way. Just dial into the service, using a PIN for verification, and you can set up your own conference, dialing out to other parties, or providing participants with dial-in info to let them join the conference at their convenience. No need to deal with operators or make reservations, although usually there is a live attendant available in case you need help.

Customer Premise Audio Conferencing Devices

Recent advances in audio conferencing (as well as a reduction in cost) have made it a practical medium for in-house business communications. Digital Signal Processor (DSP) chips and echo cancellation have improved sound quality greatly. Digital systems are clearer and echo cancellation eliminates "clipping" which causes sound to drop out when someone starts speaking from the other end of the line and cuts into the conversation. Modern audio conferencing systems are also more tolerant of unusual room acoustics, ambient noise (eliminating this via electronic "post-filtering") and the vagaries of the transmission line. Ultrasensitive hypercardioid microphones provide full room coverage, and automatic gain control enhances transmit and receive levels so you always hear clean, crisp voices. New products integrate DSP circuitry, noise reduction, echo control, have full duplex (bidirectional) transmission, and a control panel or integrated keypad, into a single device

Of course, at the bottom of the conferencing food chain, a very simple form of audio conferencing is provided by SOHO units, such as desktop speakerphones. Conference phones are practical solutions if you can get your participants into two rooms. These days nearly all conference phones require absolutely no configuration on the user's part, since they are "smart" enough to learn the acoustics of the room, cancel echo, equalize gain among participants, and control the noise level, all by themselves.

One superlative example of such a device is Polycom's super-sleek, triangular, full-duplex conference phone called the SoundStation Premier. Polycom's models provide superlative sound clarity and are perfect for instant use over standard POTS lines.

It has a 16 character display, Caller ID, a full function Infrared (IR) remote control, speed dial presets, and a 20 digit redial. The unit's automatic microphone mixing intelligently directs one of three console hypercardioid microphones (which perform up to 20db better than ordinary "speakerphones") to whomever is speaking, and can switch among the mics up to 250 times per second, so no words will be missed. The dynamic noise reduction automatically adapts to background noise.

Polycom SoundStation Premier adapts to the acoustic characteristics of any room, with a separate echo canceller for each of its three console mics (the standard Polycom SoundStation, by comparison, has one echo canceller for three mics). Polycom SoundStation Premier's optimized local speech quality has twice the volume of the standard Polycom SoundStation with extremely low distortion and high stability at high volumes. The unit's sub-band echo cancellation maintains top-notch stability, even at maximum volume.

Although most people consider the experience of desktop audio conferencing a strangely formal one, with everyone sitting around the machine in a circle much as one does a campfire or during a séance, the SoundStation Premier has an add-on unit, the Polycom SoundStation Premier Satellite, that expands the system into a more flexible, informal coverage area. The Polycom SoundStation Premier Satellite system adds two powerful rare-earth (neodymium) magnet speakers to the one built into Polycom SoundStation Premier. You can also add seven more directional microphones for broader coverage, and an optional 900MHz rechargeable wireless lavaliere microphone with built in security (it selects one of five separate channels) to prevent eavesdropping and give presenters freedom to move around the room. Other features include convenient console connectivity for a standard analog phone line, an RCA jack that allows you to record calls, and an interface to connect a telephone handset for privacy.

Polycom Audioconferencing

SoundStation Premiere

SoundStation Satellite

Also in this category is The ClearOne Conference Phone from Gentner Communications (Salt Lake City, UT – 801-975-7200, www.gentner.com) another high-quality, low-cost conferencing system for small offices and home offices. It's proprietary VTM technology provides 360-degree dynamic audio pickup, while LEDs track the direction of the speaker's voice. Full-duplex operation allows two people to speak simultaneously.

The ClearOne phone has an ingenious Express Conference button that makes conferencing easy. You just push the button and you'll be instantly linked to your friendly local Gentner conference call operator who does the rest.

Conference Bridges

For a medium sized or larger office, many office PBXs or the more sophisticated "hybrid" key telephone systems provide some teleconferencing abilities. Basic PBXs can provide around six- or eight-party conferencing, but they may limit the number of outside parties to around three.

More extensive multiparty circuit-switched audio conferencing requires the use of a separate ***conference bridge***, which is a device employed either at the customer premises or by a conferencing service that sits behind multiple telephone lines, has several (or hundreds of) "ports" into which the lines are plugged, and connects multiple (more than two) voice calls so that all participants can hear and be heard. In essence, a conference bridge unites and balances multiple voice inputs to achieve the "we're-all-in-the-same-room" effect. While most PBXs and the more advanced key systems have built-in audio conferencing capabilities, they usually don't offer the sophistication of conference bridges, so larger external units are necessary for more scaleable, robust conferences.

If you really want to have full control over the audio conferencing process and avoid high per-conference charges, you can plunk down from a few thousand dollars to $15,000 or more and own your own conference bridge. Today, the biggest reason for choosing to own a bridge over using a conferencing service is that the equipment is rapidly becoming much more affordable. In the late 1990s, for example, Autel (Wellesley, MA – 781-239-8219, www.autel.com) surprised the industry by introducing a basic bridge listing for about $14,950, down from the approximately $70,000 that similar equipment cost a few years before. Prices continue to drop.

If situated at the customer premises, the conference bridge will generally connect to the PBX through a digital T1 or multiple analog lines. Totally digital bridges having little or no trace of analog equipment tend to be better since they offer more flexibility in conference planning. With analog systems, conferences must be bridged in groups of six callers. If a company has seven participants, it must pay for 12.

A separate operator console or PC interface allows an operator to manage conferences, monitoring who's connected to each port and having the ability to add or delete conference members. To enter a conference in progress, conferees generally dial in to a certain number and then input a PIN or reservation number. Conference participants have different abilities (or permissions) depending on their role in the conference. The host (leader, or moderator) guides the conference and can assign these permissions as necessary. Permissions generally include speaking, transmission (for data), listening, and conference entrance/exit. The host may also initiate subconferences (private conversations), participant muting, or even pass the host status to another conferee. Conference bridges also provide security features such as passcodes and secured network transmission.

For example, publicly-held corporations often enjoy conducting shareholder meetings using conference bridges or conference service bureaus. Naturally, different "permissions" are granted to different participants of the meeting; those with the stock numbers having "speak" or "transmission" rights, and those wanting the stock numbers being restricted to "listening" rights. Still, this is a popular and efficient way to pass along information to a geographically diverse group of people at the same time. For such an endeavor, a rather sophisticated and port-heavy conference bridge is used.

Two popular audio conference bridges for businesses are the Confer II and Consortium from Forum Communication Systems, Inc. (Richardson, TX – 972-680 0700, www.forum-com.com).

The Confer product line of digital conference bridges is recommended by the major US PBX and key system manufacturers. Confer II bridges are installed in 45 countries worldwide and are used widely in business, government, and distance education facilities in Third World emerging nations. These products are designed to be both cost-effective and easy to install.

The basic Confer II is a six-party stand-alone unit. For larger systems, you can couple up to four six-party units together using a desktop coupler unit to build a 24 port system. You can also purchase the Confer II six-port board, which installs in an expandable 19" x 15" x 10" card cage. Using a coupler board, you can install eight of the Confer II boards in the card cage to build a 48 port system.

Conference management can be done two ways with the Confer II. The Director option is a dedicated desktop control console that gives the conference attendant / moderator the ability to monitor and control a conference call with simple push buttons and LEDs. It attaches to the Confer II through a communication cable. The Director comes in six or 12 port configurations.

The Confer II conference bridge easily connects to two-wire, loop-start central office (CO) lines, Centrex lines or analog station lines of a PBX or key system, via standard RJ11 modular connectors. The lines are usually programmed into a rotary hunt group in the PBX or CO so that only the lead number needs to be used.

The Confer II detects ringing signals on a line and answers that line. To release a line when a party hangs up, the Confer II requires a disconnect signal on the line (from the PBX or phone company central office). The preferred method is to see a loop current interrupt on the line when the calling party hangs up, a technique commonly used with many voice mail systems. Loop current interrupt is provided by virtually all COs and by most PBXs (except those of Nortel). You may need to install a "voice mail" card or an OPX card in the PBX to achieve this signal.

If a loop-current-interrupt is not provided, the Confer II can be set (by switch selection) to disconnect when North American dialtone appears on the line. This is the method used on Nortel PBXs (and some from other manufacturers) and the PBX typically must be programmed to provide dial tone to the Confer II when the caller hangs up. During one percent of all calls, a participant's voice may sound like dial tone (yes, some speakers can sound that boring!) and this will trick the system into disconnecting in the middle of a conversation. While the bridge has been designed to minimize this hazard, there remains about one percent chance that it will occur.

A third, switch selectable "last resort" disconnect is available if no disconnect supervision is provided. It requires that a line remain silent after the caller hangs up. The bridge will release all lines after two minutes of silence is determined in the conference. Again, if some kind of "coffee break" is taken during a long conference call, the system might think that the conference is over and start releasing lines, so steps must be taken to assure that some sort of sound is "heard" by the bridge during the break.

Forum's more powerful, computer-based audio conferencing system, the Consortium, has, unlike the (purely telephonic) Confer II, a GUI interface that eases the managing, scheduling, monitoring, and recording of conferences. A Windows NT-based 19" wide rackmount PC chassis powered by a Pentium III processor, it connects to the corporate LAN and can use a T1 digital interface to PBXs or the public local exchange network (LEC). Its modular design supports from two to 96 participants / conference ports. The basic system starts at 24 ports; the number of ports can be incrementally increased by plugging in additional eight-port conference cards.

Consortium's LAN Interface connects to any 10Base-T Ethernet network and communicates with desktop clients via TCP/IP. It allows for integration with both LAN messaging tools (such as MAPI compliant e-mail software applications) as well as the Web for Internet and intranet access to conference management functions.

Participants can call in to join a conference by using a touchtone phone and following simple voice prompts (or via intranet/Internet access). The conference initiator can also dial participants separately, adding them to the conference at his or her discretion. Or one can use the blast-dialing feature, which automatically dials participants simultaneously either when the conference is scheduled to start, or when instructed by the conference initiator.

Digital echo is eliminated, and full-duplex audio lets all participants interact spontaneously, without worrying about their conversations being clipped. Unwanted background noise can also be selectively muted, say, if one participant is calling from a noisy area; some minor background noise will remain, though, so you don't mistakenly think the caller has been disconnected.

TEC International (Irvine, CA – 949-250-9400, www.techinter.com) also produces conference bridges for use with PBXs, key systems and Centrex service. Their Meet-Me II (MM-II) unit can setup audio conferences either automatically or manually. In the automatic "meet me" method, callers are linked together by dialing a designated phone number, or an extension through an automated attendant. Manual conference calls, on the other hand, are set up by an operator / coordinator, who places or receives calls to or from the participants and transfers them to the conference bridge. Whichever method is used, the MM-II will answer the calls with an entry tone. The first caller into the system will hear the entry tone and if other callers are already on the bridge, the conversation may begin.

The MM-II base unit includes a primary wall or rack mounted shelf, equipped with common equipment, power supply, a four and/or a five port Line Card, and optional Switched Gain-Equalizer circuit packs for each port. The primary shelf has a capacity of four, five, or nine conference participants. The MM-II may be expanded with two additional shelves to accommodate a system of 13, 14, or 18, to a maximum of 27 participants.

Even more impressive is their MultiTelepatcher II (MT-II), a solid-state, audio teleconferencing system that connects to standard twisted copper wire telephone lines that can handle from eight to 48 lines (expandable in increments of two) in a single system. The unit is PBX / Centrex / hybrid compatible using analog POTS emulation.

The MT-II can handle eight simultaneous group conferences of any size mixture or up to 24 simultaneous two-party conversations.

MT-II can be configured as a stand-alone system, or it can be operated in conjunction with an in-house telephone switching system. The unit has a DB-25 serial port, a Parallel port, an internal analog modem, a Music input port for Music On Hold (MOH) devices, and one Auxiliary input port.

Conference calls can be arranged in advance and participants simply call into the system at a designated time. The conferees can also be contacted individually and transferred into the system when desired.

The system has a built-in automated attendant. Once a conferee has entered the MT-II, they can be greeted by a personalized, recorded digital announcement providing instructions on how to use the system. Each conferee may also be prompted with their two-digit identification code.

Conferees may elect to participate in any one of up to eight separate conferences by following the voice prompt menu. Once a caller is participating in a conference, he or she may exit that conference to join another by using their touch-tone (DTMF) keypad. Entry and Exit tones are optional.

For distance learning applications, there is a "Counselor Console" option.

The use of two-digit ID codes allows two conferees from the main conference to transfer to a private one-on-one conference. The one-on-one conferees can elect to transfer back to the main conference, if desired. Up to eight four-digit group affiliated security codes can be configured and capability for 10 digit Personal Identification Numbers (PINs) is also available.

When the MT-II is used in conjunction with a telephone switching system, callers may use the system to transfer to other extensions or to the operator. The operator can use an inexpensive DOS compatible PC to monitor the system, observe its usage, and moderate the conferences.

QuickTalk from Autel, Inc. (Wellesley, MA – 781-239-8219, www.autel.com) is a compact, lightweight unit that can support up to 24 callers and 15 simultaneous conferences via T1 connection to the PBX. When callers dial into a DNIS, QuickTalk asks the caller to enter a PIN (determined by the

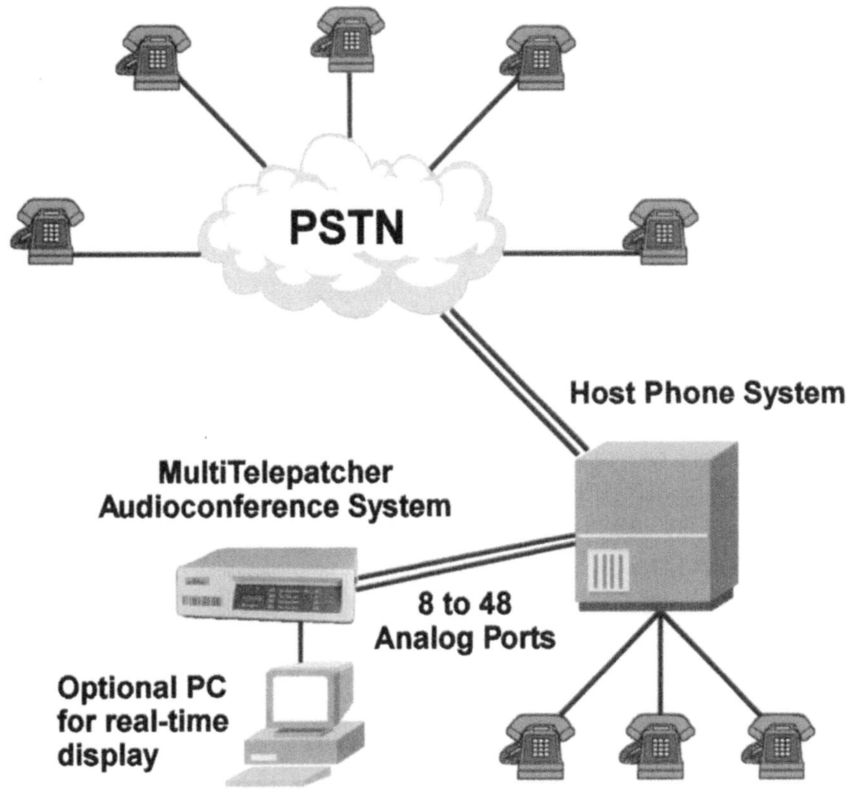

conference organizer), then confirms the correct PIN and plays a message welcoming the caller. Next, it plays a headcount message announcing the number of conference participants. This message is played every time a person joins (or leaves) the conference. Callers are added to the conference and can begin talking immediately.

The unit has few controls. The front panel hosts a series of LEDs – one set for the system, one for the trunk alarms, and one for the alarm status. Its rear panel has trunk connectors for either an RJ-45 or twin BNC jack. It also has a fuse drawer, an earth stud, an alarm port, a diagnostics port, a reset, plus a power button and connection. QuickTalk is easily deployed: connect the unit, turn it on, then conference away.

Compunetix, Inc. (Monroeville, PA – 412-373-8110, www.compunetix.com) makes a nifty line of conference bridges for small to mid-sized businesses. The Mini-CONTEX, which expands in increments of 24 ports to 120 ports, is designed for the enterprise with limited audio conferencing requirements. It supports 12 operator consoles, delivering clear conferencing with digital signal processors (DSPs) that reduce extraneous line and background interference. Operators can enhance voice quality directly from their consoles with individual gain controls. The unit provides real-time diagnostics and local or remote access to the bridge.

For larger enterprises, Compunetix offers the CONTEX 240 and CONTEX 480 which support 24 and 48 operator consoles, respectively (up to 40 in standard configurations), and 240 to 480 conferencing ports, respectively. Line interfaces include T1/E1 (D3/4 or ESF Framing Modes) and ISDN PRI. The bridges' features include preset and meet-me conferences (with digital recording and playback and database-managed reservations), subconferencing, Q&A, and voting. Similar features are available for unmonitored conferencing, including chairperson and participant functionality. If 240 or 480 conferencing ports are beyond the scope of your current needs, you can initially purchase a CONTEX with as few as 24 ports and later add more ports by simply purchasing additional plug-in circuit modules, so there's no real need to buy another CONTEX bridge.

The larger CONTEX bridges are fully adaptable to multimedia conferencing. Users can add data and video to the conferencing by simply replacing the plug-in module. Compunetix's Space Division Switching architecture provides enough bandwidth to handle multimedia conferencing.

Compunetix has also added enhanced security features for unattended conferences to its CONTEX platform. One, the Conference Level Passcode, lets conference hosts create (during the call) a second-level passcode for entry into an unattended conference. Also new is the Conference Vetting Process, which allows the host to secure the conference and permit or deny additional participants. The host can step out of the conference to meet and greet callers who attempt to join a secured conference prior to granting entry.

Gentner (Salt Lake City, UT – 801-975-7200, www.gentner.com) offers a product line that includes the AP400 and AP800 (AP means Audio Perfect).

The AP400 features an integrated telephone interface for natural sound quality between participants. It performs automatic gating of microphones, mic mixing, audio processing, and matrix routing all in one slim, rack-mountable unit. The AP400 can be expanded (using Gentner's G-Link) to accommodate apps that require up to 64 mics. The unit can also handle the audio component of video conferences.

The AP800 performs eight-channel automatic mic mixing, 12 X 12 matrix routing, distributed echo cancellation, and audio control. It expands to 32 lines and 64 mic inputs and provides DSP-based audio. The unit features six programmable presets and allows users to program and operate diagnostics from connected PC (via the front panel) or from a serial remote control. The AP800 supports audio conferencing apps or the audio portion of video conferencing.

Phone System Conferencing Add-Ons

One relatively low-cost approach to audio conferencing is by using an onsite PBX telephone system, which is either owned or leased. Initiated from the desktop, a PBX conference often requires complicated keystroke sequences and frequent returns to the operator manual. It's an inexpensive method that takes advantage of the desktop telephone, but it's seldom capable of more than a three-way conference, and sound quality can sometimes be poor.

Autel's QuickTalk takes advantage of the existing PBX, and eliminates the cost and inconvenience of third-party administrators by attaching an audio conferencing bridge to the PBX. This provides desktop-initiated teleconferencing to each and every telephone within the organization – the instant business solution to automated audio conferencing.

PBX and key system manufacturers have also gotten into devising audio conferencing device add-ons for their own line of products. To illustrate, the Meridian Integrated Conference Bridge (MICB) is a fully integrated, all-digital enterprise audio conference bridge for the Meridian 1 PBX from Nortel (Research Triangle Park, NC – 800-466-7835, www.nortel.com). It allows for plug-and-play

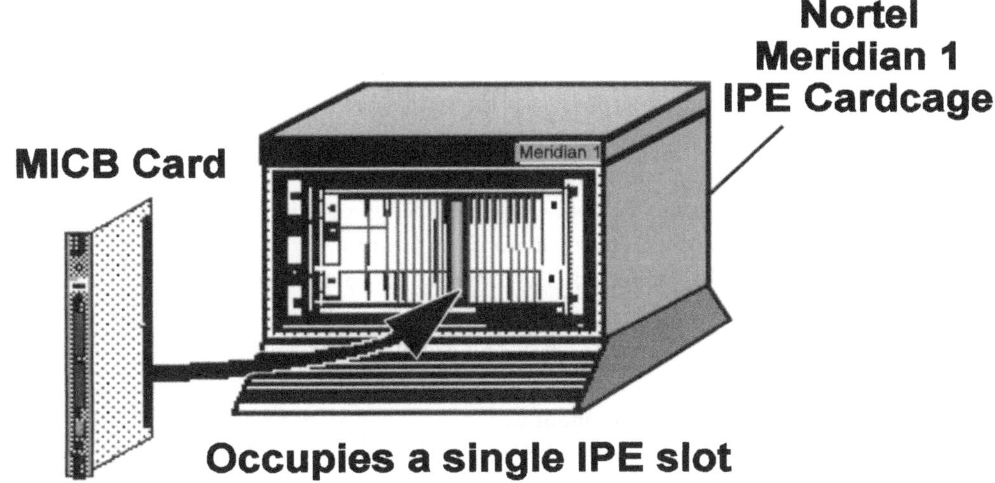

installation within the Meridian's Intelligent Peripheral Equipment (IPE) shelf. Since the MICB provides quick access to an in-house conference bridge, it obviates the need to periodically contact conference service bureaus or retrofit complex third-party conference bridge equipment.

A single MICB card supports up to 32 ports and up to 10 simultaneous conference calls. There are four MICB card capacity options available: 12, 16, 24 and 32 ports. If your conferencing requirements increase, software keycodes can activate additional ports on the MICB card to support the larger port capacities. Also, you can plug in a second MICB card in the Meridian 1 Communications System to create a 62 port system (for configurations beyond 32 ports, a Windows NT server is required).

IP-Centric Audio Conferencing

The Spectel MultiSite application from Spectel (Atlanta, GA – 770-936-9700, www.spectel.com) allows multiple conferencing platforms to work on one network that is accessed locally. It allows up to 100,000 ports distributed over 200 sites. The app uses Spectel's 700, 780 and 7000 platforms. "MultiSite participants experience the ease-of-use that is consistent with traditional conference calling, without concern for capacity limitations and costs incurred due to toll charges and port overdraft charges," said Gerard Moore, president and CEO, Spectel.

When scheduling a call using the Spectel MultiSite application, the moderator selects the global conference option on the Spectel Web Portal or an alternate scheduling tool. Participants join the conference by dialing a pre-defined access number, and are routed to a conference on their local platform. Participants join the call via the local network with a local phone number. If one site reaches capacity, excess traffic is re-routed to another site. If one site stops functioning, all callers will be routed to another site. It accommodates conference access across VoIP or PSTN.

Although Spectel is the leading provider of multimedia conferencing solutions for service bureaus and ASPs, they also wanted to make inroads in the huge broker-dealer community. To achieve this they entered into an exclusive contract with the master reseller in this area, Essential Telecommunications Corp. (Glastonbury, CT – 860-652-9370, www.essentialtel.com). Dave Harding, EssentialTel's VP of Sales, tells the author that, "Our core business is 'Hoot and Holler' and four-wire private networks (see the sidebar box on four-wire circuits) for the broker-dealer industry with customers such as Smith Barney and Prudential."

"Essentially Spectel has a bridge product," says Harding. "You can schedule all of your conference calls through Outlook or through Lotus. You can do a recording of your call. But the tremendous advantage of this Spectel-Multilink product is the versatility of its connectivity. The box can take analog in or four-wire or two-wire, digital, T1, E1, or any array of interfaces. If you have a PBX with a bunch of analog POT phones and you want to own a Spectel system, you can just take all of those trunk ports, plug analog cards in the Spectel and the lines go right in. You can dial in from the outside and participate in a Hoot and Holler circuit if you want to."

"This product is not for everybody," says Harding. "If you do even 50,000 minutes a month through a service bureau, keep using the service bureau. This system is for companies that do really high volume conference calls and want privacy. This product is for companies (such as Smith Barney) that do almost two million minutes of conference calls a month. At that rate, a half million dollar system will pay for itself in about four months."

"Voice-over-IP is hot," says Hardin "and so everybody's talking about doing 'Hoot over IP,' because the four-wire circuits are expensive. Instead of paying for a four-wire private line circuit, you can just allocate some bandwidth, use a Cisco router network and bridge the Hoot over IP packets into a Spectel using a router E&M card, thus bridging all the Hoot and Holler streams centrally

➤ Four Wire Circuits

The two principle types of circuit-switched connections are point-to-point and multipoint.

If somebody picks up the phone and calls you, the result is two phones connected by a single pair of wires. That's a point-to-point connection. Multipoint circuits allow for as many points and users to talk to each other as desired. Only one person at a time can talk but everyone can hear what's said, and anyone can talk as soon as the line clears. But for this to happen a "hybrid" (an induction coil and related circuitry) must be installed at the central office to physically isolate the transmit and receive signals, which means we must employ two pairs of wire instead of one.

This is the technology behind the "hoot and holler" squawk boxes of brokerage firms, where an analyst may scream into a phone that "Company X is folding!" and the words will emerge from hundreds of loudspeaker phones in brokers' offices around the country.

Oil and gas companies also use four wire circuit systems for maintenance and emergency communications, as do railroads. Emergency management agencies and the National Weather Service use them for statewide and nationwide communications. Auto recyclers even use them to link their salvage yards. If a recycler needs a part for his customer, he just picks up the handset on his circuit phone, presses a button, and announces what part is needed into the handset. Every salvage yard on his circuit will hear him, and any one of those recyclers can pick up the handset and provide him with price and delivery information on the auto part.

Four wire systems are generally used on leased four wire private line circuits, satellite and radio repeater systems, in-house or campus private intercom systems, small microwave networks, and even "switched 56" 56 Kbps digital data circuits if multiplexers and voice cards are used (multiplexers can't bridge the point-to-point voice circuits together into one voice circuit, but the voice circuits from the individual point to point circuits can be externally bridged into one large circuit).

where you can now manage them, control sound levels, and reroute them over the matrix if a 'line' goes dead. The box is also a two-wire bridge so if a location fails you can dial it up and bring them back up on a regular two-wire circuit."

"Roll your Own" Audio Conferencing

Most sane experts would not recommend that an end-user build a conferencing product out of board-level components. There are plenty of simple, turnkey speaker-type products and conferencing bridges to choose from. Still, if you have specialized needs and if you've got the talent (or the staff with the talent), you can build your own PC-based audio conferencing system using a voice processing board from companies such as Intel, Amtelco or Pika Technologies. These boards are installed in a server behind a PBX which can also be expanded with other computer telephony resource cards to supply the enterprise not only with audio conferencing capabilities, but also with IVR, voice mail, and fax as well.

Generally, anyone attempting to build a specialized PC-based conferencing system should be well-versed in what components they'll be using, including what operating system, what network interface, what specialized TDM bus, etc.

For example, audio conferencing boards mix sound in two different ways. In ***sample switching***, only the loudest and next loudest voices are mixed. Benefits of this approach are low noise and greater stability at higher volumes. Sample switching models the dynamics of real conversation, where perhaps two people dominate a discussion and the rest listen, until someone starts speaking sufficiently loud to displace the current speaker. Some, though, would simply call this scheme a form of ***half-duplex***, (one-way-at-a-time communication). The other approach is called ***sample summing***, where everyone's voice is mixed. This one has the major disadvantage that noise, from all inputs, gets mixed in. However, some would argue that sample summing is true ***full-duplex*** (continuous bi-directional communication). There are variations on these two approaches, and if you question vendors and service providers closely, you can find how they're doing the mixing.

If you don't already have this information sorted out, you'll certainly need to acquire it before attempting to build something incorporating the latest in switch matrix boards, etc., rather than just buying a system or using a service.

Intel's Communications Systems Products (Parsippany, NJ – 973-993-3000, www.intel.com/apac/eng/network/csp/index.htm) – in particular their Dialogic product line of Digital Signal Processor (DSP) based audio conferencing resource boards – provide for building solutions that range from small party conferences for call centers, small business applications, and chat lines to large systems including collaboration servers, Internet-based audio conferencing systems, and business audio conferencing solutions. These products use SCbus and CT Bus board-to-board signaling to interface with other Intel board products, including network interfaces.

Intel's MSI-Global Series is best for small-party conferences of three to eight, and is suitable for call center application and small business conferences. You can connect eight, 16, or 24 analog telephone devices directly to computer telephony systems using a legacy ISA slot in your PC. The boards' call conferencing supports up to 32 conferees in flexible configurations of two to eight parties per conference. Conferencing resources include broadcast, coaching (where a "coach" conferee is able to speak privately to another "pupil" conferee without any of the other conference participants hearing the coach), and dynamic additions and deletions without annoying training tones.

The boards come with application program interfaces (APIs) for UNIX and Windows NT.

For more modern PCI-bus PCs, Intel's MSI/PCI-Global Series comprise 8- or 16-port analog modular station interface boards that support up to 32 conferees in flexible configurations of three to eight parties per conference.

Finally, Intel's DCB/SC Series are high-density, DSP-based audio conferencing boards that support advanced conferencing features. Each DSP conferencing resource chip supports up to 32 conferees who can be dynamically assigned to conferences of any size up to a maximum of 32 per conference. Thus, the DCB/320SC board with its single DSP chip can handle 32 conferees. With its dual DSPs, the DCB/640SC supports 64 conferees. The DCB/960SC uses three DSPs to sustain 96 conferees.

Larger audio conferencing systems tend to be built on fault resilient CompactPCI (cPCI) computer chassis. Amtelco (McFarland, WI – 608-838-4194, www.amtelco.com) has taken a precise, carefully reasoned approach in building their world-class, high density conferencing board that's easy to use – the XDS Infinity Series H.110 512-Port Enhanced Conference Board.

Amtelco for years has offered popular, simple conferencing solutions in the 100 port range – they started out as add-ons for pre-existing Amtelco MVIP, SCSA and H.100-compliant telephony boards. They're easy to maintain and manage, and there aren't any collections of manual fancy controls that you have to endlessly tinker with that users or technicians have a difficult time controlling and getting right under all situations.

With Amtelco's cPCI conferencing board they took their basic conferencing block of circuitry that's been fully debugged for years, put more of them on a cPCI form factor, then added their standard CompactPCI kernel which gives it cPCI features such as high reliability and hot swap features. As a result, the board comes right out the box being easy to use and dead reliable. It also has a capacity that serves about 95% of conferencing needs found worldwide.

That's not to say that some neat technology is indeed on board, working behind the scenes. For example, the board's enhanced conferencing features not only allow participants to control the board with DTMF touchtones, but the "tone clamping" circuit keeps other conference participants from hearing them.

The XDS H.110 CompactPCI Enhanced Conference board comes in either a 256-port configuration, or a 512-port configuration. Conferences can include a maximum of 64 parties in a single conference, or a maximum of 168 independent conferences on the 512-port board (or 84 independent conferences on the 256-port version). Full-duplex, transmit-only, and monitor-only connections are possible. Transmit/receive attenuation adjustments are possible on a per-input basis, or globally for all inputs of a given conference.

The board has DTMF detectors on every single conference input, so that DTMF tones can be detected, filtered, and reported back to software applications on a per-input basis. Each conference input has energy detectors that monitor relative energy values of each input, and can be used to determine the loudest talker, or the average energy value of all inputs of a given conference.

There's also an analog port, accessible through a front panel audio jack, that can be used to monitor a conference, or provide music-on-hold (MOH).

The board can generate its clock signal internally, or derive it from standard clock signals on the H.110 bus. It currently runs under Windows NT 4.0, Solaris, and Linux operating systems. Software drivers are being developed for other OSes such as Windows 2000.

For a "roll your own" audio conferencing project, take a look at Amtelco's XDS Infinity Series H.110 512-Port Enhanced Conference board, which is super reliable and easy to use.

Audio Conferencing Buyer's Checklist

Whether you buy an external audio conferencing unit, enhance your PBX with an add-on, build your own system or employ an outsourced service, your audio conferencing environment has to serve your needs. While audio conferencing devices have improved, you still need to assess certain features to choose the right system.

Here's a list of tips for evaluating audio conferencing units:

- **Installation**. Is the product quick and easy to install? Complicated installation requires longer setup time and makes the unit difficult to move from one room to another.
- **Training sequence**. How does the unit "train" itself. Audio conferencing terminals that use echo cancellation technology must train themselves to the room and network environments to operate effectively. Some units use a burst of noise to train themselves, which can be annoying to participants at the other end. Others use the voices of the meeting participants to train their acoustic line-side echo cancellers.
- **Important features**. Does the product include an integrated dialpad? Systems that need an auxiliary phone for dialing require extra steps to place a call, as well as additional cabling. Does the unit also include a mute button, recorder jack or switchhook flash key to access PBX or Centrex features?
- **Stability**. Does the unit operate under all calling conditions? Some systems become unstable (echo or "howl") in "low loss" network situations such as intraPBX, short distance calls or over digital networks. Does the system remain stable, even at the maximum volume setting? Units that adhere to the CC/TT standards for audio conferencing terminals can be counted on to maintain their stability under a variety of conditions.
- **Adaptability**. Does the unit automatically adapt to changing room and network conditions without losing stability? Some audio conferencing products become unstable when: an object is placed close to the unit; the unit is moved; the speaker volume is increased; the auxiliary handset is lifted off-hook; different line conditions are encountered during calls over a multipoint bridge. Under these circumstances, some units require the user to pre-train the echo cancellers.
- **Interruptibility**. Can participants at both ends interrupt each other without clipping (loss of the beginning or end of sentences)? Do spurious noises (air conditioner, rustling papers, coughs) cause clipping?
- **Bridge compatibility**. Is the unit compatible with audio conferencing multipoint bridge equipment? Does the unit "relinquish the floor" when no one is talking but the room is noisy? Even full duplex bridges limit the number of simultaneous talkers. If an audio conference system does not give up the floor, other locations may not have an opportunity to speak.
- **Audio quality**. Is the sound emitted from the speaker clear? Can a soft voice on the far end be heard clearly? What is the maximum volume of the unit's speaker? How sensitive are the mics? Can a person walking around the conference room be heard at the other end?
- **Disconnect handling**. If you've never installed a conference bridge before, you might not realize that there are unique disconnect issues (e.g., How does the system know a port has become free and available for another conferencing party?). The thing is that many phone systems do not provide port-by-port disconnect supervision to analog extensions, or if they do, it is in the form of dialtone, reorder tone, or DTMF. In these cases, when one of the parties drops out, the conference call is disrupted, and the other participants must contend with dialtone or telco recordings that force the entire conference to collapse. Audio conferencing equipment manufacturers have solved this industry-wide problem with various kinds of conference bridge-telephone system interfaces.

- **Service and Support.** Does the vendor provide a toll-free number for users to obtain factory support? Are there reputable resellers actively supporting the product? What sort of warranty is available?

Data / Web Conferencing Gaining in Popularity

Data conferencing caught up with audio conferencing technology when the first on-line ***whiteboards*** appeared (such as the Xerox Liveboard), allowing two or more conference participants to collaboratively draw, edit, annotate and save diagrams just as if they were sharing a blank sheet of paper at the same desk. Whereas a traditional whiteboard was a modern version of a blackboard, an erasable board which allows one to draw and write in several colors on a melamine white board, the modern whiteboard is actually a "board" that exists virtually on a PC screen. It's a collaborative software application that allows two or more people to talk to one another, annotate on the "whiteboard" by using the mouse to draw or the keyboard to place text. One can also display photographs and bitmap graphical images for others to view and annotate.

When the T.120 data conferencing interoperability standard was ratified, the concept steadily improved, evolving into "application sharing," or the ability to share a document or spreadsheet and mark up changes. This was first achieved over the ordinary telephone network (the "PSTN") using modems and now happens via the Internet. Thus, "data conferencing" increasingly became "Web conferencing."

In the mid-1990s, data sharing products such as TALKShow from FutureLabs (now discontinued) and PictureTel's LiveShare (PictureTel is now part of Polycom) seemed quite exotic. Today, however, a *de facto* standard for the masses has appeared in the form of Microsoft's free NetMeeting software. There's also Web-based interactive groupware, which is software that allows those with PCs and browsers to surf up to a Web site and collaborate on documents or brainstorm ideas.

Such data collaboration tools and services are frequently – and incorrectly – referred to collectively as "video conferencing" tools, even though many low-end products don't offer video at all – unless you think an animated Powerpoint presentation counts.

Today, Web conferencing has gone from being the hot new kid in town to mainstream usage. It lets you share more information than an audio conference and is not as cumbersome to set up as a video conference. And because even the smallest SOHO is becoming dependent on an Internet connection to keep the business alive, more companies are offering audio- and data conferencing services via the Web because they know everyone has access to it. No muss, no fuss; just cruise to a Web site and you can quickly set up a conference.

Again, most companies require that you establish an account for billing purposes, but that's about as formal as the scenario gets. Essentially, you just go to the conferencing Web site, upload your data presentation (some services provide platform-independent apps for you to create your presentation in case, for some inexplicable reason, you don't have PowerPoint), click on a few buttons and you're in business. Far-end participants can cruise to the site, log in with a pass code, and watch your demonstration. In some cases you can also use a whiteboard to illustrate points in real time as the conference progresses, and even solicit suggestions from participants.

Because of the widespread accessibility of the Web and the populace's growing ease of use of Web-based products, it seems like Web conferencing is going to be around for some time. In addition to the voice and data examples already mentioned, it's also currently used for Web marketing, software application sharing, and e-learning. Areas that seem to be emerging include on-demand e-commerce customer service and ***customer relationship management*** (CRM).

One of the great pioneers in the Web conferencing field is Latitude Communications Inc. (Santa Clara, CA – 408-988-7200, www.latitude.com) which offers MeetingPlace, an enterprise-wide application for desktop voice and data conferencing. Unlike other Web conferencing applications, which are typically hosted on publicly-shared Internet servers, MeetingPlace provides customers with dedicated servers, regardless of whether they deploy MeetingPlace on premises or at Latitude's hosting facility. Some public-shared service providers have hundreds of customers sharing the same servers; a dedicated server makes it much more difficult for hackers and competitors to access meetings and meeting data.

If you purchase MeetingPlace outright, prices begin at about $1,500 for a user license depending on the options you choose. If you deploy it as a service, you pay for actual usage only, with prices beginning at about three cents a minute. The MeetingPlace conference server itself is a rack-mountable piece of hardware that connects to a LAN and the PSTN. Each server has 120 ports, but it can be daisy-chained to provide up to 960 ports.

A notification agent sends participants an e-mail notifying them of a scheduled meeting. Scheduling a meeting takes about two minutes and is a simple combination of GUI prompts. You can access conference functionality through familiar desktop applications such as the corporate Intranet, Microsoft Outlook, Office 2000, and Lotus Notes.

Once in the voice conference, there are more than 30 options controlled via DTMF. You can do things such as use the "breakout" feature and break from the multi-party conference to talk separately with any one of the parties. MeetingPlace allows nine breakouts per conference.

The data conferencing methodology is broken down into four components: whiteboard, collaborate, chat, and share. This allows writers at locations around the world to collaborate on a Word document simultaneously.

In October 2001, Latitude Communications announced the fourth generation of its Web conferencing application, MeetingPlace Web Conferencing v2001. This release provides greater security for corporations, while making Web conferencing even simpler for end users. With more organizations choosing Web conferencing as a means of increasing communication and replacing expensive and time-consuming travel, companies are realizing the need for the security and enterprise conferencing capabilities that MeetingPlace provides.

MeetingPlace Web Conferencing v2001 allows users to configure meeting security for individual meetings. Users can isolate confidential Web conferences behind the corporate firewall while allowing greater access to other meetings that require less security. This new capability, along with the fact that MeetingPlace provides customers with their own dedicated server, further establishes MeetingPlace as the only conferencing solution that meets enterprise security needs.

Does the Future Belong to Video (full multimedia) Conferencing?

True video conferencing, the most advanced (and bandwidth hungry) of conferencing technologies, provides a good facsimile for an interactive meeting, giving each conference participant a personal experience as close as possible to "being there." Many believe that we will all one day have access to some kind of video conferencing capability, even via cell phones and PDAs.

The traditional view of video conferencing was that one group of people would come together in a conference room situated in, say, New York, and would conduct a business meeting with another similarly-convened group of people sitting in a room in, say, Los Angeles. It was generally a big "videophone" connection between two points. Today, however, video conferencing is seen as a more broadly expansive multimedia communications platform, with multipoint connections among many participants, broadcasts, streaming video, and collaborative application environments making possible finer platform "granularity" so that individuals either at their desktops or traveling with their laptops can easily create or participate in information sharing, decision making and applications sharing.

Primitive video conferencing, in the form of a simple "Videophone" that could work between two points, was demonstrated at the 1964-1965 New York World's Fair. However, what we would now consider ***bona fide*** video conferencing was only really made possible when the telecommunications networks could handle the tremendous amount of video and audio information that must be transmitted back and forth between a number of systems simultaneously.

Four Sizes Fit All. . . .

Video conferencing systems come in four sizes:

Room systems are high-end, expensive systems that were among the first video conferencing systems ever used by those large corporations capable of affording them. Room systems are designed for large groups of up to 40 people sitting in a theater designed for a high-quality presentation. A "built-in" room system can reside on shelves behind a facade wall creating a stately, permanent look. Large format displays for both data and video are used along with sound systems capable of several inputs (desk microphones and/or radio microphones) and output via amplifier and speaker system. A programmable movable camera and a document camera typically complete such a system. Although the capabilities of these systems are similar to rollabouts, they often accept a greater variety of peripherals and are more customized to specific applications.

Rollabout systems are complete video conferencing packages contained in a cabinet and typically mounted on a wheeled trolley so that they can be moved from room to room. Still the most popular of corporate video conferencing systems, rollabouts are designed for an audience of 10 or 12 people, thus filling a "sweet spot" between expensive theater-sized conference systems intended for large groups and desktop systems for individuals. Typically rollabouts offer high bandwidth connectivity and have one or two large monitors housed in the cabinet, along with at least one Pan, Tilt and Zoom (PTZ) camera, a desktop microphone, the control system, a software codec, and if required, a data monitor for display of PC applications.

A rollabout's audio system consists of an echo canceller, microphones, speakers and amplifiers. The control system provides the meeting participants with control over the video images, camera orientation, audio levels and other peripherals. The camera in the rollabout can be remotely controlled to select different views of the room. Presets allow you to easily switch between commonly used viewing angles. And there's also usually a graphics or document camera, which is used to share documents, charts, maps, objects and other graphics.

Popular rollabout configurations can use up to two cameras and two monitors. Such systems can transmit and receive two video signals – a motion video signal, usually from the "people camera" at the front of the room, and a captured still image graphics signal, usually from a special graphics camera. A two-monitor system can display the live video on one monitor and the captured still image on another; while a single monitor system uses "picture-in-picture" to display both on one monitor. While the "still graphic signal" is being sent to the remote location, the motion video will briefly freeze until the still image transfer is complete.

Desktop systems are for single users. They used to consist of a camera and one or two boards inserted into PC slots, but now the USB and Firewire (also known as IEEE-1394) ports available on PCs and Apple Macs allow the easy installation and configuration of a small fixed-focus video camera, a microphone, speakers, video conferencing software and some kind of ***digital subscriber line*** (DSL) or LAN connection enables a user to video conference and share data with remote sites. Since the single PC monitor is used for all display purposes, both data and video must be accommodated on the same screen. Although DeskTop Video Conferencing (DTVC) had existed in some form prior to the PC, from 1995 on the introduction of cheap Intel Pentium and Motorola PowerPC CPUs brought sufficient processing power for video conferencing to be on nearly all desktop and portable computers.

Portable Systems. Video conferencing is appearing more and more on laptops and transportable computers. Generally, news photos of President Bush showing him hunkered down at Camp David and video conferencing with other government officials reveal a system resembling a rollabout. When on-the-go, however, Bush and/or his staff use systems from Harris (Melbourne, FL – 321-727-9207, www.harris.com) as well as the Aethra VOYAGER from Aethra (Miami, FL – 305-375-0010, www.aethra.com), a flat screen video conference system built into a briefcase that's a favorite of journalists (CNN correspondents) and people who are ***really*** on-the-go. The VOYAGER handles bandwidths of from 56 Kbps to 384 Kbps, is fully H.320 compatible and supports audio and 30 fps video over ISDN, leased line, or satellite connections. A transparent data channel supports T.120 compliant applications such as Microsoft NetMeeting or it can be used to do background file transfers. The system can be expanded for large groups by connecting additional devices to the unit's input/output jacks.

As advanced third generation (3G) mobile telephones start to appear, some of them will have a little solid state cameras on them coupled with a small full color screens, and will be capable of joining in video conferences. A bit farther out on the horizon is the adoption of SIP by the third-generation wireless community. Late in 2000, the Third Generation Partnership Project (3GPP), the organization responsible for developing technical specifications related to global 3G wireless standards, agreed upon SIP as the basis for signaling in all future, IP-based wireless networks.

With the rise of the Internet in the mid-1990s, major efforts appeared to hammer out IP based video conferencing systems, or "packet-based multimedia communications systems," as they are often called. But in spite of the vendors' acknowledgment that the move to IP is inevitable, and even with able protocols such as H.323 and SIP, not to mention gateways to interconnect legacy H.320 and H.323 systems, the move to IP has slowed over fears regarding unpredictable Quality of Service (QoS). Packet loss, jitter, and latency over the Wild and Woolly Internet are still concerns, with no magic panacea in sight.

Still, as video cameras continue to drop in price, bandwidth increases, and convergence-friendly software and operating systems (Windows XP) appear, IP-based desktop audio-, data- and video conferencing will become the norm. Reasonably good $500 camera and software packages are now running on many PCs, and the act of conferencing individuals instead of groups allows for better "granularity" and flexibility in scheduling, since a special room doesn't have to be reserved.

Yet another advantage: If a video conference participant has a prior commitment, video meetings (and audio ones too) can be recorded and reviewed later. This has been done in the past with VCRs but today video servers, CDs and DVDs can stream back video and audio sessions. Video streaming itself has become a hot topic in the conferencing world, since most employees now have desktop PCs and so can receive efficient "Webcasts" or "videocasts" from their CEO, trainer, news source etc., instead of mere e-mails (the day after the disaster, RealNetworks Inc., the biggest provider of Internet broadcasting video, said 11.4 million people accessed audio and video content delivered through its network by news organizations and companies, as compared to the typical two million users).

One might assume that the era of the big video conferencing room is over, but certain corporate cultures still tend to gather people in groups, huge "high impact" presentations are still done and technological costs have fallen sharply for big room systems too. Although market analysts such as Sarah Dickinson of Personal Technology Research (Framingham, MA – 508-875-5858, www.ptr-corp.com) long ago labeled room-based video conferencing systems "dinosaurs," even the large $100,000+ room system of just a few years ago now costs about a tenth that much to set up, so some years will pass before those big conference rooms are reallocated as office space.

Centralized and Decentralized Multipoint Conferencing

As for the actual network configuration, in the simplest case, all the terminals participating in a video conference can transmit their audio, video, data and control streams to a central MCU, which centrally manages the conference and can send processed multimedia streams back to the participants. This model originated in the early H.320 ISDN conferencing era.

One problem with centralized multipoint conferencing is that the MCU / bridge must receive a lot of video and audio information from all the endpoint terminals, mix all the video into one composite CIF image, then send that image back out to the endpoints terminals of the conference participants. Each endpoint therefore receives just one video image of CIF size that contains the reconstituted, re-encoded images from all other endpoints. On many systems the end result is a composited stream of about four participants at a time, with each smaller video image having only "grainy" QCIF resolution. Also, MCUs from some manufacturers can only support up to around 384 Kbps in a four-way IP-based conference, and if participants are using different kinds of conferencing terminals with different codecs, then the MCU must do ***transcoding***, which means that the video must be decoded, then re-encoded before transmission. Such decoding and recoding of the video increases the "latency" or time it takes for the packets to reach the endpoints, which can affect video and audio Quality of Service (QoS).

However, not all multipoint conferences require a MCU. In a ***decentralized*** multipoint conference, the participating terminals multicast their audio and video to all other participating terminals directly. This can be done in many modern systems since participant terminals now have enough computing power to simultaneously receive and process more than one video and audio channel, summing the received audio streams, and choosing one or more of the received video streams for display.

For example, the product iVisit, developed by Internet video conferencing pioneer Tim Dorcey and Eyematic (Inglewood, CA – 310-342-2944, www.eyematic.com), uses a camera and a sound card

– but doesn't need an MCU to provide integrated multipoint audio and video conferencing over the Internet within any Windows 95 / 98 / NT or Macintosh environment.

Similarly, the PowerPlay IP video conferencing package from Broadband Networks, Inc. (State College, PA – 814-237-4073, www.bnisolutions.com) has at its core a program called IPContact that allows for conferees to conduct high quality, multipoint video conferences over any IP-based network without a central MCU. IPContact can perform fully decentralized four-way continuous presence multipoint video showing all four conference members in full CIF resolution at 30 fps. It can also do eight-way continuous presence audio. PowerPlay / IPContact can even run on lower-end PCs without the help of expensive hardware acceleration cards or other assistance.

BNI's PowerPlay allows each video stream to have different bit rates, frame rates, and codecs in the same conference without having to perform transcoding the way MCUs do. If you're in a multipoint conference with three participants on a high speed LAN and one is on a lower speed connection, that person will receive lower quality video while the others can still enjoy high quality video.

"Multicasting" and Broadcast Conference Modes

You've no doubt installed RealPlayer and/or Windows MediaPlayer on your computer system at one time or other, and have enjoyed the experience of streaming audio and video, be it a song from an Internet radio station, a movie trailer from a Hollywood studio or a video clip from a major news organization. Many corporations periodically "broadcast" or "multicast" their own big-event presentations (shareholder reports, product roll-outs) to branch offices, customers, business analysts, and other interested parties.

When a 1,000 desktop users connect to a Web or FTP server and demand 1,000 different streams, the content travels from the source server in separate streams to each user. This individualized, point-to-point technique is called ***unicast*** streaming. It's a wasteful technique that causes a lot of network congestion.

If a company wants to broadcast a video/audio announcement to 10,000 employees, or do a product roll-out to impress 100,000 customers, unicasting just won't do. Delivering the same streaming content to so many users is impossible, since the server and Internet router don't have the bandwidth to serve so many users.

Instead, we must resort to some form of ***multicast*** or "broadcast conference" where the network is more efficiently transmitting ***protocol data units*** (PDUs) from one source to many destinations. To do this, multicast routing protocols are needed to achieve optimal routing paths through the vast network of routers via a system of node registration. Once such a path is set up, instead of broadcasting thousands of streams (as in the case of unicast), the server sends out just one, which is distributed among the multiple participants. Bandwidth is thus conserved at both the server and the entire network, having been reduced to whatever can handle a single stream.

A simple broadcast conference is one in which there is one transmitter of media streams and many receivers. There is no bi-directional transmission of control or media streams. Such conferences are implemented using network transport multicast facilities and under guidelines provided in a relative of H.323, ITU H.332 (formerly known as "H.loosely coupled").

Hybrid configurations of the above can also be cobbled together, such as a "broadcast panel conference" which combines a broadcast with a multipoint conference. This involves several terminals actually particpating in a bidirectional multipoint conference while other terminals just passively sit there and receive the media streams, with no ability to communicate with the multipoint terminals.

The M-Bone

Much multicast activity relies on the ***M-Bone***, or multicast backbone. This is not a separate physical network, but a virtual one, an address space laid on top of the existing ATM and optical backbones that make up networks such as the Internet. The network is linked by virtual point-to-point links called "tunnels". The tunnel endpoints are typically workstation-class machines having operating system support for IP multicast and run the "mrouted" multicast routing daemon. The M-Bone started out as an experimental concept early on in the days when the various classes of IP addresses were formulated. The M-Bone occupies what used to be called the Class D address range – that is, addresses between 224.0.0.0 and 239.255.255.255. Several addresses within the range are reserved; for example, 224.0.0.1 is reserved for all hosts connected directly to the local network, and 224.0.0.2 is designated for routers on a LAN.

Multicast sessions using the M-Bone are assigned a multicast group ID, an IP address within the Class D range. A host may join as many multicast groups as needed by placing a particular stream's Class D IP address on his or her IP terminal. The host can leave the conference at any time. There can be any number of participants and they can be anywhere on the network. All they have to do is use the same IP address – by using the same address (the group IP address) the multicast stream heads for this imaginary destination, and the multicast routers distribute it to everyone using that address. To achieve this feat, multicast routers rely on a group membership protocol, such as IGMP (Internet Group Management Protocol). When someone wants to join a group, the person's terminal sends an IGMP message to a series of routers and thence to the multicast router, which begins broadcasting the sessions requested to the member's local "subnet," the member adds the group ID address to its interface, and the stream is received by the user's terminal.

Since the arrangement of participants and routers in the network can resemble a tree, multicast uses one of two spanning-tree technologies – ***dense-mode*** or ***sparse-mode*** – to transmit streaming media.

In sparse mode, the router forwards multicast packets only if it has received a join message from a downstream router or if it has group members directly connected to this interface. It assumes that members receiving the streams are going to be in a few clusters at scattered locations, not everywhere in the network. Sparse-mode protocols also deal well with lower-bandwidth connections, such as corporate WANs. Sparse-mode protocols include CBT (Core-Based Trees) and PIM-SM (Protocol Independent Multicast Sparse Mode). Both of these protocols build routing trees by requiring the routers to participate in creating the tree. Sparse-mode routers ask to join a multicast session only when a downstream member requests admission.

In dense mode (used in massive Web broadcasts and presentations), the router forwards multicast packets until it can determine whether there are group members or downstream routers. Unlike sparse-mode, it doesn't wait for a join message to begin sending multicast packets, it just blasts information out to the "leaves" of the network "tree" in the most expedient manner. Indeed, dense-mode routers never send a join message, but they do send messages to "prune" the "branches" of the network "tree" as soon as they determine they have no members or downstream routers. When a

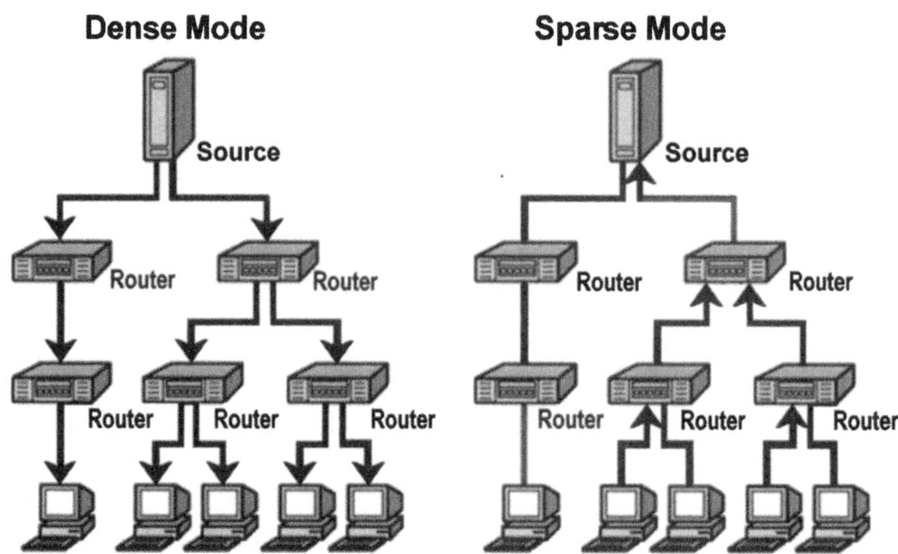

With dense-mode routing, the source broadcasts to everyone on the network until unneeded branches are "pruned." Sparse mode is initiated by the receiver sending a request to the first router that receives the multicast, or all the way back to the source to initiate the boradcast.

member leaves the multicast, the routers prunes the branch and bandwidth is conserved. Still, dense-mode operates under the assumption that there's copious bandwidth to support the broadcasts and that there are many members in dense pockets of the network who are getting the broadcast stream from the same location. Dense-mode networks employ routing protocols such as DVMRP (Distance Vector Multicast Routing Protocol), PIM-DM (Protocol Independent Multicast Dense Mode) and MOSPF (Multicast Open Shortest Path First).

Some routers can operate in what's called ***sparse-dense-mode***, which means that the router can operate in both sparse-mode and dense-mode, depending on what the other routers in the multicast group are using.

Video Conferencing Buyer's Checklist

1. High quality audio.

- Full-duplex audio lets participants interrupt without audio clipping.
- Less than 200ms delay in audio/video synchronization, so you don't get a "dubbed movie" effect.

2. High quality video.

- For lower data transfer rates, conferencing using the H.263 standard offers clear video.
- A 15 to 30fps rate makes for acceptable video. Remember, the higher the frame rate, the smoother the image motion will appear.

3. Ease of use. Systems should make it easy for participants to place calls and take calls without major setup effort or assistance from the system administrator. Some nice features that simplify conferencing include the following:

- Intuitive GUI plus remote control, so users can select phone numbers and place calls within seconds.
- An automatic voice tracking camera that lets the camera itself follow the action, and camera presets that keep the focus on the meeting, not on the electronics.
- Dual monitor support allows for near- and far-end display, or display of shared presentations.

4. Easy installation. No matter how sophisticated the system, the following features should be included for the user:

- All required and optional cables should be included in the package.
- Graphical (on screen) setup menus should be present.
- On ISDN-compatible units there should be ***AutoSPID detection*** so that the system will automatically recognize SPIDs, or Service Profile IDentifiers. When you order an ISDN line, your phone company will give you a SPID for every device you have connected to an ISDN line for circuit-switched (not packet-switched) network access. The SPID can be the 10-digit DN (Directory Number, essentially the telephone number), although it usually includes a four-digit suffix or prefix.
- Color-coded cables and a comprehensive (not to mention comprehensible) users' manual should be included.

5. Affordability. This criterion is naturally subjective. The best way to get the most for your money is to research, and then compare different models and manufacturers.

6. Data sharing and remote participation capacity. The system should have T.120 support for data and application sharing.

7. Interoperability. The system should be interoperable with other endpoints, bridges, and gateways by:

- Adhering to the H.320 standard to insure interoperability between legacy ISDN systems and newer H.323 or SIP systems. This will be less important as the whole world becomes more IP oriented.
- Conformance to the T.120 standard for application sharing between different endpoints (just about all vendors now have this).
- Conformance to the current panoply of audio and video standards.

8. Upgradeability. There will always be newer, better and snazzier features available, so make sure your system supports software upgrades.

WHAT'S NEXT?

Now that the technology, cost, and ease of use considerations have been worked out, the issue of secure conferencing and fraud protection is becoming increasingly important to customers. They want to be absolutely sure they know who is in any particular conference and they don't want people crashing conferences or using corporate bridges for personal use. Conferencing vendors are souping up these security capabilities in their products as you read this.

The other expansion will involve sophisticated audio and video conferencing over GUI-enabled cell phones, once the so-called third generation phones are in place by 2005. In 2001, for example, Octave Communications (Nashua, NH – 603-459-5200, www.octavecomm.com) a company involved in instant group communications, introduced mobility applications enabling wireless carriers and service providers to offer enterprises and consumers new subscription services for instant group communications. The applications, collectively called MeetAbout, utilize Octave conferencing hardware and software to allow any mobile phone user to instantly start speaking with multiple members of a buddy list via an "always-on, always available" conference room.

MeetAbout mobile can quickly start and attend their instant conferences from any mobile device using simple speech, Web browsers or SS7 signaling. Additionally, the applications use network presence technology so a mobile user instantly knows the online presence/availability of an individual or group member being contacted. The system can integrate with today's carrier networks and with the emerging *General Packet Radio Service* (GPRS) and 3G packet-based networks.

But aside from such exciting technical innovations and issues, it looks as if the hand of Fate in the form of terrorist activity has grabbed the adoption curve for conferencing technology and wrenched it up, thus ensuring the imminent ubiquity of electronic conferencing in all its forms.

Of course, this isn't the way it was supposed to happen. Above and beyond travel savings and the safety factor, conferencing technology reduces wear and tear on the employees themselves, empowering them to apportion more quality time to things outside of the business – to their families and themselves – so they become more satisfied and productive employees, not just simply employees "who are still alive and functioning." That, after all, has been the real goal of every enlightened management theory of the last century.

COMPANIES OFFERING VIDEO CONFERENCING PRODUCTS AND SERVICES

First Virtual Communications. First Virtual Communications (Santa Clara, CA – 408-567-7200, www.fvc.com), also known as FVC.com, was founded in 1993 by Ralph Undermann (who also founded Undermann-Bass). Specializing in broadband video applications by partnering with companies such as PictureTel and AT&T, it merged with CUseeME networks in 2001 to add Internet video conferencing to its portfolio.

First Virtual's Click to Meet is an integrated platform for multipoint conferencing designed for enterprises and service providers. It has a browser-friendly HTML interface used for placing video conferencing calls. It combines voice and video telephony, data collaboration and streaming content into a single rich communications environment. Although considered to be an H.323 platform, Click to Meet integrates various IP, ISDN, ATM, and DSL network standards.

When implementing a conference from the desktop GUI, restrictions regarding access to resources are resolved, and Click to Meet maps the requests to the necessary endpoint devices, media servers, and network infrastructure. It makes sure that all necessary resources are available, and then schedules these resources for the duration of the session.

A complete Web Development Kit is available for the Click to Meet server so developers can integrate Click to Meet capabilities into other applications, or develop their own customized user interface based on XML.

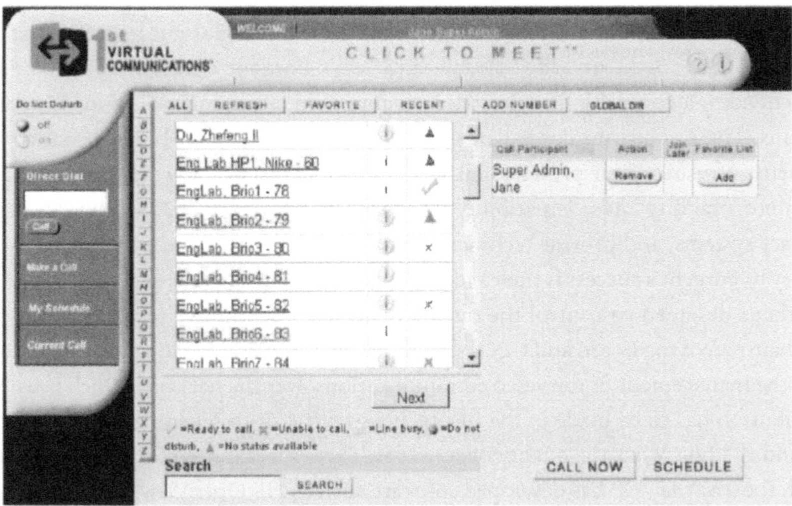

First Virtual's Click to Meet interface allows scheduling and calling for conference participants from a single location. The display shows the status of all of the registered users and their availability to meet with you. Just click on the people you wish to conference with, and either call now or schedule a call for later. Click to Meet combines voice and video telephony, data collaboration and streaming content into a single communications environment.

One of First Virtual's most impressive feats was to provide the video conferencing technology for the 2001 G8 Summit in Genoa, Italy, working with the telecom provider Wind Telecomunicazioni, which is partly owned by France Telecom.

Fabio Tessera, Managing Director of FVC.com's Italian Branch, says that "video conferencing for remote meetings was a critical aspect of the event since there were anti-globalization protesters and various political implications. With our technology people could communicate during the event without driving throughout Genoa and meeting somewhere in the city."

Wind requested First Virtual to provide three different applications: Click to Meet video conferencing (both point-to-point and multipoint), videostreaming and video-on-demand. All data was transported over a Metropolitan fiber Gigabit Ethernet network that Wind set up covering Genoa and its surroundings. "There were about 700 PCs installed in dedicated rooms and there were up to 300 LAN connections, so a total of about 1,000 journalists and G8 members could plug in their laptops and run these kind of applications," says Tessera. 30 frame per second video was supported, with each endpoint providing 768 Kbps of bandwidth over IP. For live video streaming, about 30 different servers were needed for broadcasting live TV channels.

All domestic Italian and international channels were broadcast during the event, so if journalists wanted to see CNN, the official G8 channel covering the event, or some other broadcast TV, they could log into the official G8 portal using a simple laptop or desktop PC, and get streaming video.

"Five large rackmount servers with RAID storage were used for the video on demand application," says Tessera, "They could store up to 500 hours of video in MPEG-1 format and were load balanced to appear as a single large virtual server. The software we developed for this application we call iManage." Journalists and G8 member could use video-on-demand to call up previously recorded meetings. Background movies and videos for the previous G8 event were also available.

Forgent. If you're a Fortune 1000 company and have no clue about how to get started in video conferencing, just pick up the phone and call Forgent (Austin, TX – 512-437-2700, www.forgent.com) a company that provides advanced video network solutions for enterprise communications. These solutions consist of professional services, major brand products and network management software. Their professional services people can do network planning, design rooms, provide interoperability and compatibility standards testing for IP H.323 networks as well as ISDN H.320 legacy systems, and provide technical support, maintenance and training.

One key to Forgent's success is their Video Network Platform (VNP), special network management software that's designed to control the quality of service for multi-vendor video networks. Richard Snyder, Chairman of the Board and CEO of Forgent, says that "we develop proprietary software which will allow the management of the video communications over the network, which is where we feel the real contribution has to be made in the future. It's so difficult to schedule calls, make calls, keep the calls up, and to know what the quality of service is. There's a real need for software that can manage this, much the way that HP has developed software such as HP OpenView to manage data."

"We also work with partners," says Snyder, "so that if a company wants to have bridging services, we can provide that as an outsource as well. But that's not the main focus of what we're doing."

Companies pay Forgent via professional service fees, depending upon what they want to accomplish. Forgent also responds to RFPs and RFQs. "We have a direct sales force that concentrates on the Fortune 1000," says Snyder, "but we also have a Channels Group that focuses on those people who are active in working in the video communications space."

Polycom. Perhaps best known for their triangular SoundStation audio conferencing device, Polycom (Milpitas, CA – 800-765-9266, www.polycom.com) has rapidly become the largest video conferencing equipment maker in the world.

At the low-end, their ViaVideo is an integrated, USB-based personal video system for the desktop. It enables voice, video and data collaboration, and for what it does it's pretty inexpensive.

For small conference rooms, the ViewStation SP and SP384 deliver high-performance group conferencing in an integrated system. You can also use this unit in executive suites and professional offices.

Moving on up to medium-to-large conference rooms, the ViewStation H.323 and ViewStation MP define Polycom's larger entries in this area. An embedded Web server handles diagnostics and simple software upgrades over the Net. The ViewStation MP has multipoint capability and supports four locations at 128 Kbps and three locations at 256 Kbps over H.320. These devices' Web-based presentation system eases the sharing of graphics and slides. There's also a voice tracking camera and track-to-preset function that automatically focuses on the speaker.

For the largest conference rooms, the top-notch ViewStation FX (set-top) and VS4000 (rack-mounted) systems are true enterprise solutions that provide TV-quality video. Features include the voice tracking camera, embedded streaming capabilities, and custom application support. You can connect up to four sites at 384 kbps or three sites at 512 Kbps with no outside bridging over H.320 or H.323. Full-motion video at 30 fps is supported, with TV quality beginning at 512 Kbps. An embedded Web server handles diagnostics and simple software upgrades over the Net.

Jennifer Sigmund, director of product marketing for Polycom, says: "Exciting developments are happening here. With our recent acquisition of Accord Networks – now called the Polycom Network

Systems Group – along with their network MCU and gateways, we now have the ability to offer complete end-to-end communications solutions for companies involving the convergence of voice, video and data, regardless of whether your network connectivity is ISDN, IP or ATM."

Proximity. As a kind of service provider – what you'd call a "video conferencing rooms broker," Proximity, Inc. (Burlington, VT – 800-433-2900, www.proximity.com) represents over 3,500 video conferencing sites worldwide, about half of which are in the US.

"We're like a video conferencing travel agent," says Bob Kaphan, president of Proximity. "We rent rooms wholesale and resell time in them on a retail basis. You pay the same price to us whether you work with us or walk in the room cold off of the street. We can do multipoint conferences of up to 50 or 60 sites on the same call. Our rooms are typically boardrooms that can hold six to eight people." Rooms can be reserved via an 800 number, fax, or Web form.

Proximity's other large revenue stream is their "High Impact" business, which they call Large Scale Event Productions. These are multipoint events with hundreds of thousands of people involved. They're done in hotels, conferences, and academic settings. "We'll connect several of these settings together for a big joint meeting in real-time with large projection screens," says Kaphan, "participants can see a presenter and interact with him or her, even though they're in another city. One Fortune 500 company had a 23 site multipoint conference where the chairman of the board was addressing everyone in his company. We had sites in Moscow, in Poland, in Singapore, and Chicago. We put equipment on factory floors so the workers could see it on projection screens."

Proximity's third revenue stream is corporate outsourcing. "Many Fortune 500 companies have internal video conferencing networks of anywhere from a few to 200 sites," says Kaphan, "and they use us when they want to call outside of their internal network. They rent rooms from us. We also have managed the sites of some clients, in the form of outsourced IT. We take care of the training, the maintenance, scheduling, everything that has to be done. We do it through vendors all over the world and we show up on site when necessary."

"We don't own the network," says Kaplan. "ISDN H.320 systems still abound. We all know the world is migrating to IP and H.323. Where you're seeing it now for the most part is in virtual private networks (VPNs) within enterprises. There are some public service types of packet-based things, such as Wire One's Glowpoint network. Obviously everything will gradually migrate to IP, but it's not there yet. The traditional Internet does not yet have adequate bandwidth to support high traffic and quality of service."

Kaphan says that 80% of Proximity's conferences are 30 fps, 384 Kbps H.320 ISDN transmissions. IP is on the way, but things have to be ironed out. "We were once experimenting with an IP call to a company in Philadelphia," says Kaplan, "and the packets bounced around Albany, New York six times before they left that area! Things like that can degrade service. Using a virtual private network (VPN) helps but doesn't solve everything. IP is very close, but it's not quite perfected yet."

Sonic Telecom. A company that provides global solutions for end-to-end multimedia sessions with products developed for bandwidth intensive and "bursty" applications such as video conferencing and broadcast-quality video transmission. Sonic Telecom, Ltd. (Chantilly, VA – 703-818-0057, sonictelecom.com) has over 100 points of presence worldwide connected by a global fiber optic network stretching around the globe.

Santiago Testa, VP of Corporate Communications, says: "Our Sonic Video1 service uses ISDN

At left in the photo is the Tandberg 7000 for larger settings; at right is the desktop Tandberg 1000.

to reach an ATM gateway. We have a patent pending product that sits between the video camera and the ISDN walljack. That device lets the CO know that this is a Sonic video conference and right at the switch the call pops over to our network which has an ATM backbone and a nice constant bit rate. There's no time wasted waiting for pairs of ISDN lines to bond in the ISDN cloud."

Sonic Telecom customers can also self-provision bandwidth whenever it's needed and pay according to their actual usage.

Aside from bandwidth, Sonic Telecom provisions, installs and manages custom-designed international networks created to support customer applications. Customers can assign the Quality of Service (QoS) they want for each application. They can choose Constant Bit Rate (CBR), Variable Bit Rate real-time (VBRrt) and non-real-time (VBRnrt) or Unspecified Bit Rate (UBR).

Their Sonic Video1 system can connect up to 2 Mbps, giving a near-TV video experience. "We're using H.320, but we're working on H.323," says Testa. "For H.323 in enterprise network scenarios, we can certainly do video conferencing, but our patent pending product for Sonic Video1 is a quick and easy entry level video conferencing product that anybody can use. We provide the cameras, the training for it and everything very economically."

Sonic also offers a high-end bridging service that can be controlled by a Web-based scheduling tool.

Tandberg. The second largest video conferencing manufacturer in the world (behind Polycom). Tandberg (Dallas, TX – 972-243-7572, www. tandberg.net) is based in Oslo, Norway, but most of the company's revenue (approximately $62 out of $92 million in 2000) comes from North America. They have two wholly owned US subsidiaries:

- Tandberg, Inc., the indirect sales channel for TANDBERG. They sell through resellers, and target the government, telehealthcare and distance learning markets. They also sell to enterprises through their resellers.

- NuVision, which was acquired by Tandberg in 1997, is the direct sales channel of Tandberg. NuVision provides everything to the customer, from systems to bandwidth to training.

Tandberg has a reputation for top-notch equipment. One is the Tandberg 1000, a desktop system geared towards an executive's office or small group conference room. It can be ordered as either a freestanding tabletop system or a wall-mounted solution. The system includes an LCD flat screen, codec, microphone and speaker. There are no external components or TV screen. The Tandberg 1000 can deliver up to 30 fps. System can run at transmission speeds of up to 384 Kbps on ISDN and 768 Kbps over IP.

At the other end of the spectrum, the Tandberg 7000 is the high-end group video conferencing system of the Tandberg family. Indeed, it's the highest-end product Tandberg makes. It can be configured with two 40-inch plasma screens. It comes with Tandberg's proprietary Natural Video quality technology and MulitSite functionality. The Tandberg 7000 delivers up to an amazing 60 fields per second, and can run at transmission speeds of up to 2 Mbps on ISDN, IP and other external networks.

Teleportec. The developer of a fascinating product that goes by the modest name of Teleportation Technology. Teleportec (Dallas, TX – 214 615 6555, www.teleportec.com) conferencing, or teleportation, creates an image of a person who appears to be in the same room regardless of their actual location. Thus, people in two separate locations can interact using three-dimensional images – making it seem as if the individuals are in the same room having an eye-to-eye exchange.

The network for this teleportation process uses an ISDN or T1 line and codec on each end of the transmission to transmit images across the world. Teleportec produces three products: the Teleportec Desk and Teleportec Conference Table are for an office style presentation and the Teleportec Lectern is for an audience style presentation.

Teleportec's spin on teleportation works like this: A presenter is seated at a Teleportec Desk, which has a large video display that shows an audience view at the distant location. This view has the same aspect ratio with the same line of sight that the presenter would actually see if he or she were in the same room.

Immediately in front of the presenter is a computer monitor that controls transmission operations and access to digital resource material. This monitor may be used for confidential information since it is not shared with the audience. The computer has a second monitor that displays the visual material to be shared with the audience. For one-to-one communication the remote location uses a desk.

Teleportec technology is compelling because it can be applied creatively in different industries. For distance learning, instructors and students scattered throughout the world can interact as just as if they were all in the same classroom. When used in the health care industry, individuals receiving care in remote areas are able to communicate with doctors and specialists anywhere in the world. This three-dimensional image enables doctors and specialists to actually see the patient, discuss specific symptoms and render treatment quickly.

VCON. An Israeli-based company that makes a range of networked video over IP solutions, including an interesting centralized management and administration package for multi-vendor IP video networks. VCON (Austin, TX – 512-583-7700, www.vcon.com) also offers complete meeting solutions for desktop, compact, and group conferencing over ISDN, ATM, satellite and xDSL. VCON markets its products and services through a network of reseller partners, OEMs, and value-added resellers worldwide.

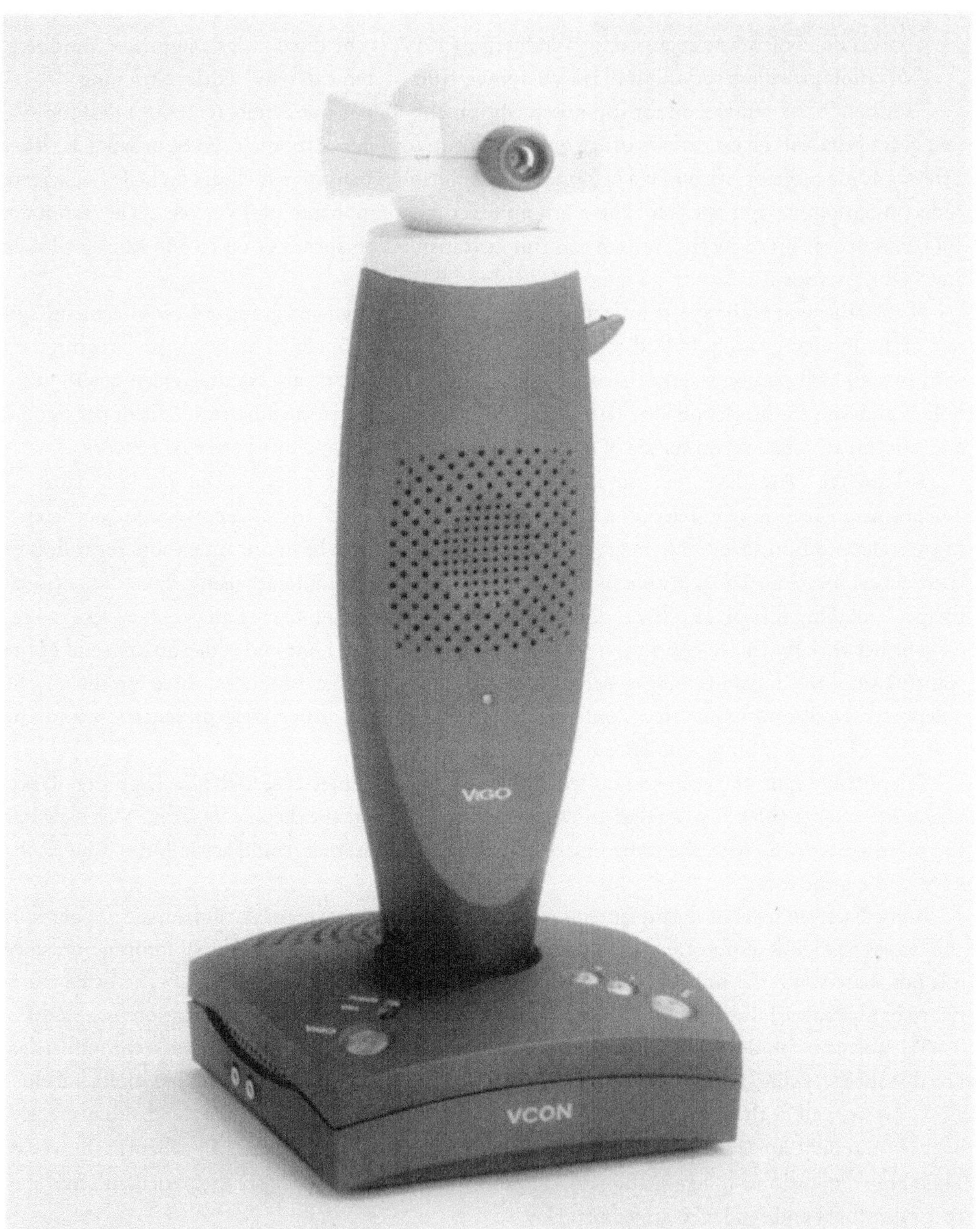

VCON's ViGO is a small, portable appliance that provides high-quality video conferencing for laptops. ViGO is seen here with its speaker tower.

Gordon Daugherty, general manager of VCON Americas, says "Video conferencing equipment is all VCON does and we're a top five vendor in the group VC arena. We offer both H.320 and H.323

systems. Our real leadership happens to be H.323 video over IP, but we still sell a decent number of H.320 ISDN based systems, especially in countries where their IP infrastructure is not as far along as it is in the US."

At the low end, VCON offers a little laptop/desktop video conferencing appliance called the ViGO. You just plug ViGO into the USB port. It can sit on a desk, and you can use a headset or a speaker tower that it comes with. Naturally, a camera attaches to the unit. One ViGO model gives you business quality business conferencing at what they believe is the "sweet spot" of $699 in the US. There are several other models in the ViGO family.

Another fascinating VCON product is an IP video PBX called the Media Xchange Manager (MXM). It also serves as VCON's management server that does bandwidth management, various policy services, QoS, online directories, all these kinds of things.

With one MXM server you can give your users telephony functionality, particularly third party call control such as Call Transfer, Call Forward. You can even do *ad hoc* conferencing, where you're in a point-to-point call, then somebody hits the conference button and invites a third participant into the call. The MXM works with any kind of network infrastructure that can run packetized video using the H.323 standard, which includes Ethernet LANs, ATM backbones, T1 backbones, satellites links, cable modems, and xDSL.

VNCI. Formerly Objective Communications, Inc., Video Network Communications Inc. (Portsmouth, NH – 603-334-6700, www.vnci.net) designs, develops, and markets video distribution systems that provide full-motion, high-resolution video networking, enabling video broadcast distribution, video-on-demand and multipoint video conferencing. Their principal product is the VNCI Video Networking System, which distributes TV-quality video, audio, and data to desktop and laptop PCs and conference rooms within an enterprise over ordinary telephone wiring (CAT 3) or LAN-capable CAT 5.

The VNCI Video Networking System is composed of three key components: the VNCI Switch (a multifunction communications center that acts as the hub of a VNCI video network), the VidModem, and the customer's VNCI Stations that can be set up anywhere there's a phone jack and a monitor or screen. Each desktop connected to the switch is fully interactive and includes a camera with microphone, stereo speakers, software and an optional video overlay card. The Java-based desktop software provides the GUI interface to control desktop settings, applications, and video service access.

VNCI uses a DSL-like modulation technique to send a signal at a higher frequency than telephone voice traffic, thus allowing both signals to travel on the same wire without interference to up to 1,500 desktops. Since the video and audio signals aren't compressed as with LAN- or ISDN-based systems, users are able to receive TV-quality images and stereo-quality sound.

Users can communicate beyond the building or campus using standards-compliant VNCI Gateways to connect to the Wide Area Network via ISDN (H.320), IP (H.323), T1 and ATM connections.

VTEL. This company makes video conferencing endpoints, as well as an H.323-based multimedia conference server. VTEL's (Austin, TX – 512.821.7000, www.vtel.com) Galaxy family, which can support both H.323 and H.320, comes in dual or single monitor versions, and includes four camera inputs, as well as support for VCR video playback and record. Integrated T.120 and support for Microsoft NetMeeting add data collaboration and application sharing features, and an Ethernet port gives you access to the LAN. Document camera with automated control and an optional interactive whiteboard are also available. Two omnidirectional mics come standard, and a 16-bit Sound Blaster is included with support for PC and DVD applications. Remote control and wireless keyboard control a Windows-based GUI that helps you easily create and manage conferences.

There is a whole suite of Galaxy video conferencing products:

- Galaxy PL – The Galaxy platform, ready to be integrated into your own custom environment. It has the same data sharing and collaboration capabilities as the Galaxy SL, RL, and XL.
- Galaxy SL – A rollabout designed for small groups, this Galaxy system includes a cabinet and features Galaxy's power document creation, annotation, and collaboration tools. There's also an integrated PC base, so you can share a file or application.
- Galaxy RL – This provides the same offerings as Galaxy SL plus a second cabinet so you can view two images at once. It's for distance learning and document collaboration sessions with medium or large groups.
- Galaxy XL – This is VTEL's top-of-the-line video conferencing system. A two-cabinet system for large groups or where full-blown data collaboration is necessary, it includes a document camera, VCR, and their SmartTrak voice-activated cameras that automatically focus on any speaker in the room.

Standard on all Galaxy systems is AppShare, application sharing software that can be run during video conferences.

VTEL's conference server, the MCS/IP, does audio transcoding to mix H.323 and POTS calls, and supports T.120 data. VTEL also offers an H.323 / H.320 gateway and a gatekeeper that can either be integrated or run on a separate NT server.

VTEL also offers their TurboCast Web streaming technology that allows you to "Webcast" events live over an internal network or the Internet, and store any multimedia content for on-demand playback.

Winnov. Providing non-proprietary and fully standards-compliant, capture card- and appliance-based communications solutions since 1992, Winnov (Sunnyvale, CA – 408-744-9777, www.winnov.com) provides high-performance products that let you capture, manage, and develop audio and video content for distribution over the Internet or company intranets. Winnov exploits audio and video over IP technology with its wide array of video conferencing and video streaming applications for such diverse markets as: Internet broadcasting, surveillance, distance learning, and telemedicine.

Winnov offers the following audio and video communications tools to help you deliver customized audio and video content:

- Capture cards. Winnov manufactures a full line of board-level audio and video capture cards that can deliver audio and video over the Web and can provide a reliable hardware platform for you if you decide to build your own applications. Every card includes drivers that enable them to run on any Windows environment desktop or notebook PC. Winnov's capture cards synchronize audio and video. A Software Development Kit (SDK) is included with every capture card. Each SDK comes with ActiveX Controls that allow developers to create their own applications by building on top of Winnov's flexible platform.
- Videoum AV PCI video and audio capture board. Built for any Windows-based PC, this board is designed for Internet video conferencing, digital video capture, Webcams and streaming Internet video. It can process 640x480 pixels @ 15 fps uncompressed.

- Encoding appliances. Winnov also sells turnkey solutions that can do streaming audio/video Web broadcasts. The devices can be rack-mounted or portable.

Wire One. A leading full-service provider of a complete range of video communications solutions, including Glowpoint, the first IP-based network dedicated to video conferencing. Wire One Technologies, Inc. (Hillside, NJ – 800-924-8596, www.wireone.com), which is the first video communications provider to receive Frost & Sullivan's Market Engineering Service Innovation Award, is a leading integrator for major video communications manufacturers, including Accord Telecommunications, Cisco Systems, PictureTel Corporation, Polycom, Inc., RADVision, VTEL and VCON. Wire One's current customer base includes more than 2,500 companies with approximately 13,000 video conferencing endpoints in the commercial, federal and state government, medical and education marketplaces nationwide and across the globe.

Yahoo. Shortly after the events of September 11, Internet portal Yahoo! Inc. (Sunnyvale, CA – www.yahoo.com) launched two new Web-conferencing packages available through the broadband services unit of its business communications division. Note that Yahoo didn't actually mention the events September 11, but did state in its announcement that with travel and meeting costs rising and the recent travel restrictions, it was introducing its Virtual Conference service and Executive Communications Center.

Virtual Conference can provide live or on-demand viewing via the Web, reaching thousands of online conference participants situated worldwide. This scalability is made possible by Yahoo streaming audio and video broadcasts from its servers to attendees who watch and listen through their Internet connections instead of traveling to a single location. Yahoo's Virtual Conference service includes features such as online registration, presentation slides, document sharing and can be customized to run a pay-per-view broadcast or include interactive tools (such as audience polling) for things such as a question-and-answer section. It also has live attendance tracking, and post event reporting on attendees; and archiving and hosting.

One of the service's first customers was Germany's Deutsche Bank which used the service to host 1,700 attendees for a conference held in New York that originally was to be held in Phoenix, Arizona, but was postponed because of the September 11 disaster.

The other service, with the full name of Executive Communication Center-Breaking News Channel, is a tool allowing companies to Webcast a time-sensitive, "breaking news" event, such as a corporate acquisition. Once clients supply the video to be conveyed, Yahoo can have a broadcast ready to go within four hours (instead of the usual 10 days with competing systems)

Zydacron. comCenter, Zydacron's (Manchester, NH – 603-647-1000, www.zydacron.com) video conferencing room system is designed for intra-company workgroups. The comCenter supports both IP and ISDN, as well as T.120 data through full integration of Microsoft NetMeeting. The system also supports streaming media capabilities, and uses a browser-based interface to access files on the Internet or intranet. Other features include 27-inch SVGA monitor, on-screen camera control with auto-tracking, and wireless keyboard and mouse. Through an OEM deal with RADVision, Zydacron offers comCenter as part of a complete solution that includes RAD's H.323-based MCU, gateway, and gatekeeper.

Zydacron has also recently released the OnWAN340 IP, a desktop IP conferencing product that delivers 30fps two-way video. OnWAN340 supports dynamic switching between ISDN and an IP-based LAN connection. The standard package includes a video/audio codec, camera, conferencing software, and Microsoft NetMeeting.

COMPANIES OFFERING GENERAL CONFERENCING PRODUCTS AND SERVICES

Act Teleconferencing. This company provides audio, video, data and Web-based conferencing products and services. ACT Teleconferencing Inc.'s (Denver, CO – 303-233-3500, www.acttel.com) ReadyConnect service allows you to make audio conference calls without advance reservations or operator assistance, or you can use ActionCall, which allows for full operator support. Conference access an be limited to select participants by issuing a passcode.

ACT's ClarionCall is also a full duplex unattended conferencing service, but it's all done over your VoIP network.

ActionView is their video conferencing service, while ActionCast is a streaming audio and ActionVideo a streaming video message/presentation service. ActionData lets you share documents or present slides over the Web while conducting an audio conference.

Centra. This company offers a fully integrated Web-based collaboration system scaleable to over 100,000 users with a common management and administration environment. Since their first product shipped in 1997, Centra (Lexington, MA – 781-861-7000, www.centra.com) has concentrated on not just business collaboration but eLearning too, having been the first company to provide a one megabyte, browser-based thin-client architecture that enables full-featured eLearning even over a low bandwidth connection. Thus, their extensible delivery platform can be seamlessly integrated with popular Microsoft platforms as well as virtually all of today's leading Learning Management Systems (LMS), learning content standards, and eCommerce systems.

Indeed, their Centra Symposium is one of the most widely used solutions for the delivery of live eLearning across the enterprise. Unlike streaming broadcast technology or static Web pages, Centra Symposium enables large groups of dispersed employees, partners, and customers to interact, collaborate, and learn – replicating typical classroom interaction – in real-time over intranets, extranets, and the Internet.

Centra Symposium offers fully integrated, multi-way, full duplex voice-over-IP audio conferencing as well as IP video conferencing that's real-time, multipoint video conferencing that enables users to see the session leader or other designated participants from within the interface. An interesting adaptive video bandwidth feature ensures the highest possible performance over connections as low as 28.8 Kbps.

Businesses, on the other hand, will be interested in Centra eMeeting, an easy-to-use virtual meeting facility enables you to easily coordinate all aspects of meeting scheduling, attendee participation, information, and content for your Web meetings. Participants only need a PC and Web browser to begin meeting you online

Another related product is Centra Conference, which "Webifies" sales, marketing, and corporate communications. Centra Conference handles live, interactive seminars, company meetings, product demonstrations, and presentations.

Centra eMeeting, Centra Conference, and Centra Symposium all share a common collaboration framework, and can be purchased either as a secure ASP service or you can install a Centra

Collaboration Server on-premises for maximum security and lowest cost of ownership.

Chorus Call. This company offers personalized, audio, video and Web-based conferencing services. Chorus Call's (Monroeville, PA – 412-373-6964, www.choruscall.com) audio and video conferencing services are built upon Compunetix audio bridges and MCUs. They also offer "Hybrid Conferencing", which allows participants to join conferences via video or, if they don't have access to a video terminal, via audio. These video conferences are increasingly popular because they integrate the features of both audio and video conferencing.

Compunetix. A maker of video and audio conferencing equipment for sale to large telecom service providers such as AT&T and Sprint (we examined their CONTEX audio conferencing bridges earlier in this chapter). Compunetix (Monroeville, PA – 800-879-4266, www.compunetix.com) produces extremely highly reliable conferencing bridges. They're one of the world's leaders in audio conferencing bridges and they also produce video conferencing bridges.

One of these is the Virtuoso Hybrid, an interesting audio/video multipoint conferencing bridge that fully integrates attended and unattended audio, video and data conferencing. It offers 24 to 120 ports of audio conferencing while simultaneously delivering eight ports of video-conferencing at speeds up to 2 Mbps via bundled ISDN channels or broadband connections. You can use it for an audio-only, video-only or mixed conference calls.

Conferences may be controlled directly from your telephone / video terminal, or from over the Internet using VirtuosoWeb (a Web-based interface), or via a LAN/WAN using the Windows 95/98/NT compatible Multimedia Windows Operator Console (MMWOC), an administrative application that enables you to set up and control any conference. The systems can support one or more MMWOCs linked to the system using TCP/IP over a LAN/WAN, direct serial connections, or dial-up modems. Several MMWOC users can view the same conference or a single MMWOC can manage multiple conferences simultaneously.

Compunetix also offers a larger system, the Orchestrator, which provides from 8 to 160 ports of attended or unattended multimedia conferencing (384 Kbps per port).

Installations of modular, scalable Compunetix equipment range from large-scale service bureau systems to small-scale corporate systems.

Genesys Conferencing. Said to be the largest independent conference services provider in the world, with a customer base of 17,000 companies, of which 800 are in the Fortune 2500. Genesys Conferencing (Bedford, MA – 781-761-6200, www.genesys.com) has a complete portfolio of services that they sell, including both automated and attended audio conferencing and room-based video conferencing services. They maintain around the world bridges that will work with H.320 compliant equipment from Polycom, PictureTel, Tandberg, etc.

Kailash Ambwani, executive VP of technology and services, told the author about Genesys' premier service called the Genesys Meeting Center. "Today, if you look at how people do Web and data conferencing," says Ambwani, "the vast majority tends to be small groups of five or ten people who go through presentations or they work on a document together. Almost 100% of those events have an audio conference call attached to them, obviously. So if you've got five or ten people in a meeting, they're sitting on a phone in an audio conference call and they're sharing a document on the Web."

The Genesys Meeting Center brings together Genesys' existing reservationless audio conferencing service called TeleMeeting and the Web conferencing and data collaboration technologies they acquired from their acquisition of the US Web conferencing firm Astound (where Ambwani was CEO), into a single platform and a single browser-controlled interface.

"With the Genesys Meeting Center," says Ambwani, "If there are five or ten of us around the world connected in an audio conference call, I'm able to see all five people in my browser and I'm able to manage the audio through the browser. If somebody's on a cell phone and there's a lot of background noise, I can, through the browser, click on that person's name and put them on Mute. Also using the browser I can click on two or three of the participants and take them into a virtual subconference room, so on the phone we'll have a side conversation without involving the rest of the conference group."

At the same time, conference participants will be able to control data and presentation conferencing. As Ambawani says, "You can deliver Powerpoint presentations or share any application on the desktop. So if the five of us are working on a budget, I can bring up a spreadsheet and we can all work on a budget together. It's a fully integrated platform which tightly integrates multi-way audio management and data / Web conferencing through a single interface."

For the audio conferencing component of this service, Genesys leases line contracts with major telcos such as MCI and Sprint, and maintains about 50,000 bridge ports deployed in 16 countries. For the data / Web component, Genesys has an IP network consisting of eight data centers in the US, Europe and Asia, each center being fault tolerant and highly scalable. When you do a data / Web conference you'll get connected to the best data center in terms of network topology. There's also live multiway replication, so if a data center goes down or is overloaded, the sessions in progress will be automatically switched over to a different data center.

The pricing strategy for the Genesys Meeting Center service is also quite eye opening. Whereas the audio conferencing part is charged on a per minute basis as is customary, the Web conferencing part is sold on an inexpensive subscription basis.

Ambawani strongly believes that in the long term, there won't be any room-based video conferencing. "You'll be able to sit at your desktop," says Ambawani. "There will be a camera on it and with a click you'll set up a face-to-face virtual video conference and have a conversation. It's not that far away. From a business point of view, video over IP is a killer app."

Gentner. This company develops, markets and distributes products and services to the professional communications, broadcast, business and consumer markets. Gentner's (Salt Lake City, UT – 800-945-7730, www.gentner.com) products and services include audio and video conferencing systems and a full suite of teleconferencing services, including full-service conference calling, Web conferencing, and Instant Access Conference Calling. Gentner is also in the conferencing services business, so if you need to use a video conferencing bridge to connect multiple locations, Gentner can help you too.

Frances Flood, president and CEO, told the author that "Our series of video conferencing systems convers everything from the lowest price point all the way up to high-end conferencing. We're also seeing a greater demand on the audio product side." Gentner video conferencing can handle both H.320 and H.323 calls. They partner with Voyant Technologies for their audio conferencing bridges.

Gentner has two conferencing brands: Gentner Conferencing Services and 1-800 Let's Meet (at 1-800-538-7633). For first time users who want to make a conference call, you can go to www.gentner.com/t4sres.html and click the "Create a New Account" button. Once you fill out your online application, you will be contacted within one business day. Existing customers can simply click the "Make a Reservation" button or call 1-800-538-7633 to contact a professional reservationist.

"There are no limits to the size of the conferences we support," says Flood. "We have handled disaster recovery calls, up to 6,000 participants at one point in time."

Gentner's TheDataPort.com is their Web conferencing service that enables you to conduct live, interactive meetings over the Internet. All you need is a connection to the Internet and your PC. Your presentation can include graphics, slides, graphs, charts, Q & A, even video and audio. Plus, your event can be saved for those who missed it live.

For example, after the WTC disaster, the National Association of Purchasing Management reported record attendance at a regional seminar held in Salt Lake City on September 13, 2001, even though its keynote speaker Wayne L. Kost was stranded in Winter Park, Florida. Using a hastily assembled, Gentner-facilitated audio and data conference, Kost addressed an audience of nearly 200 participants.

"Watching the presentation, it was like Mr. Kost was addressing us in person. The clarity of the audio was outstanding, resulting in the best teleconference in which I have ever participated," said Carol Freasier, an attendee.

Global Crossing. This company owns its own 100,000 mile fiber optic network. Global Crossing (Westminster, CO – 303-633-3000, www.globalcrossing.com) also has an Audio Conferencing Group that's made up of two key components: One is an operator assisted, high impact, event-style conference capability that's used for major announcements, investor relations, world events, product and anniversary announcements, or training initiatives.

Then there's Ready-Access an instantaneous reservationless product. To use it you just schedule your meeting, dial in, type in a PIN and you're in your meeting. Says Brad Levy, VP of Global Crossing Conferencing Products: "We're finding huge usage of Ready-Access, especially after the World Trade Center event. It influences the way people are doing business or the way people are trying to reorganize their people resources and their employees who are now stranded around the globe to just do an inventory of who is available, who is around, and who needs to be at what location."

For several weeks following the disaster, Global Crossing offered Ready-Access service at no charge to institutions who needed that kind of help. "I think that's a huge initiative on our behalf to do what's right for the business community and for people," says Levy.

Ready-Access is also tied to Global's on-demand data collaboration product, eMeeting. Aside from the audio conference call, eMeeting allows your laptop or PC to perform data collaboration too. You can share, annotate and collaborate on any document.

Global Crossing also offers video conferencing, which came with their merger with Frontier Corporation and their Frontier Video division, founded in 1993 as the first independent video conferencing bridging provider in the world, and providing services to over 60% of the Fortune 500.

"The services are H.320 and now IP," says Levy, "The world will ultimately move IP-centric type conferencing but they won't immediately abandon H.320 ISDN conferencing. ISDN may be used as a stable back-up to IP. In some countries, of course, there'll go pure IP. It's incumbent on us to provide gateway services to end users so that they don't care how calls are being transported, as long as their meeting starts on time, it's not interrupted, the quality is good, it ends on time, and it's not expensive."

iMeet. This company started out as a Web-conferencing solution introduced in March 1999 by SneakerLabs, Inc. In response to an explosion in the demand for Web-conferencing products, iMeet,

Inc. (Pittsburgh, PA – 877.244.6338, www.imeet.com) was created as a separate entity in November 2000 to provide focused sales and development support for iMeet. iMeet immediately became a premier solution for holding live, Web-based conferences and was recognized as Best Collaborative Software at the 19th Annual TeleCon Conference and Expo.

iMeet offers both corporate and "self serve" Web conferencing services. The iMeet Corporate Meeting Center is a Web-conferencing service package provided in six components (Administration, Schedule, Web-conference, Registration, Evaluation, Archive). You gather a profile of meeting attendees through the Schedule and Registration components. The Schedule module of the iMeet Corporate Meeting Center is based on a simple form that allows you to enter all meeting details in a quick four-step process (Logistics, Audio, Invite, and Content). In this one form you set the date, time, and length of your conference; provide details on the audio portion; invite attendees and provide them with all the pertinent information; and finally, select presentations, bookmarks, and polling questions to be used during your conference.

Do you prefer to have attendees decide for themselves if they should attend a specific conference? With the registration component, you can provide password-protected access to your list of scheduled meetings and allow people to register for those conferences that are of interest to them. There is quick access to a list of registered attendees, and their responses to questions that you asked in the registration process.

The iMeet evaluation component allows you to collect valuable feedback from participants. You can modify the evaluations to be appropriate to your particular audience and subject matter, and you can export the results to a database making it easy to analyze, share, and act on them.

Finally, completed meetings can be shared 24x7 using the systems' archive. Archiving assures that everyone has access to any meetings they may have missed. You provide iMeet with an audio file of your conference, and they seamlessly synchronize it to your Web presentation.

All of these components are managed using an user-friendly Account Management Console. It allows you to manage all aspects of your account including conference activity and user details. Access to this information helps in planning upcoming meetings, in understanding billing, and in managing access to your Meeting Center

iMeet realizes that every time you interact with customers, partners, and suppliers you want to maintain a consistent message, so you'll find that you'll be working hand-in-hand with iMeet's design and development teams to create a Meeting Center that is completely integrated with your corporate look, feel, and messaging.

MessageBank. A company with a unique approach to conferencing services. MessageBank (New York, NY – 212-333-9300, www.messagebank.com) doesn't own its own bridges or even employ operators. Instead, MessageBank acts as sort of an event planner for companies planning conferences as big as the customer wants, from three to 1000+ participants. MessageBank acts as the middleman between the company planning the conference and the service providers.

Arthur Kass, president of MessageBank, told the author that his customers look for the personalized service and one-stop shopping he provides by brokering out the services to different vendors at wholesale prices. "We are very consultative in our approach," Kass said, adding that he'll even help arrange the catering and transportation for the event.

Kass said another advantage he feels his company has over competitors is its Midtown

Manhattan location. Unlike many MessageBank competitors located throughout the United States, Kass can "jump on the subway and attend a customer's meeting or be in Stanford, Connecticut in an hour to help pull the event off."

In addition to providing high-quality audio conferencing services, MessageBank offers taping services, digital replay, translation and transcription, electronic Q & A and voting. The 100% fiber networks that they use can save customers 20-40% on long-distance by utilizing their own virtual network facilities for conferencing.

Pixion. This company was founded in July 1995 to provide a cross-platform, visual conferencing solution that enables the free and spontaneous exchange of visual imagery over business networks and the Internet. Pixion, Inc. (Pleasanton, CA – 925-467-5300, www.pixion.com) is the developer of PictureTalk software – a Client-Server-Client Web-based communication solution that provides collaboration, meeting and presentation tools for one-to-one communication or conferences with thousands of participants around the world. Any application or file on the presenter's computer desktop can be shown to multiple viewers in an unlimited number of locations. The patent-pending technology conveys imagery over networks in real time, continuously adapting the compression rates to suit the performance of individual systems.

PictureTalk users can present live desktop imagery from any combination of PC, Macintosh, UNIX workstations, and Java device to hundreds of users simultaneously. PictureTalk software products are both bandwidth- and platform-independent, allowing users to hold "visual meetings" in real time, over a wide range of network connection speeds and platforms. The result is that participants with slow Internet connections do not limit the speed of the overall presentation. PictureTalk allows customers to collaborate in real time, utilizing audio and any computer application in a seamless cross-platform environment.

Pixion's clients include such firms as Bank of America, Booz-Allen & Hamilton Inc., Macromedia Inc., Siebel Systems Inc., Siemens and BMC Software. Pixion offers a very flexible licensing model to fit its customer's needs, ranging from on demand meetings that are hosted by Pixion, to perpetual enterprise wide site license that can be deployed on the customers own networks

PlaceWare. This company got its start in 1990 at Xerox Palo Alto Research Center (PARC). In its first iterations, PlaceWare (Mountain View, CA —- 650-526-6100, www.placeware.com) was a groundbreaking multi-user game known as LambdaMoo. Ironically, this success led to the technology becoming the foundation for a collaborative computing system for the US Department of Defense.

Realizing the potential of this new multi-user real-time collaboration medium in business applications, PlaceWare was formed in 1996. In early 1997, PlaceWare's first product, PlaceWare Auditorium, gave users a live, Web-based presentation solution for field and customer communication. And in the late 1990s PlaceWare became one of the first companies to introduce Web conferencing as a solution for holding meetings over the Internet.

Today, more than 2000 leading companies including HP, Sun, Ingram Micro, Autodesk and Cisco use PlaceWare for marketing seminars, product launches, press and analyst tours, customer meetings, corporate announcements, and quarterly IR/earnings calls.

PlaceWare's Web conferencing solution offers the choice of two kinds of virtual environments: large auditoriums that can hold up to 2500 participants, or highly interactive online meeting rooms for smaller groups.

Unlike other Visual conferencing technologies, with PlaceWare there is no need for additional hardware or software installation and PlaceWare requires absolutely no special plug-ins for participants or attendees. All you need is a Web browser and a phone. Just log-on to your meeting and go. That's it.

Voyant. In the 1980s and 1990s, the audio conferencing field was dominated by equipment from ConferTech International, which became a subsidiary of Frontier Communications (which became part of Global Crossing) and, though a management buyout in August 1999, became an independent company called Voyant (Westminster, CO – 303-223-5000, www.voyanttech.com). Voyant has several teleconferencing products, and invents new ones at an impressive pace. Voyant tends to focus on supplying service providers.

Voyant's ReadiVoice is a reservation-less, on-demand conferencing system that catapulted Voyant into a leadership position. ReadiVoice uses Voyant's high density rackmounted hardware platform, InnoVox, which has hot-swappable cards, power supplies and fans; and is designed with other redundant and fault-tolerant features.

During 2002 ReadiVoice evolved into an IP product. "ReadiVoice IP is simply the same ReadiVoice product and application architecture with IP I/O," Richard Schute Voyant's COO told the author. "You eliminate the PSTN and plug in IP. Everything else stays the same. Many of our customers in the IP space will be using a gateway to begin with, at least until pure IP trunking becomes more prevalent."

"One exciting area we're developing concerns wireless networking and carriers," says Schulte. "Even before the WTC disaster, there was a renewed interest, in providing group conferencing capability initiated from a wireless endpoint via perhaps an Internet based app or even a DTMF interaction. It's still a sort of PSTN based application, because the wireless carriers aren't close to implementing IP. It may be based on a WAP or VoiceXML enabled phone. There are a variety of interfaces that we think the wireless carriers are going to use but it's unclear at the moment what the interface of choice will be."

V-SPAN. A leading video-oriented collaboration and conferencing services provider. The integrated V-SPAN (King of Prussia, PA – 610-382-1000, www.vspan.com) portfolio includes video conferencing, audio conferencing, Web conferencing and streaming for the enterprise and channel markets. Through Web-based self-service solutions, experienced full-service call centers, and diversified global networks, V-SPAN enables its customers to conduct electronic meetings, events and training.

WebEx. A major player in real-time communications infrastructure for business meetings on the Web, WebEx Communications, Inc. (San Jose, CA – 408-435-7000, www.Webex.com) provides Web-based carrier-class communication services that integrate voice, video and data to enable full interaction and collaboration, across geographies and platforms. These services are based on WebEx's multimedia switching platform and are deployed over a global network. WebEx's services enable end-users to share presentations, documents, applications, voice, and video spontaneously with anyone, anywhere, using Windows, Macintosh or Solaris operating systems, all through a standard Web browser. WebEx services are used across the enterprise in sales, support, training, marketing, engineering, and various other functions. With its modular framework and standards-based APIs, WebEx's real-time communications platform is the "dial-tone" for meetings on the Web.

Industry veterans Subrah S. Iyar and Min Zhu founded WebEx Communications, Inc. in 1996. WebEx has evolved into a leader in real-time, interactive multimedia communications services, with over 5,000 corporate customers and hundreds of thousands of individual users.

The WebEx platform seamlessly integrates data, audio and video. The platform enables services with features such as real-time sharing of applications, presentations, or documents as well as Web co-browsing, live chat, record and playback, remote control and file transfer. WebEx services are known for their flexibility and support for a high level of interactivity between meeting participants. For example, files and applications on any meeting participant's desktop or on the Web can be opened and shared in real time during a WebEx meeting.

WebEx provides a range of real-time multimedia business communication services and also teams with resellers, telecommunication vendors and online service providers to deliver services.

WebEx services include the following:

- Meeting Center, providing an easy-to-use interface for powerful on-line meetings and training across the enterprise.
- WebEx OnCall, giving technical support personnel the fastest and easiest way to resolve customer problems in real time
- WebEx OnStage, offering your company the ability to reach broad audiences through powerful, live online seminars
- Business Exchange, featuring outer offices and a business directory front-end to your company's meeting rooms.

CHAPTER 3

Fault Tolerant Computing

COMMUNICATIONS APPLICATIONS NEVER STOP. Whether it's a PC-based phone system in a small business, a communications server distributing calls to a large enterprise call center, a system sending packetized voice calls over data networks or a huge collection of interconnected rackmounts acting as the backbone of a telco's enhanced services division, computers are being called upon as never before to bring millions of people new and exciting communications offerings. "Computing" also implies "data storage," and of course businesses need instant access to their inventory, customer, employee and external Web databases. Most organizations operate with an ever-increasing number of business-critical systems involving communications, storage and more, such as eBusiness enabling systems, CRM (Customer Relationship Management) systems, e-commerce and e-mail.

But in a world where system downtime can mean the difference between profitability and failure, computer-controlled platforms for voice and data communications must have "high availability" or HA. Uptime is crucial because failure means lost revenue, lost profits and lost customers. This means they must be far more reliable than conventional PCs, allowing them to be used in "mission critical" systems. I'm not talking about only e-commerce, e-mail, groupware and databases, but also systems where you can now find multi-span T1 / E1 interfaces with on-board high-density media processing ("service bureaus in a slot"), SS7 components, Voice over IP (VoIP) and Fax over IP (VoIP) processing, ATM boards, Frame Relay and ISDN PRI firmware.

According to CNT (Minneapolis, MN – 763-268-6000, www.cnt.com), a company that provides global storage networking solutions, computer downtime costs US businesses $4 billion per year, chiefly through lost revenue, with an average company's hourly cost per outage being $330,000. Computer systems fail nine times per year, with four hours' downtime per failure. From 1995 to 2000, statistics indicate that business interruptions were caused by human error (34%), power problems (29%) and environmental incidents (10%).

THE NEED FOR RELIABLE COMPUTING POWER

Since September 11, 2001, it appears that disasters fabricated by the equivalent of B-movie evil masterminds have joined accidental, technical and natural disasters on the list of calamities that could befall both public and private communications systems. Everyone needs assurance that communications networks and the services that run over them will continue to function flawlessly, despite the worst efforts of Osama bin Laden, Lex Luthor, or Dr. No.

Communications Must Run 24/7/365

Service providers, call center operators, large and small enterprises and even the individual customer wouldn't have it any other way. Indeed, makers of high-end telco equipment must offer proof of tested compliance with the super-strict comprehensive electrical and mechanical criteria for telephone-related equipment, referred to as the NEBS (Network Equipment-Build System) standards, before any such computer equipment can be installed in the PSTN as an adjunct applications processor. Because many telecommunications companies use NEBS-compliance as a prerequisite for product deployment, the certification strengthens a products reputation. For example, any enterprise using mission-critical applications also like NEBs equipment, since NEBS certified machines can take shock, earthquakes, fire, electrostatic discharge (lightning strikes) – you name it. The "NEBS Criteria" were originally formulated by Bell Labs in the 1970s, further developed by Bellcore (now called Telcordia) and were made public documents in 1985. Central office equipment manufacturers were the prime target audience back then, but today NEBS compliance has also become the benchmark of excellence for "next-gen" service providers (such as CLECs and ISPs), ASPs, data centers and corporate organizations.

Of course, any "Ma and Pop" company that can bend sheet metal into the shape of a box can claim that they've got a "NEBS compliant" PC, but while many companies have NEBS compliant designs, few actually go to the trouble and expense of undergoing the rigorous process of NEBS *certification*, thus verifying that their gizmo actually meets the standard. Performing actual NEBS certification on a product costs over $100,000 and is fascinating to watch. For example, at Underwriters Laboratories Inc.'s testing facility in Research Triangle Park, North Carolina, there's a servo-hydraulic "shake table" that can simulate earthquakes measuring 7.0 on the Richter scale on samples weighing up to 5,000 pounds.

This seemingly nit-picking behavior can instantly be put into perspective if one imagines plugging a pile of power-hungry multi-port telephony cards from Intel, NMS Communications, or Brooktrout into an ordinary mail order computer built on a conventional "active motherboard" (or "baseboard" as Intel calls it) and a single disk drive; then try running a high-volume call-center, VoIP or service-bureau application on it. You're asking for big trouble. The overtaxed power supply will expire with a pop and the disk drive will thrash away until it literally flies apart.

Still, despite such grueling demands the levels of computer availability demanded by telcos and well-known communications product companies (such as Nortel, Cisco and Lucent) are truly awesome. Most traditional carrier-class equipment is expected to deliver at least "five nines" performance, or 99.999% availability, which is a total annual downtime of only five minutes. Many companies and carriers are now demanding "six nines." Obviously, crashing or rebooting is verboten in such an environment, both for the hardware and the software.

What really distinguishes a fault resilient or fault tolerant system from your $500 cross-your-fingers-and-hope-it-will-work mail order special, is component redundancy. If one component fails another takes over. Load sharing power supplies, multiple fans, multiple drive I/O channels and host controllers add to the continuous working life of your machine.

Many people then ask, "Okay, then what's the difference between a computer that's "fault tolerant" and another that's "fault resilient"? Both fault resilient and fault tolerant computer systems have redundant components, but a true fault tolerant system will also duplicate the CPU, boosting the system uptime from "four nines" to "five nines." Both CPUs will work on the same processes with some kind of polling process taking place to determine if a CPU has failed.

If a CPU board does fail, then an "automatic failover" process switches control to the healthy CPU while the bad one is replaced. A fault resilient computer has but one CPU, which makes it a lot less expensive than a true fault tolerant system. However, since CPUs only tend to fail if the fans fail, and if fans are monitored by an alarming board, then a fault resilient CPU can give you nearly the same reliability as a fault tolerant system, but it will simply cost a lot less (incidentally, it was the author who coined and popularized the term "fault resilient" for computers with such a configuration, back in the mid-1990s). During the "computer telephony" era of the 1990s, many reasonably-priced IVR systems, fax servers, and similar platforms were built using fault resilient (not fault tolerant) systems. As prices began to drop, more and more fault tolerant systems are finding their way into the business community.

Like all manufacturing industries, the fault tolerant communications platform business is based upon three major activities: Developing, producing and selling products. As we shall now see, all three areas have undergone a revolution in the world of fault tolerant converging communications.

FAULT-RESILIENT FOOTPRINTS: DIVERGENT APPLICATIONS, DIVERGENT FORMS OF FAULT RESILIENT HARDWARE.

The computer hardware used in computer telephony and communications convergence must now be as varied and scalable as the tremendous number of applications they serve. Such hardware ranges from the smallest Internet appliance to a kiosk with its glowing screen sitting in a lobby or store, to a pile of 1U high (a "U" is a vertical unit of measurement – 1U is 1.75-inch, or 44.45mm) "pizza boxes" at an ISP, to a cluster of tower servers, or to the familiar room full of 80 pound, 19-wide rackmount computers, all of them sitting in sturdy aluminum frameworks seven feet high.

➤ Levels of Availability

In 1995, prior to the author's popularizing the term "fault resilient computer", vendors were calling their machines "industrial", "fault tolerant", "ruggedized", "heavy-duty", "highly available" and other semi- or inappropriate terms. A table finally appeared (shown below) that pretty much puts the various degrees of availability in their proper perspective, along with the now-accepted terminology:

System Availability Levels	Total Availability	Approximate Total Downtime
Continuous Availability	99.999% to 99.9995%	2.628 to 5.256 min. / yr.
Commercial Fault Tolerant	99.99% to 99.995%	26.28 to 52.56 min. / yr.
Fault Resilient	99.99%	52.56 min. / yr.
General High Availability	99.9%	8.76 hr. / yr.

Thus, difference between "fault resilience" and "fault tolerance" is the difference between "four nines" (99.99%) and "five nines" (99.999%) availability.

Motherboards vs. Passive Backplanes.

One of the great dividing lines in high availability hardware is whether one uses a small ruggedized motherboard / baseboard, or a large passive backplane with up to 21 slots for additional plug-in cards. In an ordinary PC, the processor, memory, and other chips that comprise the "computer" section (everything except the peripheral cards, such as telephone interface cards, network cards, etc.) are contained on the motherboard. The motherboard also contains the connectors into which the peripheral cards are inserted.

On the other hand, in a passive backplane system, the backplane contains no active circuitry, just connectors (lots of them), hence its name. Despite their large size and number of slots, the passive backplane architecture is electrically compatible with the same desktop silicon used in desktop PCs; and is therefore compatible with off-the-shelf operating systems, software and available PC compatible plug-in I/O boards. In such passive backplane systems the board housing the CPU, its memory and various I/O ports (Ethernet, video, serial, parallel, mouse, etc.) are situated on its own plug-in card, called a "single board computer" or SBC.

Originally conceived as ready-made hardware platforms that would reduce time-to-market for OEMs, SBCs have evolved to the point where they can support complete hardware and software architectures. The first SBCs operated stand-alone or were designed to fit proprietary bus architectures. VMEbus, Multibus (I and II) and STD emerged as *de facto* standards, but were incompatible with each other. Over time, and as a result of trends in software development circles that resulted in "open architectures," OEMs began using off-the-shelf tools and software to manufacture and support SBC based hardware products. The adherence to the concepts of open architecture in the design and development of SBC products parallels a trend toward SBC compatibility with commercial hardware and software products – the so-called "desktop environment." A key reason for this is that development tools available for desktop PCs provide a simple and cost effective method for anyone to develop custom applications around the desktop environment.

Since the SBC occupies one slot (or two slots if it's very wide) you can upgrade your processor simply by swapping boards. The SBC is an independent card and not part of the backplane, which has nothing on it other than slot card connectors. Therefore, the probability of passive backplane failure is quite low.

Directly and indirectly, the considerations as to whether to use a motherboard or passive backplane come down to cost and space. A single-board computer and a passive backplane can cost three or even five times as much as a motherboard. Even though a passive backplane doesn't do anything but route signals, it alone can cost the same as a motherboard complete with a microprocessor.

In "embedded" applications, these considerations loom particularly large. An "embedded application" can be defined as one in which the computer drives a machine's functions, without making the user aware of its presence; users simply interact with what they see as the machine's interface. Traditional embedded applications include transaction terminals such as automated teller machines and kiosks.

If the computing platform must be small and energy efficient for embedded applications (some telecom / networking, industrial automation, gaming, medical and transaction terminals) then a large passive-backplane system doesn't make sense for such an application that basically needs just one CPU and one or two additional resource cards, not 8, 16 or 21.

Passive backplanes also can require more expensive upkeep. First of all, since so many cards are packed into a passive backplane system, the slots are closer together than those on a motherboard. If the system must have a reduced height, you may end up with a CPU card that's so fat you've got to depopulate a slot next to it, unless you use low-profile memory modules, which can bump up the cost. Single-board computers also tend to have smaller surface areas than motherboards, so you have to pack a lot more electronics into a smaller package, which in turn makes the smaller board more expensive to produce.

Then there's the problem of all of those cable connectors squeezed onto a small rear end of one CPU card; a motherboard spreads these out over a more manageable area.

More crowded components pose greater cooling challenges. You need bigger fans. While you often have to blow hurricanes through passive backplane systems, you can use much less expensive fans with motherboards; indeed, you can eliminate fans entirely on some motherboards and use passive heat sinks instead. This means you might be able to extend the life of the board to as long as eight years.

Also, with a motherboard you get the benefits of many companies using the same product, so you're buying something produced in high volumes with common suppliers. Many motherboard manufacturers build a large series of boards in various standard sizes, which brings down the cost. In short, with motherboards, it's all about economies of scale.

Embedded Boards

Embedded computing made its first appearance in industrial control applications (remote telemetry systems, programmable logic controllers, robotics), then disk servers (RAID array controllers, Web disk servers) and has now moved on up to Internet related devices (kiosks, vending machines, ATMs, set top boxes, Web caches, and firewalls).

Normally, when one thinks of a small "embedded computer" one thinks of really small boards such as PC/104, PC/104-Plus and EBX form factors. These may be fine for vending machines or Web phones, but they're usually too small for the more demanding embedded applications now appearing, applications that require a larger motherboard, perhaps half-size or even a full-sized one.

Such "real" motherboards appear in several distinct forms, but all of them are based around industry standard form factors such as the following:

ATX – A form factor invented by Intel in 1995 to replace the old AT form factor. It's a full sized board (12" wide x 9.6" deep, or 305 mm wide x 244 mm deep) holding about four PCI and three ISA slots

LPX – A defunct form factor. This was found in "low profile" PCs mass marketed by large manufacturers. The LPX introduced the concept of the "riser card" that is used to hold expansion slots. Unlike AT or ATX motherboards where the expansion cards plug into system bus slots right on the motherboard, LPX form factor motherboards placed the system bus on a riser card that plugs into the motherboard. Up to three expansion cards then plugged horizontally into the vertical riser card. At the time, the board was considered to be quite small (9" wide x 11" to 13" deep, or 229 mm wide x 280 mm to 330 mm deep)

Mini ATX – These boards use the same ATX form factor power supplies and cases except that they're slightly smaller (11.2" wide x 8.2" deep, or 285 cm wide x 208 mm deep).

microATX – Introduced by Intel in 1997, the microATX form factor is slightly smaller than ATX (generally a square-ish board up to 9.6" or 244 mm on each side).

FlexATX – A very low cost offshoot of microATX (an addendum to Intel's original microATX hardware specification) introduced in 1999. It's about 25 percent smaller than microATX – indeed, it's the smallest board of the Intel ATX family (9" wide x 7.5" deep, or 229 mm wide x 190 mm deep). Both the microATX and the FlexATX can use either full-size ATX power supplies or the newer, smaller, 90 watt SFX power supplies.

NLX – This successor to the LPX design was also introduced by Intel Corporation. It's about the size of a microATX (8" to 9" wide x 10" to 13.6" deep, or 203 mm to 229 mm wide x 254 mm to 356 mm deep). It allows for quick processor upgrades, faintly reminiscent of the way a CPU board can be replaced in a passive backplane system: The base NLX motherboard plugs into a vertical riser card having the slots for your expansion cards. Any cabling plugs into the riser card. The riser gives you a low-profile solution in the chassis, because the cards are mounted horizontally. The NLX form factor supports modern AGP graphics cards. The NLX motherboard and case form factor use the ATX size power supply because Intel didn't want yet another power supply form factor on the market. The ATX form factor is therefore occasionally referred to as the "ATX/NLX" form factor.

EBX – The immensely popular "Embedded Board, eXpandable" (EBX) standard is the result of a collaboration between Motorola and Ampro (San Jose, CA – 408-360-0200, www.ampro.com), to unify the embedded computing industry on a small footprint embedded single-board computer standard. Derived from the old Ampro Little Board form factor of the 1980s, EBX combines a standard footprint with open interfaces. The EBX form factor (5.76" wide x 8" deep, or 146 mm x 203 mm) is small enough for deeply embedded applications, yet large enough to contain the functions of a full embedded computer system: CPU, memory, mass storage interfaces, display controller, serial/parallel ports, and other system functions. EBX also offers the ability to add off-the-shelf expansion modules in such form factors as PC/104, PC/104-Plus and PCMCIA.

Such "real" motherboards are now being called to service in routers, small switches, firewall boxes, Web servers, and 1U and 2U rackmounts. Motherboards have also invaded some of the other traditional embedded applications, such as transaction terminals: ATMs and kiosks placed in retail stores where shoppers can access the store's Web site solely from within the store.

Companies such as AAEON (Hazlet, NJ – 732-203-9300, www.aaeon.com), American Predator Corporation (Morgan Hill, CA – 408-776-7896, www.americanpredator.com), Nexcom (Taipei, Taiwan – +886-2-2278-2215, www.nexcom.com), RadiSys (Hillsboro, OR – 503-615-1100, www.RadiSys.com) and U.S. Logic (Carlsbad, CA – 760-929-2700, www.uslogic.com) are among many who build "industrial strength" motherboards for embedded applications.

Be Careful in Choosing a Motherboard

While form factors are standardized, quality control and revision control among commercial, off-the-shelf motherboards is "iffy." Motherboards have to be tested and qualified on a regular basis for use in an application, which can be expensive in the case of some commercial motherboards from Asia, which have a PC life cycle of only around 12 months. Also, some motherboard makers tend to use underspec capacitors in their boards that can dry out at high temperatures and eventually fail. They might compromise on testing or just don't test at all. They'll even build boards and ship them straight out of the factory with no "end-of-line" test!

Another issue with off-the-shelf motherboards is that you don't get change notifications in the manufacturing process, or if you do, they're not that well controlled and disseminated. In embedded environments, it's very important to tightly control revisions of the BIOS and other rapidly changing technologies.

All these reasons are why US manufacturers have instituted their own quality control measures for motherboards. RadiSys, for example, established their Endura program to give their boards a longer liftime. RadiSys is targeting a five year lifespan for all of their motherboards. They came up with that figure first by using core technologies such as chipsets, processors, and Flash memory chosen from what Intel is calling their "embedded roadmap." The Embedded Intel Architecture group out of Chandler, Arizona, has picked certain processors, chipsets, and peripherals off of the mainstream roadmap and has said that these parts will continue to be supplied for five years. So you'll notice that quality motherboards such as those made by RadiSys don't cover every Intel chipset that's available. What you'll see, instead, are specific chipsets that come off of that proven embedded roadmap.

As for non-Intel parts, they always have to be selected carefully, which can be a bit tricky, since not all companies are in the business of guaranteeing that their products are going to be around in the embedded market for a long time. Companies such as RadiSys will generally have a process of trace certification, where customers are given plenty of notice if they have to make a change to any vendors or components on the board.

Finally, there's the problem of public fickleness, of what goes in and out of fashion. For example, the NLX board design was enthusiastically accepted by many of the "Tier 1" PC manufacturers such as HP and Compaq, but it has never really caught on with lesser Tier 2 and Tier 3 manufacturers. Many customers liked the small mechanical features of NLX, and they designed customer hardware mechanicals around that form factor. But when Intel pulled out of the stagnant NLX market, these customers found they were stuck. They couldn't retrofit their systems with a standard, larger, ATX board.

So, beware of what you're buying!

Pizza Boxes

Moving up in size, we next encounter a relative newcomer to the PC field, the very flat (1U and 2U high) rackmount or "pizza box," as it's affectionately called. Pizza boxes allow for greater "granularity" or "modularity" in the communications infrastructure, and their popularity exploded along with the number of ISP Internet access points, Web servers, e-commerce servers, applications servers, and heavy-duty kiosks and ATMs.

Such pizza boxes often use a "riser card" arrangement for accepting I/O expansion cards, which places the cards in parallel to the motherboard instead of perpendicular to it, as with passive backplanes and other common form factors. Riser card designs allow for flatter cases, but they typically offer only two or three expansion slots. By placing one card on each side of a riser card in a "butterfly" arrangement you can build some amazingly thin systems that can still hold two full-length PCI cards.

Every single fault resilient PC company has several 1U and 2U rackmounts in their catalog. CSS Labs (Irvine, CA – 949-852-8161, www.csslabs.com) even has a 1/2U rackmount, which is actually

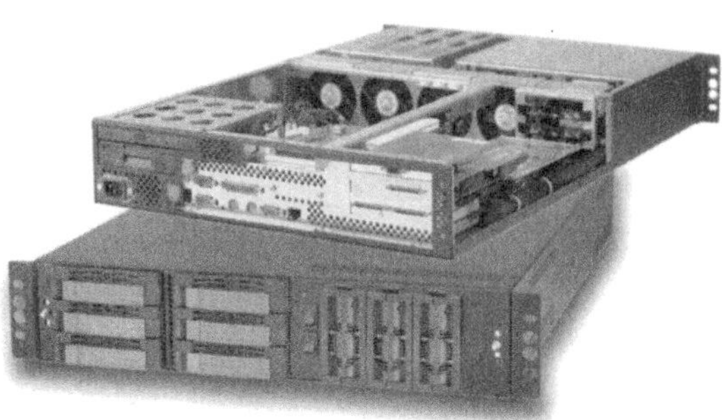

In this example of a high-end "pizza box" made by APPRO International, APPRO has somehow managed to take the storage and processing power normally found in a 4U high 19-inch rackmount and squeeze it down into this APRE-2003HX-1 2U high server. This unit sports dual 1GHz Pentium III CPUs, a 400 Watt redundant hot swappable power supply, and a hot swappable SCSI hard drive back panel plane capable of supporting up to six devices, each with up to a 160 MBps transfer rate. An internal riser card supports up to four full-length 64-bit 33 MHz PCI cards. Operating systems supported include Red Hat Linux, Windows NT Server 4.0, Windows 2000 Server and Windows 2000 Advance Server. APPRO took the lead at the dawn of the "pizza box" era and they continue to surprise the industry with their ingenious designs.

two separate motherboards sitting next to each other in a 1U high enclosure.

The space-saving pizza box design has grown so much in popularity that there are now 1U and 2U high patch panels, open-frame power supplies, flat-screen monitors and keyboard drawers.

APPRO (Milpitas, CA – 408-941-8100, www.appro.com) led the way in the late 1990s with the first high-quality 1U and 2U pizza boxes, followed by Advansor (Fremont, CA – 510-580-0338, www.advansor.com), Advantech (San Diego, CA – 858-623-0838, www.advantech.com/nc), Siliconrax-Sliger (Fremont, CA – 510-360-1600, www.siliconrax-sliger.com) and many other companies.

Shoe-Box Computers

A related space-saving design originated with Crystal Group (Hiawatha, IA – 319-378-1636, www.crystalpc.com) – the "shoe box" computer. Crystal's most refined version of this is RIA (Rackmount Integrated Applications) server, that allows up to 52 units to occu-

Crystal Group originated the idea of the "shoe-box" computer, which allows up to 50 or more PCs to occupy a standard seven foot high rack.

py a standard seven foot rack – this equals a mere .87U of rack space being occupied per server. RIA can be configured with a three-slot ISA or ISA/PCI passive backplane along with options such as Intel's latest processors, two expansion slots, non-volatile and drive storage, a programmable LCD, and a USB front port.

Since these computers are so small, it soon became apparent that if components could be swapped in and out easily, why not whole computers? Crystal Group pioneered the concept of a hot swappable computer with their QuickConnect cable management system. QuickConnect combines all cables coming from the server into a single connector, which easily slides into a fixed, self-aligning connector on the back of the rack.

Another company that specializes in shoe-box designs is ClearCube (Austin, TX – 888-266-8115, www.clearcube.com), which produces perhaps the smallest shoe-box PC of them all.

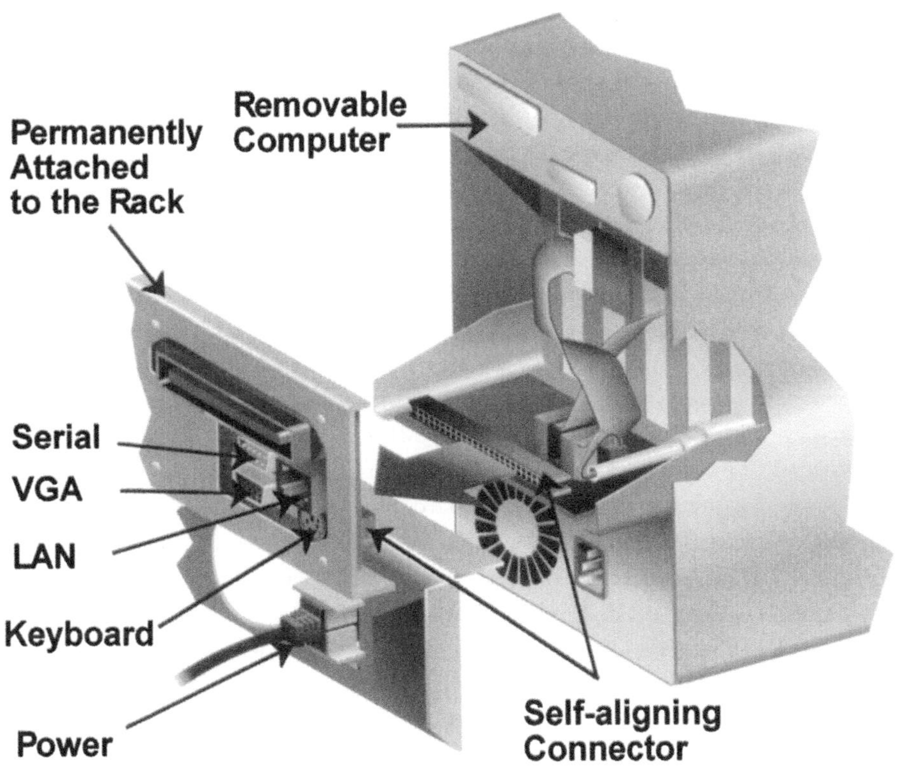

The QuickConnect feature of Crystal Group computers combines all cables coming from the computer into a single connector, which slides into a fixed, self aligning connector on the back of the rack. All cables leading from this connector are wired in place. This option virtually eliminates your cable-related failures and lets you remove and install a computer in less than 10 seconds. Costly hours of time spent tracing wires and cables, connecting and reconnecting are gone.

The shoe-box computer has made possible what the author calls "concentrated computing." This also involves the use of "extender" technology that allows hundreds of corporate office workers, teleworkers in a telecenter or call center agents to use PCs that are in fact sitting together in a seven-foot rack perhaps hundreds of feet away, locked securely in a single rack that sits in an air-conditioned, controlled room (or an armor-plated bunker, if you like) environment. We'll examine this idea further in the "maintainability" section of this chapter.

The Classic 19-inch Rackmount

Moving on up the size scale, the oldest and most recognizable form of fault resilient computer is undoubtedly the standard 19-inch wide rackmount. Sometimes the "standard" 19" rackmount is actually 23-inches wide (to accommodate certain large racks in Central Offices) and can range in height from 3U (5.25 inches or 134 mm) to 12U (21 inches or 534 mm) or more.

There are literally buildings filled with examples of the classic 19" rackmount. They're found in corporate server rooms, ASPs, ISPs, fax service bureaus, data centers and more. They can serve as softswitch-like gateways between the PSTN and the packet world, they can be found in giant storage systems for storage area networks (SNAs) and storage service providers (SSPs), and can be used as media servers and as the basis of advanced switching devices.

These computers almost always rely on passive backplane technology. While it's true that the new generation of high density telephony boards can hold eight T1 spans and can do enough packetization to serve up VoIP to a small town, designers always find that applications end up demanding just a few more resources than they expected.

Also, converging communications applications demand many other kinds of high density cards too, particularly if your system is also going to be a media server. Computer telephony applications use many add-in cards for voice, fax and digital switching. 19-inch passive backplanes provide up to about 21 ISA, PCI or cPCI expansion slots and are easier to upgrade and service than motherboard-based systems.

The "conventional" 19-inch rackmount has a PCI or PCI and ISA passive backplane and ends up being a compromise between the benefits of a standard motherboard system and what you can achieve with a CompactPCI (cPCI) or VME rackmount system (see their descriptions below). With a PCI passive backplane system your computing engine is a single board computer, which is more expensive than a motherboard. However, a conventional rackmount allows you to take advantage of the generally higher volume and lower cost of a wide variety of peripheral and I/O cards normally used in conventional desktop PCs. Also, these plug-in cards are only available in standard desktop PCI or ISA form factors, not in CompactPCI or VMEbus form factors, so you've got a lot more choices available if you decide to buy or build a conventional 19" rackmount system.

Collectively, then, the price for a PCI/ISA rackmount system will end up between the pricing of a motherboard and a cPCI-based system.

The Chassis / Blade Approach

Passive backplanes can also be segmented (and even power segmented, with each segment having it's own separate power input) so you can have up to four independent computer systems occupying one chassis, each with its own single board computer.

Classic Rackmount Design

Examples of the classic 19-inch rackmount design by Alliance Systems (Plano, TX — 972-633-3400, www.alliancesystems.com). At left is their 20 slot, 9U high I-Series 9000, which can hold eight 5.25-inch drives and offers not two but three, auto-switched, hot-swap, dual-lead, load sharing power supplies. At right is the more spacious, 11U high I-Series 11000, which has greater capacity power supplies than the 9000 (12 amps at +5 volts instead of 90).

The logical extrapolation of subdividing space in a 19-inch rackmount and putting more and more independent computer systems in the box is what's called the "blade" (not to be confused with ClearCube's small, individual "shoe-box" computers that go by the same name). A blade is a complete computer system on a board that plugs into the backplane. Think of it as a single board computer having a disk drive and communications abilities. For example, Rack Solution's (Tustin, CA – 714-368-3676, www.racksolution.com) I-Server is a compact 4U high rackmount that can house no less than 13 front-accessible single board computers (500 MHz Pentium III with up to 768 MB), each running as a separate system. Each CPU board also comes equipped with an onboard 2.5-inch, 20 GB hard drive as well as two 10/100Base-T ports. By using the I-Server architecture, over 130 systems can sit in one seven foot high rack.

The whole system only consumes about 700 Watts, which is provided by four hot-swappable load sharing 250 Watt power supplies. Each CPU board supports Linux, FreeBSD, Win 98/NT/2000 and Solaris x86. Monitor ports check on the temperature of the chassis, power supply voltages, cooling fans, and temperature of the CPU boards.

Certainly a chassis/blade approach sounds inherently more compact than the rack-and-stack pizza boxes used by others. Pushing CPU density to these limits and beyond, however, instills apprehension among some engineers, since, when you achieve incredible CPU densities on a rack, many disturbing issues arise.

Rack Solution's I-Server houses 13 complete single board computers, each running as an independent system.

According to David Medin, Crystal's Director of Technology, "we experimented with really high density racked systems beyond anything that exists now, by taking our CS500 platform and doubling up on the PICMG CPUs in each box. But we really didn't go any further with that concept because we've learned, being a manufacturer of CPUs and computers since 1991, that heat has a definite proportional impact on the life of the product. We see an extremely strong correlation between heat and product lifetime."

This is why Crystal has an internal corporate policy dictating that they won't design or sell any PC product that causes more than a 10 degree Celsius rise within the box at any point in the airflow. Telephony resource boards, after all, are very sensitive to heat. That dictum has throttled Crystal back to the well-known current density that they provide, though the company is always looking at interesting new cooling schemes (for example, the one they use in their pizza box models).

Aside from the problem of heat, simply bringing enough power to a super-dense rack may be impossible. For instance, in a co-location space ("co-lo" or "cyberhotel"), when a service provider or Competitive Local Exchange Carrier (CLEC) starts to put enough Pentium IIIs in a rack to really "up" the density, the rack as a whole starts drawing more power from the power mains than a cyberhotel can provide in a circuit, which is usually a single 120 amp circuit for a low-end rental. Every time you bring in another 20 amp circuit, they hit you with another Draconian monthly fee. So, if you keep increasing the number of CPUs in a rack, either the facility won't be able to provide you with enough power for the rack, or they simply won't allow that much power consumption in such a small space because they don't feel that they're able to cool that part of the suite sufficiently.

So, yes, it's possible to get the CPU density up in a rack or a room, but then you immediately run into problems where the facility can't cool it, they won't permit it, the power costs too much, or they simply don't have enough branch electrical circuits to supply it.

"The only way we at Crystal have been able to attack these problems is by providing extremely efficient power supplies," say Medin. "Instead of 70% efficiency, which is pretty much the median for a PC, we're looking at 85% on most of our models. To achieve that figure, we had to go out and get special power supplies. We try to give our customers a 15% edge. With our RIA system we can achieve a bit over 50 units in a rack, and we can still give you two slots per unit. We continue to look at these problems, none of which will simply go away."

Perhaps some readers out there remember ChatCom, a company now defunct. In its day, ChatCom attempted to push CPU density to where no system had gone before. Their 30-inch tall ChatPower Plus server used industry standard motherboards that piggybacked onto an interface board, and made available lots of PCI/ISA slots. For those more interested in CPU density, the backplane supported up to 14 front-mounted, hot-swappable server blades. Each blade had its own RAM, CPU and hard drive.

The ChatPower Plus server also supported Sun SPARC server blades, so you could mix and match blades within the same chassis. The chassis could be chained together in groups of three, in which case only one box needed a "community" hot-swappable power supply. The other units drew their power over an interconnect cable. hot-swappable power supplies. The power came from four load-balancing, N+1 fault tolerant power supplies.

One recent challenger in the super-dense rack-mounted server arena is code-named "Razor," from RLX (The Woodlands, TX – 281-863-2100, www.rlxtechnologies.com), a start-up founded by former Compaq executives who were originally responsible for Compaq's server business.

RLX is going after the ASP / ISP end of the high density computing scenario. If you think racks at the customer premise have a space problem, consider the plight of Internet data centers. More than 14 million square feet of data center space will be built during 2001, which is expected to not be enough to keep up with projected demand. Web hosting companies and soon ASPs will desperately want to pack more computing power into less space.

RLX claims that Razor systems can achieve eight times as much processing power in the same space as Intel-based rackmounts, while using a quarter of the power. They are to achieve this miracle by using Transmeta's super-low power Crusoe processors, which will be running Linux, Microsoft Windows 2000 Server and other leading operating systems. Each Razor server takes up three units of rack space and holds up to 24 removable Crusoe-powered blades, or about 150 blades per full-sized rack. RLX's software lets an administrator manage each blade individually from one computer.

The Crusoe processor is interesting in that it stays cool by monitoring the application and determining from moment to moment how much processing power is needed. If not much processing is needed, then the Crusoe chip can drop down from high to low power operation very quickly.

Some industry experts believe that one fly in Razor's ointment is that while the Crusoe chip can shift quickly in response to processing demands, it can't move instantaneously. When you're in a 24x7 application you actually may lose throughput because the processor is spending time shifting from low power states to higher power states. And before it can get to a higher power state, your application, which may have very limited peak processing loads, has already moved on and thus won't get the full attention from the processor.

Intel has also been working in this area, with their "Mobile" line of processors addressing this more esoteric part of the market. Processors such as the Mobile Pentium III are power optimized, so your application can shut off parts of the processor you don't need and you can literally throttle the processing power up or down. But trying to achieve this in such a way so the whole process won't sap power from your 24x7 application is going to be quite a challenge (and this also presumes that Intel will supply such nonstandard processors for more than just a short six-to-nine month stretch, as they've done with some of their other chips).

The Centauri product line from Centauri NetSystems, (Richardson, TX – 469-330-4998, www.centaurinetsystems.com) also has an architecture that follows the idea of processing blades plugging into chassis, but Centauri's blade-oriented architecture departs from others in that they're not using Intel processing solutions, nor the Crusoe chip from Transmeta (as is RLX). Instead, they're using G4 PowerPC technology.

BUS WARS

Carriers, high-end service providers and multinational enterprises dealing in real-time convergence applications have demanded up to "six nines" (99.9999%) reliability from their switching and computing systems. This is why early PCs, with their microprocessor, non-redundant circuitry and mass-produced ambiance, took so long to become a player in the stringent operating environments found at the top of the convergence chain.

The intense reliability requirements for high-end NEBS certified, carrier-class equipment led to a "big iron" view of what could and couldn't work in a central office or multinational corporation, and for many years high-end equipment consisted of dedicated, proprietary switching devices that took years to pass regression testing. Such equipment was inevitably late to market, inefficient, and generally didn't offer many enhanced features.

The appearance of high availability rackmount PCs caused some concern to those mainframe or minicomputer-trained technicians who tended to look down condescendingly at the PC to begin with. Still, the high availability PC makers persevered.

It was felt that fault resilient PCs still needed to run well-known software, but should somehow be running with "tougher" hardware.

CompactPCI

In early 1994, "RuggedPCI" by Ziatech (now part of Intel) was conceived of as a way to bring relatively inexpensive, open architecture, highly efficient, reliable, and easily upgradeable equipment to both circuit-and packet-switched networks. A gentleman named Joe Pavlat – then running a company called Pro-Log (now the Motorola Computer Group), coined the name "CompactPCI" (cPCI) in September of 1994 at a PCI Industrial Computer Manufacturers Group (PICMG) meeting when Dennis Aldridge, a marketing fellow from Texas Microsystems (now part of RadiSys) complained about the name "RuggedPCI," saying that his passive backplane products were already rugged. The new name stuck.

CompactPCI, initially promoted by Ziatech (San Luis Obispo, CA – 805-541-0488, www.ziatech.com), now a part of Intel, is electrically similar to the PCI bus found in a desktop computer, so it can run the same software and operating systems. But cPCI gives cost effective PCI circuitry a rugged home on Eurostyle cards and connectors (originally popularized by VMEbus). Whereas a desktop PCI card uses a card edge connector at the bottom of the board and the I/O connectors are on the side, a CompactPCI card generally uses a 3U (100 x 160 mm) or 6U (233 x 160 mm) board size with a pin-in socket connector at the bottom and the face plate and I/O connectors on the top. The 6U cards can have up to five connectors on the rear of each card. The CompactPCI bus itself uses two of the five connectors, with the other three providing up to 315 pins for user-specified I/O connections. CompactPCI cards are firmly held in position by their connector, and there are card guides on both sides and a face plate which solidly screws into the card cage.

Thus, the high quality 2 mm metric pin-and-socket connector of the CompactPCI card (one of Ziatech's chief contributions to cPCI) is a lot more reliable than a regular PCI card edge connector. Moveover, the power and signal pins on the cPCI connector are staged to support the most tantalizing feature of all – hot swapping, which, along with redundancy, lies at the heart of fault tolerant and fault resilient systems, and which isn't possible on standard PCI.

CompactPCI has many advantages over the conventional PC architecture for applications demanding a high mean time between failure (MTBF) and a low meant time to repair (MTTR).

Because CompactPCI cards are accessible from the front and I/O cables from the rear (cPCI can support front and/or rear I/O cabling), cards can be replaced very quickly. And under the standards brought forth from PICMG, "hot swappable" or "live insertion" boards have appeared, which can be powered down and replaced while a system continues to operate, thus satisfying the quick maintenance requirements of telcos, service providers, data centers, e-commerce operations or any enterprise that needs to maintain continuous operation.

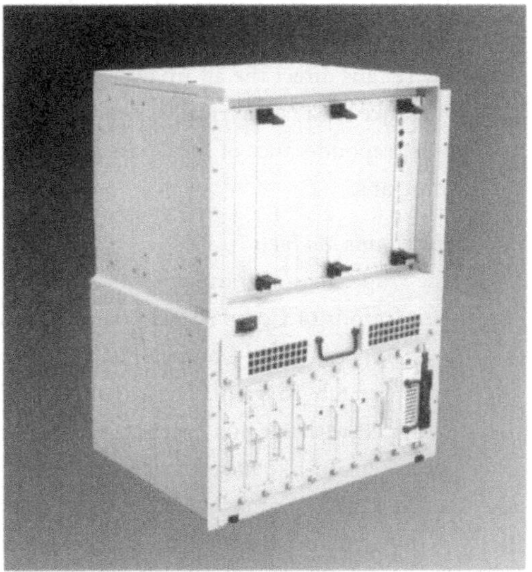

An example of a CompactPCI system: the FTC620 from Diversified Technology (Ridgeland, MS — 601-856-4121, www.dtims.com). This system is a hefty one, even for a 19" rackmount, having a height of 14U or 24.5" (622 mm) — add another 3.5" (89 mm) for optional air plenums) and a depth of 14.5". The FTC620's cPCI card cage is slightly over 20 slots wide. The standard backplane that comes with the FTC620 has slots for a processor module, two PCI-to-PCI bridge modules, and 16 cPCI expansion cards. The processor slot area is two inches wide to allow CPU boards with large heat sinks to be inserted into the slot without bumping into any of the adjacent boards. The expansion slots are all 6U high. The FTC620 supports a myriad of drive configurations. All the drives are mounted on carriers which plug into a mid-panel mounted in the chassis, which allows all types of drives to be serviced without having to pen the system to access cables.

Modern, "convergence-friendly" options for cPCI include -48 volt power supplies to make cPCI compatible with other central office equipment – most of the -48V cPCI technology suppliers offer dual or even triple, N+1 redundant power supply configurations. There was also the adoption of the ECTF's H.110 CT bus for computer telephony resource and I/O boards. H.110 provides an intra-chassis time-division multiplexed digital telephony highway with 4,096 timeslots so that communications resource boards can communicate with each other over a "private" bus of 4,096 multiple 64 Kbps voice channels, not the PCI bus. The equivalent version of this bus for regular PCI rackmounts is the H.100 bus.

A number of manufacturers have placed the Sun's famous UltraSPARC 64-bit processor, the PowerPC and even DEC's 64-bit much respected (and dearly departed) Alpha processor on CompactPCI cards, running more exotic software such as Solaris and real-time operating systems.

This matter of PCI compliance not only provides a wide choice of hardware, but also enables the free choice of operating systems from NT to UNIX to Linux to Solaris to real-time OSes. Meanwhile, the core PCI standard has added features important to communications, including Maximum Completion Time to reduce transaction latency and Message Signaled Interrupts which provides for a virtually infinite number of peer-to-peer mailbox interrupts so that voice cards can process data packets without host intervention.

CompactPCI uses a vertical passive backplane and vertical card orientation for good cooling, which is a must for today's power hungry telephony DSP-powered cards. Cards are loaded from the front and rear, simplifying maintenance. Air is generally brought in the front-bottom of the chassis, where fans direct the air up over the cards and out the rear.

Since cPCI is electrically similar to regular, mass-produced PCI desktop systems, there will always be a preponderance of Intel-based software solutions for cPCI versus PowerPC and SPARC platform solutions.

Mezzanine Madness

The reader may at some point notice that some additional mystifying acronyms have crept into the descriptions of CompactPCI boards. These relate to what are called "mezzanine modules," another by-product of PCI / CompactPCI's rapid evolution. One no longer has to design totally new boards from scratch to gain certain functionality, since there's now a preponderance of cPCI boards equipped with various "sites" for add-on mezzanine cards. Need T1 access? There's a best-of-breed daughtercard-like mezzanine module for you. One side of the board connects to the main carrier board while the other side allows I/O signals to enter/exit via a front panel.

Even if your quest for the perfect mezzanine board comes up empty-handed, it's still easier (and faster) for somebody to design a mezzanine board and market it than a new main carrier board. You can thus "personalize" your I/O functions, tailoring it specifically to the needs of your application.

Like everything else in this industry, there's a bewildering array of mezzanine form factors, though some standards have emerged: The M-Module is used mostly in Europe for industrial control applications, while the new and slim PC-MIP module bus has electrical and logical layers similar to PCI and allows modules to have components on both sides. Then there's the communications industry favorite – the PCI Mezzanine Card (PMC). Even Sun Microsystems, which used to team so-called S-bus modules with their SPARC CPU boards, recently switched over to PMC technology.

As you'd might expect, the standards body PCI Industrial Computer Manufacturers Group (PICMG) has a mezzanine specification. PICMG 2.15 enhances CompactPCI as a user configurable platform for telecom and datacom oriented I/O controllers in a mezzanine form factor. PICMG 2.15 describes what's called the PCI Telecom Mezzanine Card (PTMC). Based on the standard four connector PMC module, the PTMC can coexist with older 32-bit PMC signaling but also interoperates with newer 64-bit PCI devices. PTMC is thus not a replacement for PMC but will coexist with it, supporting common industry standard telecom bus interfaces.

For example, the MEN D3 board from MEN Micro (Carrollton, TX – 972-939-2675, www.menmicro.com) is a one-slot 6U CompactPCI single board computer that's available in three different standard configurations to accommodate three different types of mezzanine card I/O: M-Module I/O, PC-MIP I/O, or PMC I/O. The board can have three M-Module, three PC-MIP or two PMC mezzanine sites (the latter model, the D3C, is popular among convergence developers).

To summarize, it was felt that CompactPCI equipment could be made sufficiently fault tolerant so that the $170 billion carrier-class telecom market would finally whole-heartedly adopt PC technology, replacing large, expensive, proprietary switching platforms with less expensive, open standards-based, programmable PC switches. It would be the last and greatest hurrah for microprocessor-based computer telephony, which had appeared 15 years before in simple IVR systems and add-on voice mail servers.

But as CompactPCI finally "jelled" (it's taken years for all of the specifications to be approved) it became apparent that cPCI still suffered from some limitations, as we shall see.

VMEbus

Long before the current mania over CompactPCI, the first heavy-duty computer bus to appear that became popular was the Versa Module Europa (VME) bus. It was the only game in town when it came to high-end telecom, military, industrial, and other real-time computing systems. VME, which dates all the way back to 1981, has a huge installed base. It's an open-ended, flexible computer backplane bus with a 32-bit wide data path, built upon the Eurocard standard (typical card sizes are 160 x 216 mm and 160 x 100 mm).

Since it has been around for so long, VME supports a huge number of protocols which allows newer, faster products to be added to a system while at the same time supporting older boards. At last count, VME also supports an incredible number of real-time operating systems. It has an efficient interrupt scheme and it lets you put a full 21 slots on a backplane.

Despite spurious predictions by industry experts of the rapid demise of VME, many high-end systems are built on VME / UNIX platforms, and it has taken some considerable lab data and demonstrations to persuade the "Powers That Be" to switch over to later CompactPCI / Windows platforms.

Most companies use VME64, which has a theoretical maximum transfer rate of about 80 MBps (320 MBps for some proprietary implementations), but in practice is about 40 MBps.

Somewhere along the line, however, VME got "fat, dumb and happy." In the late 1990s, CompactPCI, with its hot-swap card capability and its faster bus (264 MBps in theory, about 110 MBps in practice) stole the show, taking over the telecom market in the process.

In January 2002, however, "The Great VMEbus Revival" was launched at the annual Bus & Board Conference (www.busandboard.com) in Long Beach, California. There were a startling series of announcements from various companies belonging to VITA, the VME International Trade Association.

Leading the charge is the Motorola Computer Group (Tempe, AZ – 602-438-3000, www.mcg.mot.com), one of the inventors of VME – but also known, ironically, for their fault tolerant CompactPCI platforms.

Motorola calls their new VME roadmap "the VME Renaissance."

Today, VME supports 3U and 6U high cards that are 160 mm. deep. When mezzanine add-on cards are used, they tend to be based on the 64-bit wide, 66 MHz PCI local bus. Chip-to-chip communications on the cards are also predominantly PCI.

The current "control plane" or the actual VME bus itself tends to be VME64. To bridge from the local PCI bus on the board to the VME bus control plane you need a VME64 compliant bridging chip such as the Universe II from Tundra Semiconductor (Kanata, Ontario, Canada – 613-592-0714, www.tundra.com).

The control plane can't do everything, however. Demanding applications such as medical imaging, radar, sonar, or any kind of "heavy lifting" in terms of data movement can't be done over the regular VME control bus, so computer makers added another "data plane" bus in addition to the VME control plane. This data plane can be proprietary or one of the "quasi-standard" interconnects now available such as RACEway from Mercury Computer Systems (Chelmsford, MA – 978-256-1300, www.mc.com) or SKYchannel from SKY Computers (Chelmsford, MA – 978-250-1920, www.skycomputers.com). Typically these data planes are parallel switched interconnect buses.

When these additional buses are used, the regular VME bus is relegated to doing synchronizing and coordination kinds of events: Setting things up and tearing them down – what we call "signaling" in the telephony domain. But it's in the data plane where you'll see some pretty slick technologies eventually adopted such as serial switched interconnects (e.g., InfiniBand, see below).

Motorola's plan is as follows: in 2002, the PCI-X local bus will be added to the mezzanine interconnects, usurping regular PCI. PCI-X is a souped-up version of PCI, a 64-bit bus running at 133 MHz that can deliver burst transfer rates above 1 GBps. It will appear in servers to satisfy the bandwidth demands of Gigabit Ethernet and 160 MBps Ultra3 SCSI cards. The PCI-X bus is faster than other local buses (such as regular PCI) but still allows you to be "processor agnostic" (some technology forces mezzanine chips to connect directly to the processor bus, restricting their use to that particular kind of processor).

In the control plane, rather than using VME64, products will use a protocol technology called 2eSST (2X Source Synchronous Transfer). Approved as a VME standard by VITA in 1999, it's a signal modulation technique that can transmit two bits per processor cycle along the bus, giving 2eSST an 8X bus bandwidth performance improvement (640 MBps theoretical, 320 MBps in practice). Texas Instruments (TI) has developed transceivers to achieve this modulation, based on a technology called Incident Wave Switching.

What's been missing up to this point, is a bridging chip that connects the new PCI-X enabled VME cards to the VME bus itself running the 2eSST protocol. Motorola's answer is a chip that's code-named Tempe. It's a PCI-X to 2eSST chip.

Amazingly, Tempe and the TI transceivers allow the new super-fast VME cards to be backward compatible with the huge existing base of VME backplanes and cards. Tempe-equipped cards can talk at the new 2eSST speeds but they can also talk at the regular speeds as well. You can thus plug old cards and Tempe-enabled cards together on the same bus, and you can mix and match them and they'll all work together nicely.

Individual data transactions between cards are going to take place between two cards at the speed of the slower card. So if you have a 2eSST card and an older card and they want to talk, they'll communicate with each other at the older, slower speed. But if you also have two 2eSST cards, they'll talk at the faster 2eSST rates. This is like a Montessori school where everybody interacts as fast as they're able to. And this is unlike CompactPCI, which is more like a public school – the whole bus runs only as fast as the slowest card plugged into it.

Whether Motorola and other companies are about to bring forth a new Golden Age of VME remains to be seen, but it does give CompactPCI a run for its money in the high-end, ruggedized computing business.

CompactPCI vs. VME

CompactPCI's development was not all sweetness and light. VME engineers love to trash cPCI for various reasons. It's taken a while to get the cPCI specs nailed down. One of the challenges faced was taking the PCI bus architecture, which wasn't designed for hot swap or high availability, and making it ruggedized.

Adding value to a known architecture such as PCI is a different approach than just herding a bunch of engineers into a room and getting them to design a "perfect bus" from the ground up, which is what happened in the case of the failed computer architecture called FutureBus. FutureBus was a fun utopian vision of what a computer bus should be, and it even held a lot of promise, but in reality it never convinced a critical mass of suppliers that it could ever be viable, probably because such totally new concepts always start out as merely abstract ideas and not working hardware. There's often just too much time elapsing between the initial announcement of a new concept and the actual deployment of systems based on it.

This is why in the case of cPCI the engineers tried to bypass that problem by taking a popular architecture (PCI) already in existence and morphing it over time into a solution (cPCI) with the characteristics needed to succeed in its intended higher-end market. This ultimately involves taking a large existing PCI supplier base and converting it over gradually to cPCI with slow steady persuasion rather than trying to take a single huge step into brand new technology, something that's very difficult to do.

Why, it's even *easy* to design a board using PCI circuitry since there are scads of tools, many custom Application Specific Integrated Chips (ASICs) for the PCI environment, and other such readily available items that make it easier to design and build a PCI based board than, say, a VMEbus board.

So the barriers to designing a cPCI board are actually less than the barriers to entry in the market, at least when compared to an established, competing bus such as VMEbus.

The other thing to keep in mind is that PCI has become a near-universal standard at the local bus level. Ziatech (now part of Intel) had the same problem when designing boards for their older STD 32 architecture. They were almost forced to use PCI because all of the chipsets were PCI compliant and in order to be able to use them they had to create a layer of logic circuitry to convert signals from PCI to Ziatech's own STD 32 architecture. The same thing is true with VME. If a manufacturer is going to build a VME CPU board the PCI bus will be lurking somewhere in the solution. It's just hard to avoid. And to hook up a VMEbus and a local PCI bus, VME engineers need to devise bridges and generic chips. You may be able to eliminate PCI from your design, but these days it's pretty hard to build something that doesn't have PCI in it at all.

Adding a layer of logic to convert back to an older bus such as SDT 32 or VME is not all bad, since such a scheme allows a system to be backward compatible to the earlier, legacy technology. But it's pretty hard for consumers of such a product to pay a higher purchase price for additional pieces of silicon (such as VMEbus to PCI bridging chips) that they may not even know are running in the background anyway. It's difficult for VME CPU board vendors to offer a cheaper solution than cPCI, because they have to burden their solution with PCI / VME translation logic. CompactPCI has less cost and complexity. And any time you have fewer transistors in the design, it's going to improve your overall reliability.

At this point the VME camp counters that VME component pricing was greater at one time only because VME vendors were used to higher margins – having the military as a customer can be quite lucrative. After all, VME and CompactPCI look very much alike and can be built on similar assembly lines.

As for PCI (and hence cPCI) technology being inherently cheaper, it's true that you can go to your friendly local computer store and buy a PCI parallel port card for $30, while a VMEbus version can cost you $300 or more. Therefore, everybody thought cPCI was going to be really cheap because it has inexpensive PCI bus interface parts, but the reality is that it isn't a parts cost issue, it's an economies of scale issue. Anybody can make a $30 PCI board if they send it to China and have 100,000 units a month made, which actually happens. Build 100,000 VME cards a month and they'll come down in price too.

What it all boils down to is that, given similar production volumes, CompactPCI and VME cards can be sold at pretty much the same price, with a slight advantage in cPCI's favor. To get the kind of savings from economy of scale that you'd expect from a derivative of mass-market PCI such as cPCI, you need real high volume production to occur.

This is why cPCI makers targeted telecom applications right from the beginning – telecom companies are capable of acquiring a particular chassis or board in tremendous volumes, tens or hundreds of thousands or more pieces per year, while a "large" sale of a particular VME board for industry or the military can be 1,000 or fewer pieces per year. It thus may come as a surprise that most non-telecom companies (and even some smaller telecom companies) that buy VME or cPCI systems are buying less than 1,000 systems a year. In fact, the vast majority are buying much less than that, just a few hundred systems or even less. Of course, when you add up all of these little guys you get "the big market."

In any case, it was felt that the large-volume production of cPCI boards and chassis needed to satisfy telecom's insatiable appetite would keep cPCI's components far less expensive than VME's.

In this respect CompactPCI came along at exactly the right time, and the cPCI promoters were very lucky in many ways. CompactPCI was introduced at the moment when telecom was being deregulated and all sorts of new and exciting pieces of computer equipment were being evaluated in terms of reworking the world's telecom infrastructure. Many people were using comparatively unreliable cheap PCs and lackluster rackmounts for telephony applications, and suffering because of it. CompactPCI garnered a fair amount of business from those entities, becoming an alternative solution, not only to the PC, but to VME and even conventional PCI bus rackmount machines.

VME had been around for decades, but people looked at it in "Internet time" and squawked that "it's more than three years old so it must be obsolete!" In fact VME works just fine for just about any application. Moreover, some VME aficionados ironically point out that although cPCI is indeed gaining popularity with large telecom companies, these companies tend to build much of their own equipment. Did you know that the largest user of power supplies in the US is Lucent Technologies? And guess who builds many of the power supplies for them? Lucent, that's who. Various divisions of companies like Motorola and Siemens are big buyers of their own company's products. They want to keep their factories busy and they actually make more money that way.

"As many operations and manufacturing specialist will tell you, the breakpoint, or the point at which it makes sense to bring a board's manufacture in-house, is somewhere between 700 and 1,000 pieces a year," says Wade Peterson, author of ***The VME Handbook*** and president of Silicore Corp. (Corcoran, MN – 612-478-3567, www.silicore.net) a company that builds very high density RISC microcontroller cores and "system-on-a-chip" devices.

Peterson told the author that when the real heavyweights in the industry (the Lucents, Ericssons, and so forth) need a small number of a certain type of board ("stray dog and cat boards", as he calls them) they just go out and buy them. "But if these companies need a half million boards, they'll build it themselves or farm out the manufacture to companies in places like China," says Peterson.

"Typically, when a manufacturer is building small quantities of a board, the rule of thumb is to take the parts cost and multiply it by five and that's the board's end selling cost," says Peterson.

Peterson also says that "CompactPCI does have a future. What it comes down to is this: There will always be the need for a front-loading PC architecture. Front loading boards are strictly a mechanical issue. If you want to fix a PC that's running a large telecom application or a robotic assembly line in an automobile plant and nothing is working, you've got to fix things as soon as possible, since you're losing thousands of dollars a minute. You don't want to be pulling rackmounts (or entire racks) out and fiddling around with conventional PCs, you want a front-loadable card that can be swapped out immediately. At Control Data Corporation they used to call that 'Easter egging' since you'd have to find the Easter egg, or faulty board. Live insertion or 'hot swap' is another feature of interest to telcos. Ironically, many real-time systems don't rely on hot swap. People just want a convenient way to fix things."

Sun-based Platforms

Long ago, Sun Microsystems (Palo Alto, CA – 650-960-1300, www.sun.com) targeted computer telephony and converging communications and sought to develop fault tolerant hardware.

Companies such as GNP Computers (Monrovia, CA – 626-305-8484, www.gnp.com), Integrix (Newbury Park, CA – 800-300-8288, www.integrix.com) and Continuous Computing (San Diego, CA – 858-882-8800, www.ccpu.com) specialize in building fault tolerant NEBS-compliant systems based on Sun's UltraSPARC and Solaris technology for media servers, SS7 gateways, call processors, and announcement and conference servers. One of Continuous Computing's more interesting telco building blocks is the Telco Protocol Engine (TPE) for CompactPCI, a SPARC-based, NEBS-compliant, 19-inch rackmountable unit that integrates two completely independent computing nodes into a single chassis.

PICMG 3.X / AdvancedTCA

AdvancedTCA – the Advanced Telecom Computing Architecture – is the umbrella term for the PICMG 3.x series of specifications for next-generation telecommunications equipment that was announced in January 2002.

AdvancedTCA is aimed specifically at the narrow, super-high-end telecom market. It's for "CO in a box" systems, not the desktop. AdvancedTCA is not just an extension of CompactPCI – indeed, although an AdvancedTCA box looks like a CompactPCI machine on the outside, there isn't any PCI (or any other kind of parallel bus) in it at all. Instead, the AdvancedTCA form factor is meant for housing the new switching fabrics such as packet-switched backplanes, InfiniBand, and StarGen's StarFabric (these are discussed later in the "Automatic Failover" section).

At the end of 2001, PICMG's AdvancedTCA committee voted to increase the cPCI board size (from 6U x 160mm to 8U x 280 mm) so that it can hold four mezzanine modules and have more room for I/O connectors, wider slot spacing (1.2" pitch), more cooling, high speed connectors, redundant -48 volt power distribution and more coherent system management. These will be defined in the core 3.0 spec along with a 16 slot "square mesh" backplane.

The mesh provides eight high speed (up to 5 GHz) differential pairs between each board and every other board, with a capacity up to 2.5 terabits per second (Tbps) per shelf. So, instead of a T1, quadspan or octal board, imagine a super-high density OC-768 board. Such a system should be able to handle terabit routing and deliver video on demand, 100 Mbps Ethernet and fiber services to the home.

Subsequent 3.x specs will define how the different switching fabrics map onto the 3.0 core backplane. Added to the AdvancedTCA tree are PICMG 3.1 that will define a Gigabit Ethernet fabric, PICMG 3.2 for InfiniBand, and PICMG 3.3 for StarFabric. Presumably, there will be some future version for Intel's Arapahoe/3GIO. All of these fabrics should in theory be able to use the same backplane, connectors, chassis, etc. Only the protocol being used over the wires will differ.

PICMG 3.x/AdvancedTCA won't displace the existing (and complimentary) CompactPCI form factor, which will instead find itself competing with faster versions of VMEbus equipment.

STRATEGIES FOR FAULT-RESILIENCE: TEMPERATURE CONTROL, MONITORING, PREDICTION, FAILOVER, MAINTAINABILITY

Over the years the author has made much of the term ***redundancy*** when it comes to fault resilient or high availability computers. Redundancy in an HA system means that "single points of failure" can be eliminated by duplicating components: There are at least two power supplies, disk drives and CPU boards in a system.

Heat is the Enemy

One of my favorite stories about 19" rackmounts concerned a frantic call from the Pentagon. They had a rackmount (one of the best on the market) configured to be a big telephone conferencing bridge, and so it was stuffed with 20 high-capacity telephony conferencing board (from a manufacturer that will remain nameless for reasons that we shall soon discover). Each board was so high and wide that there wasn't much space between them – or anywhere else, for that matter – which in turn severely restricted the air flow in the computer. Thus, the cooling fans (or "blowers" as they're called) couldn't move enough air through the system to keep it sufficiently cool – there was too much "back pressure." The system got so hot that it stopped working. What to do?

As things turned out, the solution was to take off each of the boards' electromagnetic shielding. Slimmer boards = more space between them = more air flow (and the board manufacturer immediately redesigned that model of conferencing board).

High availability computers can hold many more boards than ordinary desktop computers, and can hold many more hard disk drives and other devices in its drive bays as well. These components give off heat, as do the power supplies that drive the system, so cooling such a mass of equipment (all of it confined in a metal box, no less) becomes an increasingly difficult problem.

Hot Power Supply, Hot Potato

As more and higher-density (higher capacity) components are crammed into a computer, larger power supplies are necessary. Modern CPU boards and other circuitry use increasingly lower voltages as the miniaturization of components increases. To maintain power output at lower voltage, current has to increase. As faster and faster microprocessors are developed, more current must be directed to the chips. Actually, current demands rise even higher still, because of an increase in the number of components. More components and larger power supplies both result in higher temperatures generated in the PC enclosure.

Heat must be properly handled or power supplies will fail, causing system wide failure. Also, communications equipment is sensitive to heat, which is a particular problem in mission-critical telecom networks today.

Thus, heat is the Number One Problem in computer design.

There are three ways of dealing with heat:

1. Build more electrically efficient components that generate less heat,
2. Move the power supply outside of the rackmount enclosure.
3. Dissipate heat and remove it from the system, usually with fans (blowers).

Some companies are taking the first route. Take for example Broadband TelCom Power (Santa Ana, CA – 714-259-4888, www.btcpower.com), known as BTCPower. Their solution is to go to one of the major sources of the heat problem, and devise a new power supply transformer design that can dish out power at both low voltages and high amperages.

BTCPower's view is that the problem with conventional transformers in power supplies is that they develop a single hot spot at the center of the transformer core. Instead of a typical single transformer core having multiple windings, BTCPower developed a flat transformer composed of multiple cores with a single winding. This yields faster switching time, lower copper losses, and reduced stress on ancillary parts. Power supplies designed using flat transformer architecture quickly distribute heat, thus maintaining a lower internal temperature.

What Power Supply?

The second route for dealing with heat, which we listed previously, "move the power supply outside of the rackmount enclosure" has led to a totally new design paradigm for rackmount computers.

To review, until now your options for achieving high density with rackmount-based systems fell into three basic categories. At one extreme, you could pile 1U and 2U pizza boxes on top of each other. At the other extreme, you could adopt the "blade" paradigm, where a compact 4U high rackmount houses 13 or so front-accessible single board computers, each running as a separate system, complete with its own onboard 2.5-inch drive and Ethernet ports. Somewhere in the middle is the "swappable shoebox computer" approach pioneered by Crystal Group and imitated by others.

Now, however, Tracewell Systems (Westerville, OH – 614-846-6175, www.tracewellsystems.com) has a fourth alternative to achieve high density – the "Multi-Node" architecture. Tracewell's new system chassis employs ingenious interconnect boards that can integrate a full-size PCI Mezzanine Card (PMC), two external PMC sites, and one embedded PMC/PCI slot in an 1U or sub-1U chassis. These packages allow you to create some of the most space-efficient rackmount computer packages ever for convergence applications, without resorting to using space-robbing PMC carrier modules or extenders.

The new Multi-Node chassis gives you the cost benefits of existing PCI technology, the robustness and CT Bus connectivity of CompactPCI, and 20% higher computing density than traditional packaging strategies. How? The secret of this new architecture is the segregation of CPU modules, power supplies and cooling into separate, more efficient, dedicated units.

Essentially, the Multi-Node concept consists of a 10U, 19"-wide, rackmount chassis that's subdivided into a computer module section and a separate, centralized power / cooling section. Systems are built in 10U segments, each of which can contain up to 12 complete computer modules; each computer module is a separate, removable sub-chassis that can house PCI and PMC cards, custom circuitry, and I/O.

With Multi-Node, system integrators now have tremendous latitude to choose the most effective combination of PCI processing and peripherals for an application, while the Multi-Node chassis provides the structural integrity and hot swap capability of CompactPCI. Also, Tracewell's micro-

Here we see 12 of Tracewell's new Multi-Node chassis stacked upon each other, all of which share the three N+1 power supply / cooling modules at right.

processor-based Chassis Monitoring and Management (CMM) module can be added to provide remote control, remote monitoring, data logging, power cycling, and event recording.

Multi-Node has redundant power and cooling. A centralized array of industrial power supplies (AC or DC) are installed in a redundant N+1 configuration. Similarly, redundant system cooling is supplied via a centralized array of fans.

This unique segregation of power supplies and cooling from the CPU modules means that the individual computer modules can be simplified and made smaller and less expensive because power supplies and fans are not duplicated in every module. And since the power supplies are kept away from the CPU modules, so is the heat they generate, thus maintaining more consistent temperatures for enhanced reliability.

Now that all of these components no longer have to be designed to exist in the same chassis or even on the same board, further cost savings are realized through the refinement and standardization of each type of component.

For smaller applications, Tracewell offers a "Single-Node" system, which takes the same Multi-Node interconnect hardware and adapts it to a standalone, self-contained 1U chassis. The 1U version contains a 175W power supply, a cooling system, and up to four sites for PMC, PCI, and PCMCIA cards. A Single-Node system provides many of the benefits of the split-chassis Multi-Node design, including high density, remote monitoring/control capability, and cost effectiveness.

Since Multi-Node is incorporated into systems during the design phase, the pricing depends on user specifications.

Blow Ye Wind

As we've seen, increasing heat generation is causing power supplies to slowly undergo dramatic design transformations.

Of course, power supplies are not the only source of heat. Packetization and telecom I/O boards keep doubling in capacity and are gulping power as never before, putting whole systems in danger. This is why nearly all fault resilient rackmount PCs take the third route listed above for dealing with excess heat – removing it. High availability computers still rely on forced air convection methods to keep all of their components cool, a technology which is still based on little fans, or "blowers" as they're called. Each computer has several fans in a plenum, circulating cooling air throughout the card cage, then vented out the back of the chassis. By definition a fault tolerant airflow subsystem has multiple fans and allows the system as a whole to operate indefinitely with a single fan failure.

To give you an idea of how much air it takes to cool a large system to an acceptable temperature, some rackmounts by Alliance Systems can completely replace the interior volume of air in just seven seconds.

The NEBS-certified, hot swappable, intelligent RiCool blower from Rittal. Here we see two of them in action cooling off a Compact-I chassis.

The amount of airflow (Q) in Cubic Feet per Minute (CFM) needed to cool an enclosure can be calculated if you know the total amount of heat to be dissipated (W) and the allowable temperature rise (T) in degrees centigrade. The formula $Q = (1.76 \times W)/T$ gives the amount of volumetric airflow required to cool a system under free air conditions without accounting for what's called "pressure losses."

Pressure losses are interesting in that as you keep adding components to a system, the airflow becomes more and more restricted, and actually resists attempts by the poor little fan to cool it. The typical electronic enclosures experience a "static back pressure" equivalent to the weight of 0.15 inch of water caused by resistance to airflow.

Before CompactPCI, back in the heyday of VMEbus computers, it was vary rare to have any I/O go out of the back of a telco computer. Because static back pressure was very low, and boards consumed only about 30 watts per slot, this allowed unexceptional "muffin" fans to be used in the rear of the subrack to keep things cool (a subrack is a fancier engineering term for the system chassis which houses printed circuit boards, power supplies, and other plug-in modules; fabricated from aluminum extrusions and stampings, a subrack can frequently be referred to as a chassis, shelf or card cage).

With the coming of CompactPCI, however, power consumption now averages about 45 watts per slot, and the density of the boards has doubled over what was found in VME boxes. This means that you now have a lot more static pressure build-up within the subrack that will resist all efforts at cooling. And since cPCI popularized the idea of crowding lots of I/O cabling at the rear of the system, you no longer can put plain old muffin fans back there. The static pressure also prevents muffin fans from being placed at the bottom of a subrack.

One of the best solutions to this problem is the NEBS certified, 1U high RiCool blower from Rittal (Springfield, OH – 937-399-0500, www.rittal-corp.com). This is a self-contained cooling module. Two of them fit neatly in a tray at the top of a 19" cPCI Rittal COMPACT-I system. The RiCool blowers pull 220 CFM of air from the unit's bottom front air intake, draw it between the cPCI boards, and then blow it out the back. There's no need to use extra ducting with RiCool blowers, since the powerful backward-curved impeller blowers can turn the airflow by 90 degrees.

RiCool blowers are intelligent, having an onboard CPU chip and circuit board. They can understand input from heat sensors, and can count the rate at which fan blades are passing by. If the fan blades slow down and stop, the RiCool will send an alarm signal.

The blowers have variable speed control. By running the fans at half speed, you can extend the life of the blower. By using either internal or external thermistors, you can determine at what temperature you want to run the blower at half speed, and at what temperature you want to bring the blower up to full speed. For systems running at 500 or 600 watts, both fans can be run redundantly at half speed. If one blower fails, the other can detect a rise in temperature and automatically be brought up to full speed, cooling the system by itself. For very hot systems consuming 1,200 watts or more, however, both blowers should run at full capacity.

RiCool blowers have a Mean Time Between Failures (MTFB) of about 60,000 hours, and a Mean Time to Repair (MTTR) of about 15 seconds. They easily slide into a Rittal subrack because when Rittal designed their subracks they designed a special housing area that the tray and fans can slide into.

The RiCool blower tray also offers an alarm output via a fan speed sensor, speed control, easy access, and hot-swap capability.

If you happen to be using different blowers but want to have some sophisticated control over the fans, take a look at the new B024-684 integrated Fan Controller and Temperature Monitoring Board from Hybricon (Ayer, MA – 978-772-5422, www.hybricon.com). Measuring at just 1.5 x 7.3 inches, it has fan speed control, fan faults (tachometer and locked rotor detection) and temperature monitoring are available. Up to four fans and three temperature sensor can be accommodated.

The temperature and fan fault signals are available as pass/fail signals (over the utility interface on J19 and J20 connectors) or as individual flags (over the parallel interface on J13 and J14). The utility interfaces (J19, J20) also provide the ability to drive external LEDs, relays, etc. with no additional hardware.

The board runs off a single 12 V power supply via a standard PC style disk drive power connector. A variety of interface options are available, including: Opto Isolated, Parallel and Open Collector. The Fan Controller is compatible with Hybricon's other monitoring modules.

But having powerful blowers alone may not handle all of the heating issues in the enclosure. In regards to the thermal properties of a subrack, Rittal and other companies have also developed airflow blockers – also called "slot blockers" or "air dams" – to help direct the cooling airflow. A slot blocker is a small plastic module that snaps right into an existing card guide. For example, if you have boards in slots 1, 2, 3 and 6, 7, 8 and you want to block the air going through empty slot 4, you just snap a slot blocker in an empty slot and it will force the air to pass only over the boards that you actually have installed. The airflow blockers help you control the airflow path and force the air to go where it's really needed.

Hybricon's airflow blockers can also be placed to block the airflow in and out of an empty slot, preventing "chimney effects" where too much airflow is traveling where it does really cool many components. Hybricon's front panels provide the system with an EMI seal (in both CompactPCI and VMEx versions) and contain provisions for mounting internal hard drives. They even have VME P1 connector jumpers that daisy chain signals to allow bypassing of the vacant slot. The blockers come in sizes from 3U to 9U high.

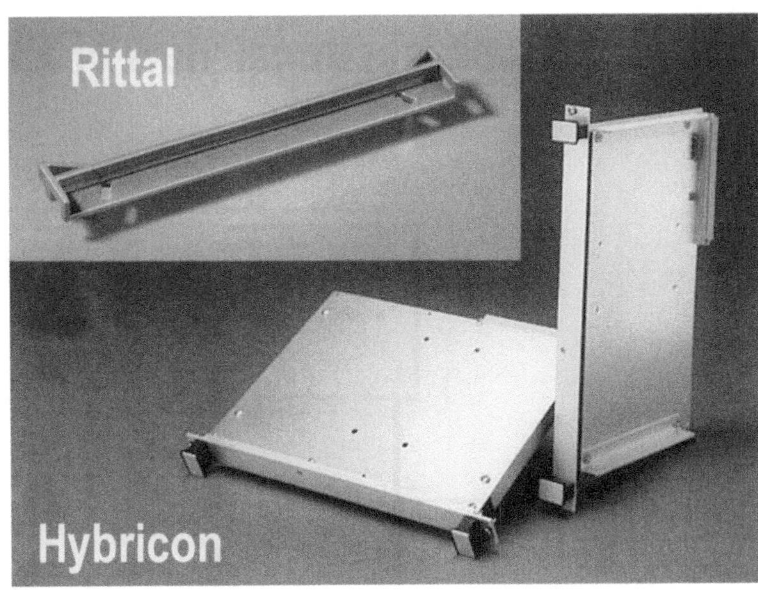

Airflow blockers such as these from Rittal and Hybricon help direct an airflow path within a computer to cool components.

APW Electronic Solutions (Waukesha, Wisconsin – 262-523-7600, www.apw.com) has come up with an interesting cooling system design that is inherently fault tolerant and also highly efficient without the need for fan control or conventional airflow blockers.

Since the design of most enclosure systems require air to intake in the front and exhaust in the rear, air to flow is forced in an S-shaped pattern making 90 degree bends in the bottom and top of the enclosure. Thermodynamic work is required to change the airflow's direction and in all cases irreversible velocity and cooling losses occur. APW's fan arrangement uses multiple fans to direct the airflow in a 90 degree bend without the need for additional "steering" devices such as baffles. This is done by angling each stage of fans from 0 to 90 degrees (see diagram on next page). The angle and location of each fan can be optimized for a given chassis layout. High cooling efficiency can be achieved with this fan arrangement because the entire bottom of the fan module comprises the air intake whereas the air intake of conventional fan trays are limited to the cross sectional area of a single fan.

Still, the cooling of high thermal loads in system enclosures has always been a challenge with traditional fans and cooling strategies, and becomes even more difficult with today's state-of-the-art circuit cards and increased packaging densities. The obvious approach, but one providing diminishing effectiveness, is to simply add larger, more powerful fans, or to increase the number of fans throughout the chassis. However, the resulting increase in cooling air turbulence frequently results in dead spots or recirculation loops that can increase heat density across circuit boards and components by as much as 10 degrees Celsius or more. Devices located in such areas can experience shortened life or degraded performance as compared to properly cooled devices.

Research in this area has led to yet another approach to system cooling: Advanced Vector Controlled Air Flow (AVCAF) technology, developed by Raytheon and popularized in high avail-

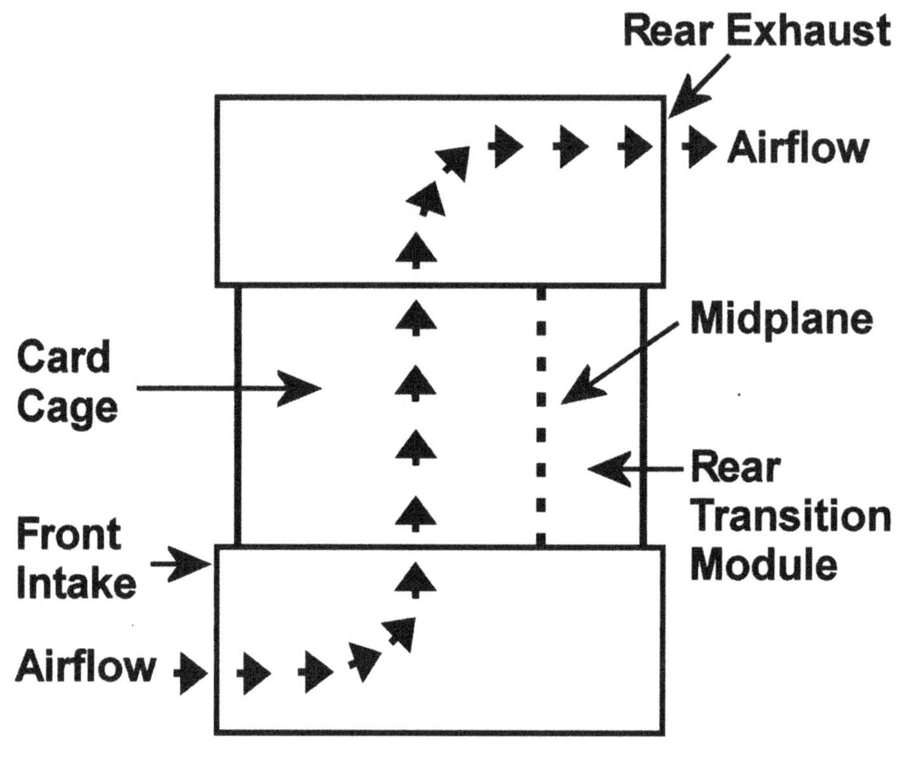
Cross-sectional view of airflow path
Rear Exhaust
Airflow
Midplane
Card
Cage
Rear
Transition
Module
Front
Intake
Airflow

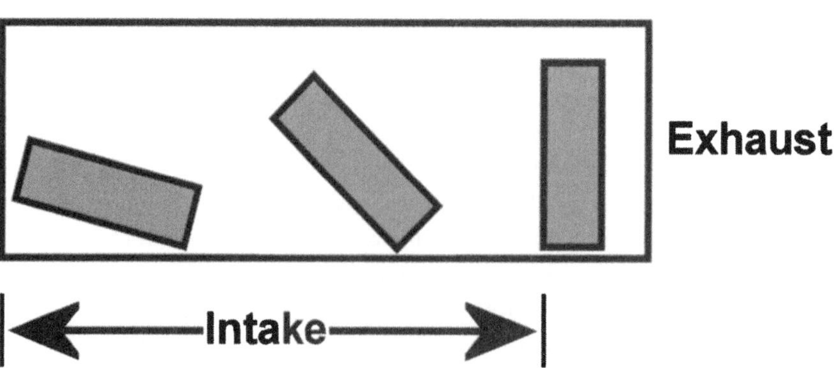
Side view of module showing
APW fan arrangement
Exhaust
Intake

ability PCs by Tracewell Systems, Inc. (Westerville, OH – 614-846-6175, www.tracewellsystems.com). AVCAF is a radical new design method that provides superior cooling and reliability of VME, VXI, cPCI, proprietary bus systems, and the latest high-power switch fabric architectures. AVCAF delivers a highly uniform flow of cooling air within each slot in a system chassis, without creating a substantial increase in air flow resistance. This results in more consistent temperatures across all circuit cards and components. AVCAF has already been proven in rigorous military and aerospace applications, and is thus suitable for mission critical telecom and datacom systems.

A Tracewell rackmount PC System incorporating Advanced Vector Cooling Air Flow (AVCAF) technology.

AVCAF relies on subtle adjustments in airflow direction to significantly improve the flow balance across system boards, without excessive pressure loss. AVCAF can allow for power dissipation densities up to a factor of four – greater in a given enclosure volume than conventional cooling techniques. In a typical measurement of air flow versus chassis slot position versus depth along the card, tests indicate that air velocity is made dramatically more uniform with AVCAF. The average velocity in each test was approximately 1000 linear feet per minute (lfm). The difference in entrance velocity diminished from +/- 1000 lfm without AVCAF, to +/- 400 lfm with AVCAF. Similarly, the difference in exit velocity dropped from approximately +/- 400 lfm, to +/- 100 lfm.

Heat Pumps, Heat Pipes

As we've seen, as more transistors are crammed into smaller areas, and cycled on and off at higher speeds, more heat is produced. CMOS devices are now pushing the limits of conventional forced air cooling. As thermal densities continue to increase, more advanced (or perhaps simply more desperate) cooling methods are starting to appear.

For example, Network Engines (Canton, MA – 781-332-1000, www.networkengines.com) makes a 1U high pizza box (they call it an Internet server appliance) called WebEngine Sierra. It runs NT or Linux, and supports large Web-based applications for Web hosting and service providers. These little servers are easily clustered and users can combine up to 42 of them in a single rack. This much hardware in a small space generates some significant heat, so, aside from their hot swappable disks and fans, Network Engines designed their unit with a unique cooling system.

Essentially, they borrowed some technology from the laptop world to keep each WebEngine Sierra cool. Network Engines uses what's called "heat pipes" or "heat pumps" – copper tubes with a circulating liquid inside – to dissipate heat. These heat pumps transfer heat from the chips to

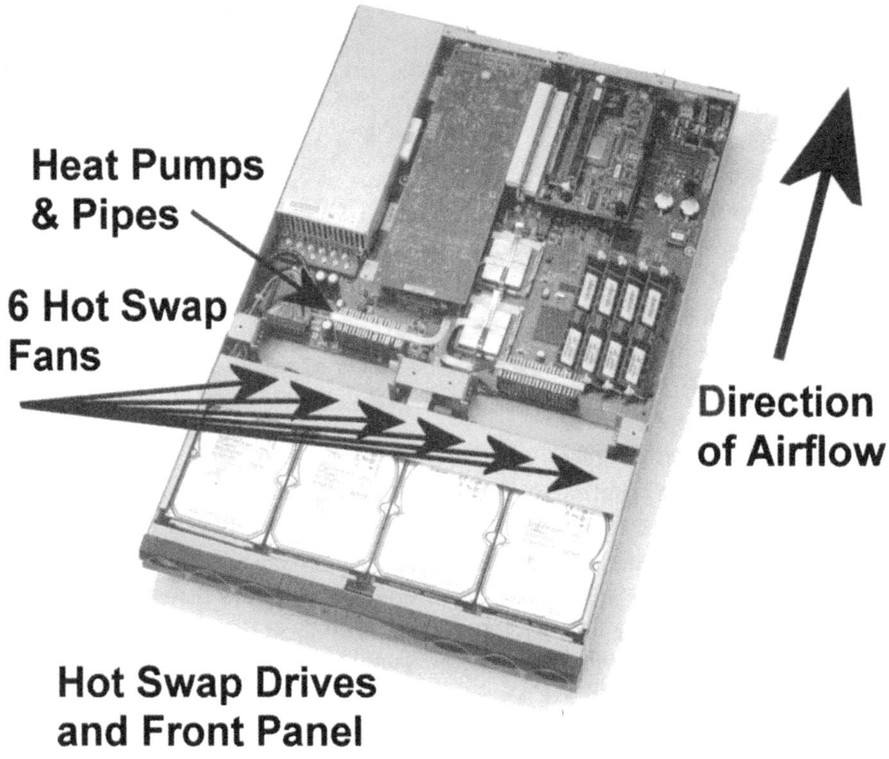

some cooling fins. Each fin is cooled by three fans, which means that the failure of any one of them won't cripple the system.

The classic heat pipe consists of a vacuum tight envelope, a wick structure and a working fluid such as ammonia, water, acetone, or methanol. A partial vacuum is formed in the heat pipe, which is then back-filled with just enough working fluid to saturate the wick. As heat enters at the pipe's hot end, or "evaporator," vapor is generated at a slight pressure. This higher pressure vapor travels to the condenser end where the slightly lower temperatures cause the vapor to condense, giving up its latent heat of vaporization. The condensed fluid is then pumped back to the evaporator by the capillary forces developed in the wick structure.

One major manufacturer of heat pipes is Thermacore (Lancaster, PA – 717-569-6551, www.thermacore.com).

In the future, more and more computers will be using liquid cooling in one form or another, perhaps unbeknownst to most employees. Don't worry, IT people will soon become all too familiar with them as they attempt to "dry swap" boards in an emergency!

Monitoring and Alarm Boards

No computer system can run perfectly forever. Some component must ultimately break down. This is why the most important component in a fault resilient system is the alarm, whether it be a simple watchdog timer that sends timed pulses though a system, or an elaborate battery-backed independent alarming board.

Top-of-the-line alarm boards generally either fit into one chassis expansion slot or exist as a module that doesn't take up a slot at all. The latter type can be positioned freely so it can be accessed from the front of a chassis. Doing so also makes it easy to provide audible, visible, and remote alarm notification of system failures.

The best alarm boards generally monitor power supply status (over and under voltages), user programmable system temperatures at several points in a chassis, the rotational speed of system cooling fans, an optional modem daughter card for dial-in (interrogation, maintenance, and set-up) and paging (during alarm conditions), or a LAN or WAN connection to a remote maintenance center. Some boards have a backup battery to ensure alarm delivery in case of a power failure.

Crystal Group has perhaps the best remote monitoring and management subsystem in the industry, the platform-independent DarkSite board. DarkSite monitors server parameters including internal chassis and CPU temperature and external room temperatures, fan rotation speeds, and power supply voltage. DarkSite can also monitor CPU load, RAM usage, disk drive usage, network connections, and host server "heartbeat" based on user-defined parameters.

When problems are detected DarkSite sends alerts to any e-mail-enabled device. The board will allow you to do remote resets, reboots, or shutdowns of the host server. DarkSite is independent of the host system, using only an ISA or PCI expansion slot for power.

The better boards can also do remote "out-of-band" monitoring or alarming so that information and management control is still available even when the host CPU isn't working.

Some alarm boards communicate with the outside world via the Telcordia Man-Machine Language (MML) protocol, others may use the LON bus architecture found in heavy-duty industrial applications. Still others can signal remotely through the Signaling Network Management Protocol (SNMP) in which case their software may include a friendly GUI interface with graphical representations of all monitored parameters, thus providing an immediate visual indicator of alarm conditions.

If a sufficient number of variables are being examined by a monitoring / alarming subsystem, then it becomes possible to note such things as a sudden decline in the RPM of a cooling fan or a change in the temperature of a chassis or the fluctuation in the output of a power supply. Such information can be used to give a rough prediction as to when an actual component failure will occur, or will at least alert the technical staff to go ahead and perform a replacement of a would-be troublesome component.

The IP-based SNMP was for a time the basis of most network management software – indeed, the phrase "managed device" used to generally imply SNMP compliance, though other standards such as CMIP (Common Management Information Protocol), DMI (Desktop Management Interface) and WMI (Windows Management Interface) have also waxed and waned in popularity. A consortium of companies led by Intel is now heavily pushing its Intelligent Platform Management Interface (IPMI) specification, which has led to the PICMG 2.9 Secondary System Management Bus for CPCI Subcommittee and the subsequent PICMG 2.9 specification / standard.

To prove that their Windows NT-based communications server (called the Enterprise Interaction Center) really did support automatic failover in a two-node cluster, Interactive Intelligence (Indianapolis, IN — 317-872-3000, www.interactiveintelligence.com) resorted to an extreme demonstration. They sent one of their employees onto the stage of the 1998 CT Demo & Expo show in New York, who, in front of a live audience, took out a sledgehammer and smashed one of two PCs. The first one took a licking, but the other one took over the call processing tasks. It was, as they say, the "hit" of the show.

IPMI defines common interfaces to the "intelligent" hardware and can be used to monitor server physical health characteristics, such as temperature, voltage, fans, power supplies and chassis. IPMI gives IT managers access to platform management information that is said to allow more accurate prediction of hardware failures, diagnoses of hardware problems and initiation of recovery actions. Such proactive advice of any potential problems within the system enables advanced servicing and, ultimately, reduced down-time.

Generally speaking, however, some major system failures can be quite abrupt with no warning, so one simply has to rely on the alarm board to alert the staff when they occur.

Automatic Failover

Knowing that a component has failed is good, but since the chances of a technician standing nearby at that moment may be slim, it's also important that some kind of "automatic failover" can occur. This means that a spare component — or even an entire spare computer — can instantly take up the processing reins when something goes amiss. Without the redundancy of components, any computer system walks a tightrope, teetering on the brink of failure.

Of course, merely duplicating components in a computer is not enough. There must be a way for an application that finds itself dealing with a "dead" component to automatically failover to another, ready-and-waiting component. Under the classic view of redundancy, when a primary (operating) component fails, a secondary component takes over. Failover policies for some subsystems are well-established. For example, power sources are no problem, since the concept of load sharing power supplies was perfected years ago. Also, the theory behind redundant arrays of independent disk drives (RAID) goes back to 1987.

Single board computers, on the other hand, are new to the automatic failover game. What complicates matters in the case of processors is that the processor is executing software — indeed a whole "process" must be preserved as automatic failover occurs. But there is no industry standard way of doing this. Various scenarios allow for automatic failover to occur, with varying degrees of data / transaction loss, and they can occur at varying time intervals, from several seconds to less than a millisecond:

"Loosely coupled" components can be spread across geographical regions and can take minutes to failover.

Hot standby servers (e.g., clustering) can take seconds. Clustering has become quite popular, since it's a less complicated, though somewhat more expensive way of achieving high availability. A cluster is two or more independent servers (sometimes called nodes) that are interconnected to work together as a single system. Since client software interacts with a cluster as though it were a single server, clusters can provide higher availability, higher scalability or both.

If one node in the cluster fails, the clustering software distributes the workload of the failed node to other nodes, and users see no interruption in service. Clustering can be implemented at different levels of a system, including hardware, operating systems, middleware, systems management and applications. The more layers that embody clustering, the more reliable, scalable and manageable the cluster.

As systems became much larger, consisting of multiple boxes distributed about in various locations, the idea of clustering becomes attractive.

I-Bus/Phoenix (San Diego, CA – 800-382-4229, www.ibus.com), for example, offers the 14U high GS1077 High Availability Sun SPARC Cluster System. Part of the G2077 family of cluster systems, the NEBS-tested "five nines" reliable GS1077 ships with two independent system boards (Sun CP1500 SBCs with up to a 440 MHz UltraSparc IIi and 1 GB of ECC RAM per board), both configured as a Cluster Server. The system comes integrated with dual hot-swap, Intraserver ITI-8241C-S quad Ethernet, dual Ultra 2 SCSI and a 4 MB video board, as well as Sun Cluster 2.2 software or the Veritas HA Foundation Suite installed.

Clustering services are now appearing in popular operating systems. For example, Microsoft Cluster Service (MCS) is a component for Microsoft Windows 2000 Advanced Server, Datacenter server and Windows NT 4.0 EE.

When a group of Linux machines is connected in a cluster, it's called a "Beowulf" system, a name arbitrarily chosen by Dr. Thomas Sterling of Center of Excellence in Space Data and Information Science (CESDIS), who developed the first Linux cluster of 16 486DX machines. Sterling's original Beowulf project led to "Extreme Linux" software, now based upon the RedHat Linux distribution. Extreme Linux includes all the common tools and libraries for clustering multiple Linux boxes for a single application.

High availability systems "tightly coupled" with hardware on the same backplane (e.g., multiple CPU boards or controllers) can achieve failover in milliseconds or less.

Certainly telecom applications demand as fast a failover time as possible, since much of their traffic is in real-time. Also, communications applications include critical billing and other transaction-based functions that must be preserved during the failover process. If switching to the backup component or system takes too long, users will notice this disruption. For mission-critical applications, this might not be acceptable.

One Nortel engineer offers a definition of high availability that's a little facetious but is in fact fairly useful: "Equipment is of high availability if the customer thinks it's working." In other words, if you can recover from an error before somebody gets mad and picks up the phone and calls you, it's HA.

For years it was thought that the fastest, most "tightly coupled" automatic failover systems that could be built were those where all of the components were in the same box and shared physical connections on shared buses. A bus can best be described as a circuit "roadway" for transferring data

Differing granularity of failover

2N Full Redundancy

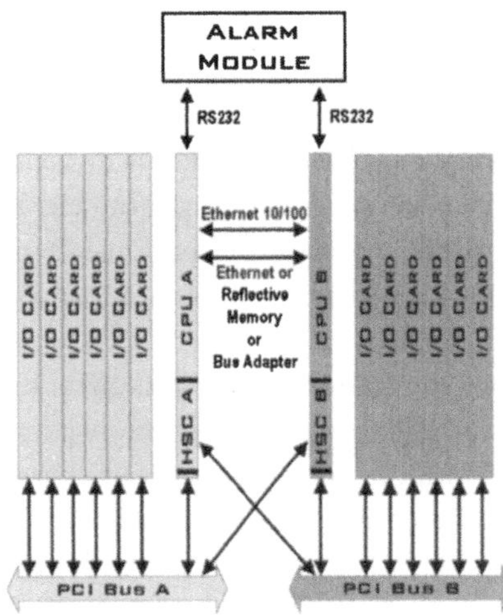

N+1 I/O Cards

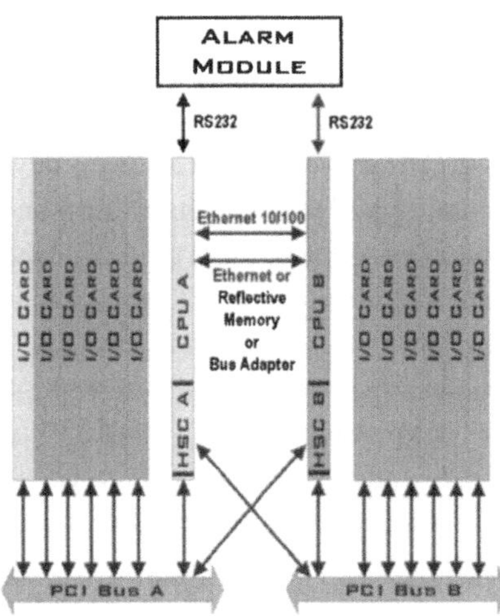

Automatic failover in a CompactPCI system (electrically similar to PCI) can take on differing forms of "granularity." In each chassis pictured above, there are two cPCI buses, an alarm module, and two CPU boards, each controlling six I/O or resource boards. The system on the left has full "2N" redundancy, but very "chunky" granularity — one segment fails over to the next in its entirety. However, a system could be configured where each I/O module or nonsystem processor module could instead be matched with an identical module on the other bus. Thus, paired modules can assume an active/passive arrangement or a load-sharing arrangement in which each carries half of the load of a single module. The system on the right is in an N+1 arrangement where multiple modules can be backed up more economically by one or more spares. For example, a single passive nonsystem I/O module can be used to back up five others, or the I/O and CPU boards could be pooled so that processing could be shared by all the surviving boards.

between a CPU and peripherals. Most computer makers today have abandoned the old ISA bus, with its 16-bit data path, 8MHz clock and paltry 5MBps data transfer rate, and now use the Peripheral Component Interconnect (PCI) Bus Specification, developed by Intel and others.

The PCI bus started out merely as a means for fast communication between chips on PC motherboards, and then grew to become the world's most popular way for PC peripheral devices on slot cards to communicate with CPUs. As we've seen, PCI's more rugged brother, CompactPCI, is electrically similar to PCI (so the software is compatible) but uses more rugged Eurocard style pin connectors and serves as the basis for many heavy-duty computer telephony and data systems.

Some early "fault resilient" CompactPCI systems divided the backplane into two mirrored seven slot segments. Each segment housed identical sets of communications I/O resource boards, and both segments were controlled by one single board computer. This allowed for "redundant I/O domains" so that if one I/O board failed or was extracted from its slot, the system host processor automatically called upon its "twin" board in the redundant I/O domain. But since there's only one single board computer in such a system, it's a single point of possible failure that could bring the whole system to a halt.

Obviously, the single board computer in the system slot should be made redundant. Each segment of I/O cards should have its own resident CPU board. Ethernet and/or serial connections on each CPU board can allow for the two CPUs to exchange data and to maintain a loosely coupled relationship by passing so-called "heartbeat" and "checkpoint" protocols between them. The heartbeat protocol resembles a watchdog timer. For example, an active CPU can send a periodic message to a passive CPU to signal that it's alive and kicking. The passive CPU can respond that's it's okay too and is ready to take over if anything runs amiss.

Checkpoint protocols do things like copy a block of memory from an active CPU board to a corresponding passive, backup board. Checkpoints generally provide a "snapshot" of the application or system program state data, so that in the event of an active CPU failure, the passive CPU is able to begin service in nearly (or perhaps even exactly) the same state as the state the active CPU failed in.

Allowing for redundant single board computers leads to systems which can be built with differing levels of "granularity" for automatic failover. For example, two CompactPCI segments, each with its own SBC, can be configured into an "all or nothing" full redundancy system. If something goes wrong with an SBC or I/O card on one segment, the entire segment shuts down and the second segment takes over. A finer level of granularity would entail an N+1 type system where the failure of any individual board (either an I/O or CPU board) is compensated by the corresponding board in the other CompactPCI segment.

Before the CompactPCI industry as a whole attempted to standardize automatic failover for CPU boards, two companies had already developed and deployed true fault tolerant CompactPCI systems: The Motorola Computer Group (Tempe, AZ – www.mcg.mot.com) and Ziatech (San Luis Obispo, CA – 805-541-0488, www.ziatech.com), now part of Intel.

Motorola's CPX8216 has been out in the field for the longest of all the fault tolerant PC systems, since October of 1998. It's a 16 slot CompactPCI system split into two independent eight-slot cPCI buses. One slot in each bus is dedicated to a system processor, and another for a Hot-Swap Controller (HSC) module. This leaves six slots on each bus / domain to support I/O devices or nonsystem processors.

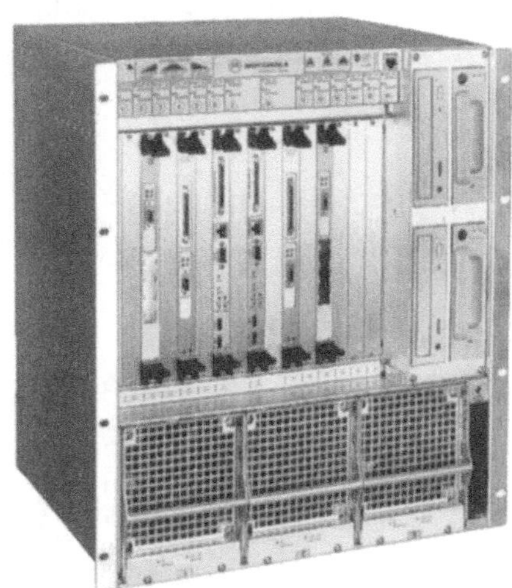

Motorola's CPX8216 was one of the first true fault tolerant CompactPCI computers with automatic failover capability.

Each system processor has direct access to its local bus through an onboard PCI-to-PCI (P2P) bridge. Each domain can support an HSC module that contains its own P2P bridge. Thus, in a fully redundant configuration, there are two bridges that have access to each of the I/O buses – one associated with the CPU and one with the HSC. Only one of the bridges may be active at a time, however.

Aside from providing bridges to the remote I/O buses, the HSC provides the services necessary to hot swap I/O and CPU boards and, also, controls the system alarm panel, fans, and power supplies.

In the fully redundant configuration, a CPU board on, say, the left system slot of Domain A is associated with HSC A that sits in an HSC slot on the Domain B bus. There is a local connection between each CPU-HSC pair that allows the CPU in one domain to control the other domain through its HSC.

Motorola's CPX8216 is a remarkably flexible system, allowing for three possible processor / control configurations:

A simplex system containing a single CPU-HSC pair controlling both I/O domains. For applications not needing true fault tolerance, the CPX8216 can be configured as a simplex (single CPU), 16-slot system. This configuration still gives you the benefits of redundant power supplies and the system monitoring capabilities of the fully redundant configuration.

Ziatech's ZT 5083 was the second fault tolerant CompactPCI system to appear.

An active/passive configuration. This is like the simplex configuration in that one CPU manages all 12 I/O slots, but a second CPU board serves as a warm standby ready to take over and run the system in case of a failure on the active system.

An active/active or load-sharing configuration. Here each CPU runs a single domain while also serving as a backup to the other CPU. Each CPU manages six of the 12 I/O slots, much like a dual eight-slot system, but with the added benefit of one CPU being able to control all 12 I/O slots if the other CPU fails. The total processing power needed in a load sharing system must never exceed the capabilities of a single CPU board, since either one of the CPU boards must be ready to take over the load carried by the other if one fails.

Ziatech followed Motorola's lead with its ZT 5083 High Availability System, similarly engineered for 99.999 percent fault-tolerant availability.

Ziatech's CompactPCI system also employs redundant, hot stand-by system CPUs; has 12 peripheral slots, and is configured with the usual array of redundant system components. Like the Motorola system, the ZT 5083 provides extremely fast CPU switchover, which makes the system ideal for central office and other mission-critical communications network applications.

The ZT 5083 used Ziatech's ZT 5550 System Master single board computer and ZT 5541 Peripheral Master SBC blades for multicomputing applications. Available with support for both Windows NT and VxWorks, the ZT 5083 can be configured with dual redundant ZT 5550 System Master CPU boards with Intel Pentium III processors and an array of I/O and media options. The system also includes rear panel I/O support.

Bill Potter, a product manager at Intel, told the author that "Ziatech got into this automatic failover business early on in CompactPCI when a manufacturer's chief option was to do multi-segment or multichassis clustering configurations for what could be called enhanced availability systems. But some telcos weren't particularly satisfied with the kind of failover times you get with clustering, though it's the clustering software that's the real limiting factor as opposed to the cluster architecture itself."

Potter continues: "In any case, since clustering technology and its multi-second or multi-minute failover times weren't very conducive to maintaining 'five nines' uptime, we then decided to develop the CompactPCI redundant system slot architecture that we ship today in the form of the ZT 5083 chassis and the ZT 5550 CPU board. These give us automatic failover times of around 10 or 15 milliseconds, which is about state of the art for a quick yet cost effective switchover system."

Pete Holmes, manager of strategic product planning at Intel, agreed, adding: "The ZT 5083 has been very well received. We've got a lot of Tier 1 and some Tier 2 and Tier 3 telco customers who are real happy with it. That architecture's only drawback, as is the case with all our competitors, is that there's no industry-wide automatic failover standard. Therefore, there's no vision of multivendor support or software standardization. For customers hurrying to get to market and who need the availability it's acceptable, but for broad industry acceptance, certainly some kind of standardization would be a huge advance for all of us."

Around this time, the same idea was popping up at Motorola.

The PICMG 2.13 Standard

Joe Pavlat, former director of the Motorola Computer Group (MCG) and president of the PCI Industrial Computer Manufacturers Group (PICMG), recalls what then happened.

"Many of the Motorola engineers have real fault tolerant backgrounds," says Pavlat. "Motorola for years has shipped a special product called the FT – the Fault Tolerant computer – that we've sold to companies such as Nortel."

"But our technology was very proprietary like all of the fault tolerant systems were and are," says Pavlat. "If you go to Tandem and Stratus or whoever, their systems are all different and they're all proprietary. They're true '2N' systems where you've got two of everything and they're expensive. Since they're proprietary it's hard to get people to write software for them."

With the advent of the open architecture of CompactPCI, some of the MCG engineers reasoned that although it's extremely difficult to devise a system that's completely fault tolerant – which means no transaction losses whatsoever, such as a banking or a billing system or a stock trading system – you could, however, build a system sufficiently fault tolerant to satisfy the needs of many modern telephony and telco applications.

"You obviously don't want to drop a call in progress," says Pavlat, "but maybe it's acceptable if you can't immediately complete a call. So perhaps there's something one step back from absolute fault tolerance that makes sense, something that the telcos want. And that turned out to be the case. We were really encouraged by Nortel and other people at Motorola to create a product with these characteristics."

"So we set off to take some of our fault tolerant knowledge and develop some technology using cPCI as the core, to build a 'five nines' HA system," says Pavlat. "We did that, and we got a lot of early customer acceptance and have shipped thousands of the CPX 8216 machines. But before we were finished with the CPX 8216, one of our large customers announced that it wanted an open standard – what it really wanted was a second source for our technology. At the time we were only part way through engineering the product; we hadn't even finished it yet, much less delivered examples and learned from our experiences."

"Still, they were a big customer so we went along with the idea," says Pavlat. "I put on my PICMG hat and we started the PICMG 2.13 technical committee, which is still striving to achieve a standard specification making CompactPCI system slot functions redundant and hot swappable."

"In the meantime, the 8216 product solidified and we shipped it with great success," says Pavlat. "We built a lot of software around it. At the same time the committee now had some good talent from companies such as Force Computers and Sun Microsystems and they were providing new input for the committee."

"The predictable thing happens when you start a standards effort before the cement is dry," says Pavlat. "You can in theory take something that's a *de facto* standard, run it through a standards committee and, boom, you've got a standard. But when you're developing a new technology such as HA, things get bog down. Motorola was already down a successful path of having made a huge investment in a product that customers love, while the committee was off in a dark room deliberating. And guess what, Motorola and the 2.13 committee diverged."

Pavlat muses: "Some of our competitors have beat us up for not being compliant to the spec effort that we started in the first place. But that was a Motorola business decision, to stick with something we knew worked, and not a technical decision."

"I've got some issues with 2.13," says Pavlat. "Wearing my PICMG hat, I've even tried to squash the whole effort a couple of times, because standards efforts generally work best when there's a general agreement on how a problem should be solved. HA is so new and there's so many different ideas and so many different opinions on what it should be and how it should be done, as a technology it's really still in its infancy."

The major stumbling block in building fault tolerant automatic failover systems complaint with PICMG 2.13 is that it's built on PCI, which can itself be a single point of failure.

The PCI bus is a funny duck. It didn't start out as a true local bus; instead, it was originally categorized at an intermediate level between the CPU local bus (processor / memory / cache subsystem) and the standard expansion buses of the 1980s and 1990s (ISA, EISA, Micro Channel). Thus, it was originally meant to supplement rather than replace the traditional I/O bus, but its high bandwidth capabilities and popularity forced it to fill some pretty big electronic shoes. But the fact remains that PCI was not originally designed to be a failover-friendly architecture.

Indeed, the PCI bus is actually quite fragile, which is why the latest revisions of the 2.13 specification still have some limitations. For example, if you have several boards plugged on a bus and one board fails or there's an electrical short in a slot, the whole bus fails. Or let's say that you hot swap a board and happen to bend a pin in the board's connector or cause some other mechanical failure. No matter what your software is or what kind of high availability board drivers you have, your whole system is going to expire because of the electrical limitations of the PCI / cPCI bus.

These limitations, however, can be overcome, either with additional hardware or by rethinking the whole concept of a computer bus. This is all being done in response to what carriers and companies such as Nortel are demanding.

"Carriers are asking for systems that are even more reliable than five nines," says Stephane Dubois, Systems Solutions Product Manager at Applicom (Boisbriand, QC, Canada – 450-437-5682, www.teknor.com), a Kontron Company. "Therefore we must forge ahead, beyond what's given in the PICMG 2.13 specification. Our future systems will be different than present multiprocessing systems, but they'll achieve a level of fault tolerance that will absolutely satisfy everybody."

As Brian Carr, Telecom Product Manager for Blue Wave Systems Ltd. (Carrollton, TX – 972-277-4600, www.bluews.com) now a part of Motorola, says: "If failover at the CPU level is a problem, then that problem is magnified many times by management of multiple DSPs and DSP resource boards. The most effective solution for development and deployment is to run a loosely coupled system where the user application requests an appropriate resource from a pool of resources and then uses that resource for a communications processing task. If there is a resource failure, the user application can request the system to assign a new resource and carry on where it left off."

This method is similar to clustering, and is employed by Blue Wave Systems' FACT software to manage multiple DSPs across multiple DSP boards. FACT can operate using N+1 redundancy from a DSP board level right down to an individual resource level (which guards against possible software failure – however unlikely).

Carr says that "within a CompactPCI shelf, one has to consider the failure of the system controller – which PICMG 2.13 tries to solve. But one needs to look at the real effect of any failure. If you just consider basic power, backplane management, clocks etc. – then a failover is pretty simple to handle. For control and state-maintaining user programs, that has to be solved whether control is local or off-platform, so that cannot be a specific criticism of PICMG 2.13. In such a case a loosely coupled control approach (using a reliable transport) once again works well."

But until recently, it was thought that clustering, with its long history of both theoretical and product development, was too slow for tightly coupled telecom applications involving single board computers. Clustering is a simple way to achieve some measure of fault resilience, though it tends to be expensive, since you must have two or more independent servers or "nodes" that are interconnected to work together as a single system.

PICMG 2.16 Packet Switched Backplanes

A novel way of clustering CPU boards in a single fast failover "pool" inside a single box, made possible by a new generation of packet switched backplanes (particularly Ethernet backplanes), is about to take the industry by storm. The least complicated way of explaining this idea is to think of a sort of "LAN on a backplane" where each "node" system is sitting on its own board on a CompactPCI backplane, and they're all interacting via Ethernet packets traveling over signal pathways or "traces" on the backplane rather than Category 5 twisted pair cabling. The standard for this technology is the PICMG 2.16 Compact Packet-Switching Backplane (cPSB) specification, initially proposed by system developer and integrator, Performance Technologies Inc. (Rochester, NY – 716-256-0200, www.pt.com), also known as PTI.

Packet-switched backplane aficionados like to use the term Embedded System Area Network (ESAN) to describe how cPSB can leverage well-known Ethernet and IP network technology by overlaying it onto the CompactPCI backplane. This allows system design to occur at the higher layers of the Open Systems Interface protocol stack.

PTI's got the PICMG 2.16 technology ball rolling with their CPC4400, a Layer 3, 6U high cPCI switch for high availability communications applications. Instead of using a shared bus architecture such as H.110 to connect the different boards on the system, the CPC4400 uses point-to-point switched Ethernet signals; these are sent through the cPCI midplane at ten Mbps, 100 Mbps, or even at gigabit speeds (the switching fabric can potentially handle up to 9 Gbps, far beyond cPCI's normal bus bandwidth of 530 MBps when running at 66 MHz).

Where the regular PCI bus uses the J1 connector, and cPCI's H.110 uses J4, PTI can use 20 pins per slot on J3 or J5, the user-definable cPCI connectors. This vastly increases the interboard communication capabilities of sub-systems in a chassis by moving system traffic from the shared bus to an abstract, fault-tolerant, packet-switched backplane (PSB).

The CPC4400 has 24 switched 10 Base-T/100 Base-TX ports, two 1000 Base-TX/FX Gigabit ports to uplink to the Internet, and one out-of-band (management) 10 Base-T/100 Base-TX port. The switched and out-of-band 10/100M Ethernet ports are accessed from the user-defined pins on the cPCI J3 and J5 connectors. The Gigabit ports are accessed either from the front panel or the cPCI midplane.

Other cPCI-based cards may be connected to the CPC4400 via Ethernet over the midplane, removing the often confusing and unreliable mass of cabling found in most "rack area network" solutions. Alternatively, a rear-panel I/O line interface module (LIM) can bring some (or all) of the Ethernet interfaces out to RJ-45 connectors.

The two Gigabit Ethernet uplink ports come in optical and copper options: You can have two GB fiber ports at the front panel, with either single-mode or multi-mode transceivers (1000 Base-Sx or 1000 Base-Lx Fiber-optic); or else copper connections can be accessed from the back panel I/O via an optional connector card (100/1000 Base-T twisted pair Ethernet physical connections).

By creating a new switching matrix for a CompactPCI backplane instead of relying on the bus, system throughput can be doubled at a lower cost and in a radically smaller footprint – about a fifth the typical size. And PTI's scheme doesn't interfere with components that are already interoperating over H.110.

The CPC4400 is thus able to replace enterprise switches in central office environments or provide a PSB that doubles overall system throughput, while cutting the cost and space associated with the cables, cabinets, and cooling fans it renders unnecessary.

You can put two cPCI CPC4400 switches in a chassis and then the other 21 slots that are available can now be fully populated in the chassis of standalone systems. Depending upon the size of the switch fabric that one populates the card with – new cards are being developed – one can operate all 21 cards or more in other chassis. One can link switch-to-switch, or chassis-to-chassis using these Ethernet connections. It's just like a gigabit LAN connection between the cards and chassis.

And instead of having to integrate a system at the operating system and driver level – where everything has to interoperate – you can put just one card running Linux in Slot 1, another card running NT in Slot 2 – whatever OS makes the most sense for the system on a card. You can pick the best-of-breed products and tie them all together using commodity Ethernet as the ***lingua franca*** with which they all communicate.

Targeted applications for the CPC4400 include all IP switching tasks associated with today's voice/fax-over-IP media gateways, signaling gateways, integrated access devices, DSL concentrators, multimedia gateway controllers, or any other next-generation network element requiring prioritized handling of Ethernet traffic. Additional applications include high-availability systems for industrial or military applications.

Essentially PTI has not just created a new bus channel, but a fault tolerant channel. You can run two switched gigabit (1,000 Mbps) ports to each slot. Each is full duplex, so you can have an active channel and a passive "hot standby" channel. Or you can have both channels active and then after a failure drop back to half bandwidth.

Besides fewer cables, PTI's system allows for much easier setups, since you don't need technicians who may plug the wrong cables into the wrong positions. PTI has also carried over to the CPC4400 some tricks they learned in the HA switching arena.

For example, they've embedded a trivial file transfer protocol (TFTP) and a dynamic host configuration protocol (DHCP) server on the switch, along with a Flash-based file system and some IP address allocation services. Let's say you've got a DSP card in Slot 5, and that card fails. A technician does a hot-swap of that card – he pulls it out and puts a new card in without taking down the system. Since we know that Slot 5 always gets a specific setup and configuration file fed to it for installation, we can send that information directly to the card from a flash module in another card in the system through TFTP and DHCP. The replacement card can thus finish the installation and boot up sequence by itself.

Similarly, two CPC4400 Ethernet cards in the same system can share setup information with each other, so if one fails the replacement card will clone its setup from the survivor.

Another refreshing aspect of PTI's system is that, after hearing for years how IP is going to be the Master of the Network, it's fun to see IP packets in a VoIP gateway get encapsulated in Ethernet packets. This is done with the help of another PTI product, the PT-CPC395 Dual-Channel T-3/H.110 TDM Switch. This is a board that can allocate circuits from the PSTN through a TDM switch to H.110-based DSP boards such as those from Intel, NMS Communications, or Audiocodes. The DSPs convert the PSTN circuit into IP frames. The IP frames are then collected and aggregated into the CPC4400's high-speed Ethernet uplink to be sent over the Internet.

PTI's board can act as an aggregation layer switch. PTI's first design win was with an IP telephony equipment manufacturer in California who took some DS-1 and DS-3 switch cards and used them as the interface to clients for an IP telephony gateway. Those switched-circuits were switched onto the H.110 bus and sent to Audiocodes DSP cards for packetization. The circuit traffic was then converted into Ethernet packets, connecting them with PTI's switching board via the midplane. Then the traffic can be switched either back to another DSP card for a call that's going to be locally terminated or, since the board is a Layer 2 or 3 switch, the traffic can be aggregated and sent up to a large router, a SONET ring, or some other mode of long distance conveyance.

The system can also act as a router between other chassis or other systems that are not chassis-based. This is possible because everybody uses Ethernet. All of this makes perfect sense for a number of different applications: media gateways, signaling gateways, IP telephony gateways. The bandwidth that PTI's product offers by relying upon Ethernet is going to be very attractive to people as they deploy higher and higher bandwidth consuming applications, such as streaming video.

In fact, there's a growing consensus that Ethernet, not IP, may eventually displace ATM. If you think about the amount of data traffic that's being carried, what sense does it make to start with 1,500-byte Ethernet packets, convert them to 53-byte ATM cells, then reassemble them back into 1,500-byte packets at the destination, particularly if you can do wave-division multiplexed Ethernet that can travel 50 or 60 kilometers without a repeater?

During the 2000 to 2002 time frame, there was a tidal wave of PICMG 2.16 compliant products:

For example, APW Electronic Solutions (Waukesha, Wisconsin – 262 523 7600, www.apw.com) was called upon by PTI to develop a 15-Slot CompactPCI EtherPlane Backplane (EPB) to their specifications. The EPB extends the CompactPCI specification with a packet-based (Ethernet) switching architecture to create an ESAN, one which integrates a scaleable LAN into the embedded systems environment along with hot swap capability and high availability.

Also, Bustronic (Fremont, CA – 510-490-7388, www.bustronic.com) offers a cPSB backplane based on the PICMG 2.16 draft specification. It comes in 16-slots, with up to four 47-pin power supply connectors, and distributed power plugs for voltage I/O, ground (GND), and +5V, -5V. By using stripline design technology, the backplane has virtually zero crosstalk. Bustronic has also teamed up with its European sister company, TreNew, also known for their high-speed backplane designs, to develop switch fabric and high-speed backplanes. TreNew (now part of Elma), also has recently released one of the first cPSB backplanes on the market, an 18-slot version, also based on PICMG 2.16.

Like the APW board, the Bustronic cPSB board increases system performance by moving data traffic off the shared bus and onto an embedded switched Ethernet network fabric (10/100/1000 Mbps), accessed via the CompactPCI J3 connector. System MTBF and reliability is also improved because cPSB reduces the requirement of cables and connectors. The specification currently reserves pins for optional H.110 telephony bus implementation. cPSB is also fully backwards-compatible to present CompactPCI technology.

Furthermore, the appearance of the PXP platform from the Motorola Computer Group (Tempe, AZ – www.mcg.mot.com) and the Centellis CO 21000-12U series from both Force Computers (San Jose, CA – 408-369-6000, www.forcecomputers.com) and ZNYX Networks (Fremont, CA – 510-249-0800, www.znyx.com) suggest that we'll be seeing both new hardware and new development environments springing up to take advantage of this packet-switched backplane technology, and that PICMG 2.16 compliant equipment may, in many circumstances, overshadow PICMG 2.13 compliant systems.

There's been so much hubbub about the packet-switched backplanes that one may not be aware that there are also boards now available that can be plugged into them. The MTN5300 from Mapletree Networks, Inc. (Norwood, MA – 781-751-2400, www.mapletree.com) is a CompactPCI board designed for carrier-class access and gateway systems capable of sending Ethernet signals across the backplane or out through conventional RJ-45 connectors. Also, the HighWire400c/M from SBE (San Ramon, CA – 925-355-2000, www.sbei.com) is an advanced PowerPC based core processing platform with dual Ethernet ports and cPSB support.

"The traditional idea of a long PCI bus that serves as the backbone of a whole computer system is about to bite the dust," says Eric Jobidon general manager of CML Versatel (Hull, Quebec, Canada – 819-771-0011, www.cmlversatel.com). "In our high-density programmable switching platform we don't use the PCI bus to share information between the CPU and the high-density communications resource cards. Instead, we're using a redundant Ethernet bus. Since card control is exclusively done using redundant packet links, this also allows for the deployment of both centralized and distributed systems."

In CML Versatel's system, dual Ethernet links are used to send control information to the cards, but the media sharing is done using another standard interface, the H.110 bus. Although the H.110 bus is not covered in the PICMG standard *per se* (it's an ECTF standard), it also constitutes a single point of failure that must be addressed. To that end, CML Versatel has devised redundant access to the H.110 bus to protect against potential bus-level failures on the media side.

"We've solved that problem by splitting the bus in half, and on all our cards we have two interfaces to the H.110 bus," says Jobidon. "Then, instead of having 4,000 timeslots running together with no redundancy, you can use the first 2,000 timeslots for communications ports and the second 2,000 timeslots as a backup. If there's a problem with a timeslot then all of them are switched over to use the second group."

"We've also implemented hot swappable boards, in accordance with the PICMG 2.1 spec," says Jobidon. "But again, instead of using the PCI bus to share state changes with other cards in the system, the hotswap capabilities are handled through the Ethernet interfaces. The behavior is exactly the same, but without the possible single point of failure of the PCI bus."

As Carr of Blue Wave Systems says: "This packetized approach is also valid at the higher subsystem and even system level – loosely coupled entities use multipath reliable transport methods (such as TCP/IP) for command and control – for example the Megaco / H.248 interface for media gateways. Again, this abstracts the control mechanism away from low level detail allowing effective redundancy strategies."

"The logical extreme of this approach is the 'no controller' system where everything is a 'processor blade' or even a standalone box [often termed "appliance"] attached only via IP," speculates Carr. "For some medium density or processor-only applications (such as ISP servers) this works well. But for high-density and I/O intensive systems, it doesn't take proper remote management and diagnostics into account. If your only attachment internally in a shelf is via IP then what do you do if a failure wipes out the IP link, or a blade fails to start its IP link or starts blasting the IP link to the point that nothing can get through?"

Carr is ready with the answer: "Well, you can resort to using the CompactPCI bus to actively take down, probe, load diagnostics, reprogram FLASH, etc. – and you can do all this remotely without scheduling a maintenance visit to hook up a terminal."

But as Philippe Muraglia, VP of Automation, North America for Kontron Embedded Computers (Boisbriand, QC, Canada – 450-437-5682, www.kontron.com/en/) told the author, "These are all different issues. 2.13 talks about doing failover on a system slot in a CompactPCI system, 2.16 defines a switch fabric that integrates Ethernet on the backplane. You can use 2.16 to do a loosely-coupled (or even a tightly-coupled) multiprocessing system, but it doesn't remove the necessity in some applications to use 2.13."

"Most people envision 2.13 as a classic scenario where a 'primary' processor board fails over to another one, but there are various ways of doing it," says Muraglia. "You could have two processor boards working in sync, and one takes over the other. The problem with this configuration is that you need some failover logic to decide what board has failed and what hasn't. Or you could set up a hot standby, that is sort of following what's happening in the system, and can step in and take over processing relatively rapidly."

Wolfgang Eisenbarth, VP of business development for CompactPCI at Kontron Embedded Computers, agrees: "PICMG 2.16 is really a packaging issue. You put everything in a small package with a small footprint and remove what was the cabling between boxes. But 2.13 is something completely different. In an application where you've got a lot of I/O cards, you could under 2.16 simply duplicate the system components with backups and then say, 'this is your solution.' But that solution may behave just like a loosely-coupled multiprocessing system, and the performance and costs involved in doing that may not be acceptable. That's where PICMG 2.13 steps in. A typical CO application where you have many line cards being used with each CPU is where you'd like to see a 2.13 type of solution, or let's say some kind of CPU failover architecture."

"But the key issue here is that, depending on the application, the failover time must be specified and the technology must be able to failover in the alloted time," says Eisenbarth. "There are applications that can tolerate a one second failover, others can tolerate a 10 second failover and some can't even tolerate a millisecond of failover. Right now as it stands, PICMG 2.13 does not necessarily address all of these issues. The failover time is not specified."

David McKinley, director of engineering, Telecommunications Division, at RadiSys (Hillsboro, OR – 503-615-1100, www.RadiSys.com) says: "We're clearly among the group that thinks 2.13 has some significant deficiencies. We put a stake in the ground and told our customers that we're not going to pursue 2.13 technology, and instead we're putting our high availability development issues and resources and strategy and everything else into a system interconnect, which is, at this point, more like the PICMG 2.16. To the extent that we're still shipping systems that look like CompactPCI systems, we are, but in as many cases as not, there may not even be a PCI bus on that backplane."

Mark Overgaard, president of Pigeon Point Systems (Scotts Valley, CA – 831-438-1565, www.pigeonpoint.com) told the author that he thinks there are many benefits to the 2.16 approach, but the thing to keep in mind is that there is no hardware independent infrastructure yet defined for managing the boards in a 2.16 system.

"In the case of the PCI bus slots," says Overgaard, "there are well-defined mechanisms for PCI boards to identify themselves, for boards to supply things like serial numbers and all that. With PCI systems all of those interactions with the boards can be done in a board independent way, by software that doesn't know about any particular board, and so the software will work with hardware from different manufacturers."

Indeed, the whole plug-and-play notion of cPCI is founded on this idea, where you take components from Vendor A and you mix them from components from Vendor B and Vendor C and your put them all together in the system and you make them work.

"PICMG 2.16 only accommodates this idea to the extent that you're limited by what you can do with the TCP," says Overgaard. "You can hook some regular PCI boxes and CompactPCI boxes together, and you may or may not get the tight integration and really fast failover that a single system with PCI slots will allow. The real issue is what's missing: There's a whole set of conventions such as those for finding out in which slot is a given board situated. If you want to tell a technician that the board in Slot 5 of Chassis 21 is busted, so go and replace it, there is as yet no board-independent infrastructure specification that will allow your 2.16 system to do that. PCI systems do have that infrastructure, as well as well-defined notions of how a board can communicate its recognition of you wanting to hot swap it out of a system, such as turning on or off a blue LED."

So the big message here is that there's lots of good things about a PICMG 2.16 based computer world, but there are also many new challenges that must be faced so that an arbitrary collection of components following the spec can be smoothly integrated into a single 2.16 compliant system.

Bye-Bye Bus? Serial and Pseudoserial Switch Fabrics Appear

The PICMG 2.16 packet-switched backplane concept lays bare the deficiencies of ordinary computer buses and points the way toward technologies that will make them obsolete.

From the 1940s, when IBM created the first computers, through about 1990, computer system throughput has always been bound by the power of the CPU. The I/O interfaces and their buses could always run much faster than the CPU. More data could be delivered to the CPU than the CPU could process. That was a big problem with mainframes, which led to the multiprocessing movement.

Around 1990, however, CPU performance soared dramatically. Suddenly, the I/O interfaces became the bottleneck. It's taken ten years to understand the underlying physics to get the I/O bus up off its duff. But I/O is still not fast enough to feed the faster and faster processors that are now appearing almost on a monthly basis.

PC manufacturers have been coping reasonably well for years, of course: The aging computer buses we known and love, such as ISA, PCI, CompactPCI, VMEbus, STD-32, etc., provide an internal information highway for the boards we insert in our respective PC's slots. But like all buses, even a newer robust one such as CompactPCI, has its limits. cPCI systems can easily range up to eight slots for 64-bit, 66MHz throughput speed with a total bus bandwidth of about 530 MBps. To build systems with 16 slots you need some fancy bridging circuitry, and you normally can't put 21 slots in a single enclosure the same way you can with cPCI's competitor, VMEbus.

Also, PCI (and other traditional, parallel computer buses) suffer as bandwidth demands increase. A futuristic system with 16 or so boards, each with 10Gbps of I/O capacity, demands over 2,000 signal pathways or "traces" on a PCI backplane. High-speed parallel buses cannot send such signals over more than an inch before the energy is radiated away. If you lower the speed of a parallel PCI bus so the signals will make it from one end of a 19" rackmount backplane to the other, your bandwidth carrying capacity is reduced proportionately, unless you start increasing the number of signal pathways – that is, increase the parallelism of the bus. Each signal pathway or "trace" is associated with a pin on each board, so what you end up with are processor and I/O boards having tens of thousands of pins!

Also, cPCI-based CT resource boards are limited to communicating with each other across the backplane via the H.110 "CT Bus" (H.100 is the equivalent for regular PCI boards and buses, which uses a ribbon cable). But it's becoming glaring evident that these time division multiplexed (TDM) buses with their paltry number of time slots won't be able to handle the kind of high density, high bandwidth multiprocessing applications that are hurriedly being developed. How do 21 different boards communicate among themselves in a rackmount, with each board running under a different operating system? What do you do with rooms filled with thousands of 1U and 2U high "pizza boxes" that must be tied together with some new kind of data transfer architecture?

The answer to these problems rapidly appears to involve dumping the traditional parallel TDM bus structure and using instead packetized backplanes between the boards or one of the new scalable and flexible high-speed serial and pseudoserial switching fabrics such as InfiniBand, RapidIO, StarFabric or PLX.

Buses start running into physics problems long before high speed serial fabrics do. The big problem with parallel buses is the "skew" that occurs between rows of bits as they move along their respective lines on the bus, like rows of soldiers marching forward in a parade.

In a 16 or 32-bit bus, Bit 1 may get to a circuit long before Bit 2 or Bit 10 does. That adds latency to the system, because if you know that everything is always out of skew to some degree, then a developer must specify a "hold time," which is when the system does a "Read" to detect a group of bits but has to wait for about 10 nanoseconds, since the system "knows" that during that time segment all of the out-of-skew bits will finally arrive. Such inherent delays place a big damper on scalability.

Another delay in buses has to do with so-called registers in computers, which used to be called "accumulators" long ago. The concept is also used in the structure of buses when addressing certain items on the bus. For example, in the Firewire high performance serial bus for peripherals, input and output plug control registers determine the properties of plugs (a DV camcorder, for example, could have two logical output plugs, one for camera out and one for tape out, plus a logical input plug for DV dubbing, but only one physical connector, thanks to the use of registers). Even DVD-Video players use registers, with 24 system registers used for information such as language code, audio and subpicture settings, and parental level.

Moreover, for mission critical telco applications with "five nines" (99.999%) uptime, buses don't fare well without some considerable engineering work. Buses just don't like to be forced into a fault tolerant environment.

"Buses are not inherently fault tolerant and never will be," says Ray Alderman, executive director of VITA, the VMEbus International Trade Association. "To build a fault tolerant system with fault tolerant software for conventional buses becomes an absolute nightmare. All kinds of things can go wrong, and each of those possibilities must be taken care of."

"But if you're working with a high-speed packet-thrashing machine," says Alderman, "moving packets around, then the software problem simplifies. You can do a five-nines system because redundancy is a function of the way the system scales, and you're using fewer wires and pins, so there's fewer things that can go wrong. Your system is looking at packets and not registers. It all makes a whole lot more sense."

Watching the communications infrastructure move their computing elements to serial switching fabrics is somewhat amusing, since telcos deal in serial information anyway. They take their serial information, bring it into a box, translate it into parallel bits so they can look at packet headers and what-not, then reserialize the data and send it on its way.

Of course, they did this because, until recently, no one had any serial processors that were fast enough. Now we've all got clock cycles to burn, and companies such as Intel have interesting packet processors (such as the IXP 2000) that can deal with these issues.

In short, the tools that we've needed – fast packet processing processors and high speed serial interfaces – have finally developed to the point where a major transition is about to take place, one that will allow companies to deploy super high density converged applications running at fantastic bandwidths and with absolute continuous availability.

Ironically, these dramatically new systems still look like CompactPCI or VME boxes on the outside and inside. The telco providers and system developers are not just strictly using cPCI as the bus architecture. They're simply using the mechanical architecture of cPCI as a means of achieving reliability and getting a lot of board pins in a small area, pins that can be used with new technology such as packet-switched backplanes, the StarFabric and InfiniBand. So they're not necessarily using the cPCI bus at all.

Recently however, you may have noticed that larger, even more expansive (and complicated) architectonic switching fabrics are nearing completion.

Some major switching fabric contenders are as follows:

InfiniBand

InfiniBand is a pseudoserial switch fabric for box-to-box system-area networks (SANs) and cluster farms. This new architecture will expand the performance of entry-level servers through high-end data-center class solutions using interoperable links based on aggregate bandwidths of 500 MBps, 2 GBps, and 6 GBps employing a 2.5 Gigabit per second wire signaling rate over either copper or optical cabling. InfiniBand supports both switch and router functions, and it can be deployed at the intrasystem, subsystem, and chassis levels.

InfiniBand is backed by the Heavyweights: Intel, Compaq, Dell, Hewlett-Packard, IBM, Microsoft and Sun Microsystems are all backing InfiniBand. Unfortunately, InfiniBand tends to be a bit "heavyweight" itself in terms of hardware and software, and may not be entirely suitable for real-time systems.

At first it was thought that InfiniBand would take longer to come to market than some of the other competing technologies, but it now appears that quite a bit of "behind-the-scenes" work has taken place, so expect the first InfiniBand products Real Soon Now. Check out the trade association at www.InfiniBandta.com.

As the alleged "heir apparent" of future computer architectures, InfiniBand will give the server world a much needed boost in terms of bandwidth capability and fault tolerance. Unlike PCI, InfiniBand's links can come out of the box, enabling flexible network connections that can scale easily and yet provide fault tolerance.

The author spoke with Kevin Deierling, VP of product marketing for Mellanox Technologies, Inc. (Santa Clara, CA – 408-970-3400, www.mellanox.com), a company that's taken the first steps toward linking the InfiniBand Future with the PCI Past. Mellanox' InfiniBridge technology was released in January 2001.

A board with an InfiniBridge chip looks like a standard PCI or cPCI card and indeed it interfaces to the circuitry of the legacy PCI bus, but it can support 10 Gbps links traveling over copper cables up to 17 meters (55.7 feet) in length. The 10 Gbps link is sent as four 2.5 Gbps pipes using what's called "byte-striping" – the first data byte is sent on the first twisted wire pair and the next byte on the next pair, and so on. It's reminiscent of time division multiplexing, except that the bytes are stripped across the channel in parallel instead of in sequence.

Dierling says that Mellanox was the first company to send 10 Gpbs over copper, "and copper is one tenth the price of the equivalent 10 Gbps fiber technology," he says.

InfiniBridge doesn't alter backplane signal trace construction, since the signals run through cable cables and connectors. "We've done designs where the signals actually run over a backplane connector," says Deierling, "but adding cables is quite easy to do and doesn't have to deal with any specific, peculiar backplane characteristics."

Competing fiber technologies are really meant for long haul networks," says Deierling, "and that doesn't address the data center or communications center where you just want to bridge from board to board and chassis to chassis instead of connecting across a metro area network. That's where we come in."

Another new Mellanox technology is called InfiniPCI.

"Our InfiniPCI chips have the ability to map the PCI semantics (the memory, the I/O cycles and the configuration cycles) to the InfiniBand semantics," says Deierling, "and at the reverse side we do an inverse mapping. This is transparent to the OS, the device drivers and even the BIOS. This means that we can boot up Linux or Windows 2000 across an InfiniBand fabric where the boot disk is on the other side of an InfiniBand fabric in another machine but the software on this side thinks it's just dealing with a local PCI-to-PCI bridge on the backplane or motherboard. So people can use their existing software and still exploit the capabilities of InfiniBand today.

"Tomorrow, as InfiniBand becomes embedded in all of the major operating systems, users will be able to do a software upgrade and suddenly change the performance characteristics and the relationships between their I/O devices and the CPU devices," says Deierling. "But even today you can deploy the technology without any changes in your software – if your application runs out of slots you can expand to another chassis, just by plugging a card in each machine, which gives you an invisible 10 Gbps InfiniBand connection between the PCI boxes."

Mellanox also offers their Nitro platform which brings the InfiniBand architecture to server blade design and allows you to construct a system that enables switch, server and I/O blades to all occupy a single 7" tall (4U) chassis. The Nitro platform consists of Mellanox switch blades, diskless server blades, and InfiniBand architecture passive backplane and chassis.

But no matter what mysterious signaling goes on in the dark recesses of CompactPCI equipment, the physical structure of PCI / cPCI will be around for some years to come. You can see how painstaking care is being taken to make future advances in cPCI compatible with older PCI circuitry. One company, Flextel (Ivrea, Italy – +39-0125-235311, San Jose, CA – 408-298-8587, www.flextel.it) has even gone so far as to build a system (the netVision 5000 Multi-Service Platform) that allows CompactPCI, PCI and ISA boards to amicably toil together in the same chassis.

"PCI / cPCI has got a long life ahead of it," says Deierling, "and some of the things we've done with our InfiniPCI technology will actually help extend the life of PCI as well as overcome some of the limitations of PCI and cPCI chassis. We've got a long-term roadmap that includes additional new devices to support PCI as the transition occurs from PCI to the new switching fabrics."

Still, as the phenomenon of convergence continues, new and more demanding applications will appear, testing the mettle of CompactPCI as never before.

Arapahoe / 3GIO

Although Intel is developing and backing InfiniBand for high bandwidth box-to-box communication, it favors Arapahoe (later code-named 3GIO) as a replacement for the local PCI bus to do chip-to-chip and possibly board-to-board communications. In early 2002 Intel decided to reformulate 3GIO, and is rumored to be switching its emphasis from telecom-related I/O to multiprocessing considerations.

HyperTransport

Developed by AMD, HyperTransport technology (formerly known as Lightning Data Transport) is a high-speed, high-performance, point-to-point interconnect technology for integrated circuits. HyperTransport provides a universal connection scheme that's designed to reduce the number of buses within the system, provide a high-performance link for embedded applications, and enable highly scalable multiprocessing systems. It was developed to enable the chips inside of PCs, networking and communications devices to communicate with each other up to 48 times faster than with existing technologies, running at up to 12.8 GBps per second.

HyperTransport runs at 800MHz and is "double pumped," which means that the data is sent on the upswing and the downswing of the processor timing clock, doubling the effective bus speed to 1600 MHz. HyperTransport can use whatever data width the bus needs – it can be 2, 4, 8, 16 or 32 bits wide. And it can send and receive data at the same time (duplex). It also enjoys low latency.

The first commercial implementation of HyperTransport was in the AMD-processor powered nForce platform from nVidia (Santa Clara, CA – 408-486-2000, www.nvidia.com).

Although one would expect AMD's HyperTransport to compete with Intel's Aprapahoe / 3GIO, Cisco is a supporter of HyperTransport, so for Intel's products to be interoperable with Cisco's, they will probably have to support HyperTransport in some way or make their own fabric compatible with it.

RapidIO

Backed by Motorola and others, this is more of a lightweight "in-the-box interconnect" pseudoserial switch fabric, though it also supports interconnects at the processor, board, and chassis-level.

RapidIO was still an eight-bit wide high speed parallel bus until two weeks after their presentation at the Bus & Board conference in San Diego during January 2001, when the RapidIO people announced that they were going to "go serial." Then, interestingly, Intel joined the RapidIO group and began adapting some of the InfiniBand technology to RapidIO.

Says Ray Alderman of VITA: "For three years I've urged Intel to develop an 'InfiniBand Lite' for board-to-board and chip-to-chip communication, because InfiniBand was way too ungainly. Rather than design such a thing themselves, Intel has apparently adapted the transceiver of InfiniBand into RapidIO as an electrical layer for the high-speed serial connection, so now the RapidIO guys can go ahead and build the chip-to-chip and board-to-board protocols stack. They can thus build a bridge between both InfiniBand and RapidIO technologies."

RapidIO thus had to play catch up. But they do have the "front end," the electrical end, and now they just have to get their protocol engine straightened out.

The initial RapidIO specification defines physical layer technology suitable for chip-to-chip and board-to-board communications across standard printed circuit board technology at throughputs exceeding 10 Gbps utilizing low voltage differential signaling (LVDS) technology. Unlike other next-generation I/O technologies, RapidIO technology is transparent to application software, and does not require special device drivers. Additionally, it has no impact on operating system software. The RapidIO Interconnect can also be a bridge to other bus technologies such as PCI, PCI-X, and system area networks built with other switching fabrics such as InfiniBand.

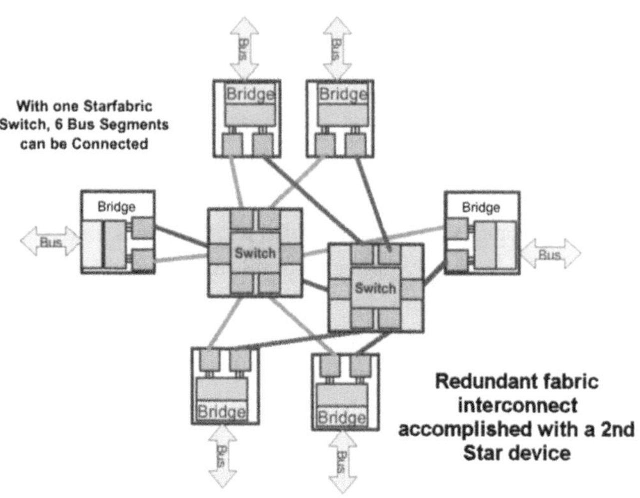

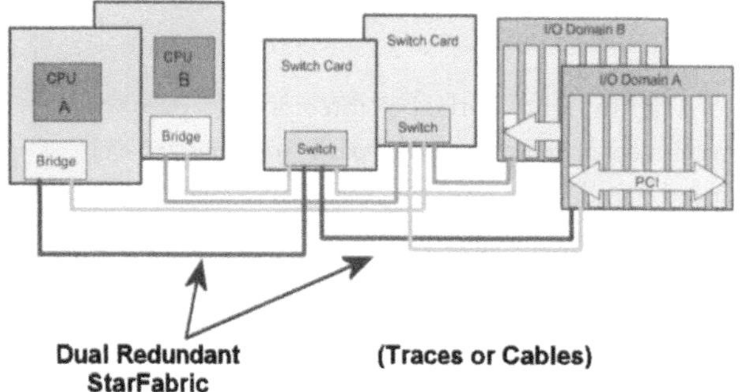

StarGen's StarFabric is compatible with PCI / CompactPCI software, can connect processing nodes at 3.2 Gigabits per second via CAT5 cable or etched traces. It can use existing CPUs, I/O cards, and chassis, and its very low overhead protocol doesn't impact on BIOS, drivers or applications. It allows Quality of Service and Isochronous data types for Time Division Multiplexing data and it can be configured as a redundant, fault tolerant system.

The RapidIO standard is a packet-switched interconnect architecture conceptually similar to internet protocol (IP). However, the RapidIO architecture is designed to be used for the processor and peripheral interface where high bandwidth and low latency are crucial. The RapidIO architecture is partitioned into a three-layer hierarchy of logical, transport, and physical specifications, which allows scalability and future enhancements while maintaining compatibility.

For more information visit the RapidIO Web site (www.rapidio.org).

StarGen's StarFabric

StarGen, Inc. (Marlborough, MA– 508-786-9950, www.stargen.com) is a semiconductor company developing an interesting pseudoserial, high-speed switch fabric called StarFabric. StarFabric supports board- and chassis-level interconnects. StarGen's StarFabric should have a bright future if only because it's supposed to be totally software compatible with regular PCI-based software. So, unlike InfiniBand, you won't have to rewrite all of your code, which is the biggest software headache imaginable.

Some companies announced that they've developed software to run on InfiniBand, but you're going to have to pay a fee to use it. With StarGen, everything should be plug-and-play with your existing PCI code since it acts essentially as a virtual PCI bus. No doubt there will be some exception cases that must be handled, but that's yet to be seen.

StarFabric can handle simultaneous packet, cell, and voice traffic and provides high availability features such as fault detection and isolation, hardware fail-over, and hot swappability. StarFabric also provides a smooth migration path from existing bus-based architectures like PCI, H.110, and the ATM Forum's Universal Test & Operation Physical Interface (UTOPIA) specification, primarily because it's 100% backwards compatible to existing PCI and CompactPCI technology. This means that developers and integrators don't have to craft new drivers and software.

StarGen's StarFabric was architected for supporting loosely coupled processors in a distributed fashion, but in a very efficient way. It's well suited for both distributed computing and high availability.

What "loosely coupled" really means at StarGen is autonomous 64-bit address spaces that can be mapped at every endpoint, so a fabric can have as many as a 100 or 1,000 nodes with a protected communications capability between all of those processors. Then there's StarFabric's ability to distribute everything from driver services to interrupts and semaphore communications. It's a fairly complex set of capabilities but it's all there, having been built into the technology in anticipation of enabling high-bandwidth, high-availability distributed applications.

Stargen's StarFabric, like Infiniband, RapidIO and PLX, competes with PICMG 2.16 Ethernet backplanes. Some people ask, "why not just use Ethernet forever inside the box?" Well, as it turns out, that's not a particularly efficient way to provide in-the-box communication at extremely high bandwidths. A "LAN in a box" paradigm like PICMG 2.16 uses networking technology that was developed to support longer range communications. Such networking – Ethernet and ATM – suffers from a lot of overhead in processing the network stack software to get data to move from one processing element to another."

At slower speeds, in the realm of 10/100 Ethernet, the gains in processor utilization turns out to be not all that significant. But as you scale up to gigabit transfer rates, you're chewing up a significant portion of your processor cycles just to handle the network stack. That's why you see such dedicated network processor chips as the IXP 1200. The stack isn't really needed to just move data from one board to the next within a system – it was put there to help move data from a system here to a system across the country.

With StarFabric, this technology is not intended to resemble a LAN. It's not intended to provide communications across miles, but rather within a room or a box. So this is more of a PCI-type of model, where memory read-write kinds of communications replace stack processing to move data around. You can get orders of magnitude better efficiency, since instead of managing the network stack, the processors can now actually run the application you wanted them to in the first place.

Aurora Technologies (Brockton, MA – 781-290-4800, www.auroratech.com) a Carlo Gavazzi Group Company, demonstrated to the author some "PCI extender" technology based on the advanced StarFabric switching fabric technology from StarGen. A card plugs into a PCI slot, then a CAT-5 cable snakes over to a "pizza box" containing three more PCI slots. A CAT-5 cable can then be connected to a whole series of boxes, allowing you to create a configuration that acts as a single system with around 220 PCI slots.

ITOX (East Brunswick, NJ – 732-390-2815, www.itox.com) also recently showed me a PC system with 42 expansion slots (6 ISA and 36 PCI slots), an example of its Star Cluster concept, also based on the StarFabric.

Although the ITOX system consists of three rack-mounted chassis that can be up to five meters apart, it functions as a single PC with common PCI and H.100 buses. The system operates as a standard Intel architecture PC and uses the standard PCI card addressing built into the operating system to provide 42 card slots with no alterations to the BIOS, the OS, drivers or application software.

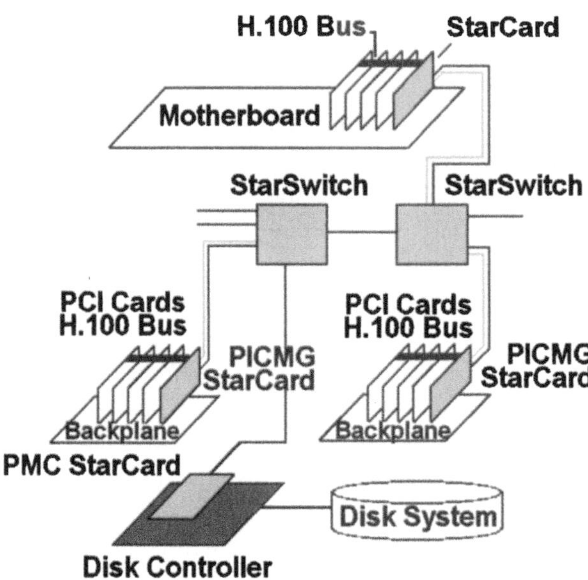

This is but one example of the ITOX Star Cluster concept, which can be expanded to support over 1,000 expansion cards and multiple processing clusters that are highly reliable, with optional enhancements for computer telephony.

To be specific, the ITOX StarCluster system allows a PC to be expanded from a motherboard to an additional passive backplane chassis or to harness the combined processing power of multiple PCs working together. If you need more slots, you just follow these five steps:

1. Turn the power off.
2. Plug an ITOX StarCard into the motherboard.
3. Plug an ITOX PICMG StarCard into a passive backplane, or another StarCard into a second PC.
4. Connect them with standard LAN cables.
5. Turn the power on.

If you're just expanding a system, that's it. There are no software implications. When you power up the system, the operating system software uses Plug-n-Play to first find the slots in the motherboard and then those in the backplane. Then the operating system works just as though all of those slots were on the motherboard. If you are connecting regular PCs together, you will need some additional software.

Expanding a system having H.100 "CT Bus" signals demands a few additional considerations:

- Both StarCards must have H.100 capability.
- The StarCards must be plugged into an ITOX Star Switch box.
- You need two sets of LAN cables.
- An H.100 software driver must be installed.

By using mezzanine form factor Star Cluster cards, both regular PCI and cPCI systems can be incorporated into the same configuration.

StarFabric plays a role in yet another PICMG specification, PICMG 2.17. The PICMG 2.17 spec may in fact solve all of the automatic failover conundrums we've been agonizing over in this chapter. Unfortunately, the 2.17 spec is not as developed as 2.16. Indeed, 2.17 doesn't exist at the time of this book's publication (early 2002), but there are some things we can project about what it will be able to do from what's known publicly about StarFabric.

StarFabric's immediate competition, the PICMG 2.16 Ethernet backplanes, are based on a familiar centralized model of a redundant star. Links come off every node card and go to one switch card or possibly a redundant switch card.

"We envision supporting that kind of a topology in PICMG 2.17," says Tim Miller of StarGen, "but we can also do a distributed topology that supports switching elements on each node card. This

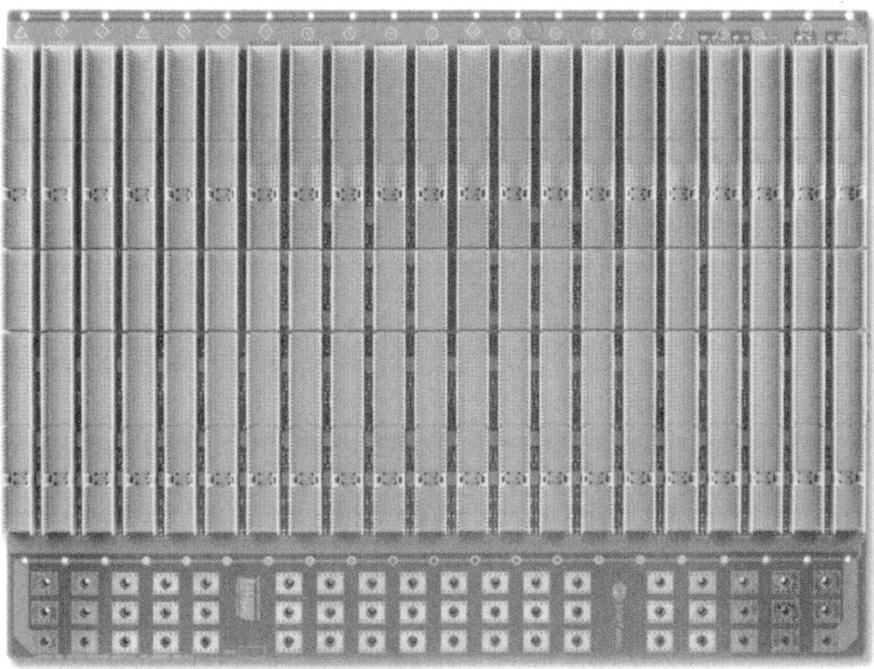

StarGen and Bustronic teamed up to create what looks like a 21-slot CompactPCI backplane, but it is, in fact, home to StarFabric, StarGen's new pseudo-serial switching fabric. The 21-slot StarFabric hybrid backplane comes in a 7U high form factor, compatible with standard 6U cards, with 1U of power connections on the bottom. Featuring a 12-layer controlled impedance stripline design, the outside layers are grounded for EMI protection and suppression. The signal layers are alternated with power and ground layers to minimize crosstalk and control impedance. Power and ground planes use two ounces of copper for enhanced power distribution. The differential wire pairs are carefully routed as close together as possible and kept on the same board layer.

means you can start with a relatively small configuration and not be required to have a dedicated switch card in your box. Platform vendors tell us that when you're dealing with little mini-chassis of five or fewer slots, a distributed model makes more sense than a centralized one. This will resemble a sort of crossbar switch, or actually halfway between a star and a fully interconnected mesh, where the node cards are connected to a set of node card neighbors, and those neighbors are in turn connected to other node cards. You can interconnect an entire fabric in this way without a dedicated switch card. This allows you to once again have efficient use of node slots and also add cards incrementally. The whole idea is to provide system level specifications so both platform vendors and board vendors have a common set of specs upon which to build their products."

StarFabric can address some of the issues of TDM traffic on legacy H.100 and H.110 CT Buses, something that Ethernet doesn't do very well. You can certainly packetize all of your calls and send the voice packets from board to board (or box to box) over Ethernet, but you've got to make some QoS compromises in the process.

Says Miller: "Ethernet packets are too large to support genuine, real-time QoS, so a lot of the

PICMG 2.16 Ethernet-type implementations actually have to overlay a TDM plane of some kind to take care of that. With StarFabric, however, you can eliminate those redundant interconnections. A single interconnect supports both TDM and packet-oriented traffic. And since we don't rely on processor-mediated communication, we can support nodes in the fabric that don't have a processor at all. Compare that to an Ethernet environment, where you need a processor at every endpoint to process the network stack."

Mark Overgaard says that with StarFabric "you don't have to make those compromises. You can take your H.110 traffic or other TDM traffic and put it over the StarFabric, fully expecting that the traffic of a hundred or more separate H.110 buses will get along fine and that you can still do switching among them with no problem. One way to look at 2.17 is as a complement to 2.16. If you're committed to building a system based on 2.16, then 2.17 will fill one of the 'holes' that 2.16 still has, which is how to handle the TDM traffic."

PICMG 2.17 systems will address the same kinds of architectures that the 2.13 Redundant System Slot architecture was aiming at, but you'll be able to do it in a more generic, realizable and scalable fashion. Under the 2.17 spec you can incorporate legacy boards that you don't have to throw away and replace with boards having 2.16 interfaces.

Says Overgaard: "You could define your system in such a way that you still have some CompactPCI buses with H.110 on the P4 connector and PCI buses on P1 and P2 and then you can use StarFabric to link these segments together in a way that allows you much more scalability and many other good properties compared to doing it the way Motorola and Ziatech did, or the way 2.13 is trying to do. And it still allows you to use your existing resource boards."

PLX's Adaptive Switch Fabric Architecture

PLX Technology (Sunnyvale, CA – 408-774-9060, www.plxtech.com) a leading supplier of high-speed silicon and software solutions for the networking and communications industries, is the mastermind behind what it calls the PLX Adaptive Switch Fabric Architecture, which is also PCI friendly. And unlike some of its competitors, it's actually starting to ship product.

PLX's GigaBridge GBP32 was the first controller based on PLX's switching fabric. PLX expects that the controller and development tools will be the first PCI-based switch fabric silicon and software on the market.

PLX is based on a redundant ring bus topology, which means that you can cut it anywhere and it will still keep working. Buses can't do that. You get one transceiver locking up on the bus and the whole system will stop. PLX's architecture lets designers incorporate 224 PCI bus segments, aggregate tens of gigabits per second and link between PCI devices up to five meters away.

In the PLX PCI ring, PCI signals carrying data between either I/O subsystems or the system controller get converted to a form suitable for a fabric connection. The ring then sends it to its destination where it's converted back to a PCI format. In this way system designers can retain all of their legacy PCI line cards, and they won't have to write any new software or invest in any new form factor or connectors.

Several hundred controllers using the architecture form a cell-based fabric of dual counter-rotating rings, which appear to the communications system as a huge network of PCI bus segments. Each controller drives up to eight CompactPCI slots (or four PCI slots) and interpolates with other controllers as "nodes" on the ring. Each node is connected on the ring via two 16-bit-wide LVDS links operating at 400 MHz.

Each controller has a protocol engine that transparently manages the routing of cells from the PCI source to PCI destination. The controller automatically converts PCI transactions to fabric transactions and back, and manages all error-correction or retry functions.

Some say that what PLX has done with its redundant ring topology is what CompactPCI should have done, because, technically speaking, it does what CompactPCI says it can do but can't.

Of course, the telcos have already run into these "high density, high availability switching" issues on a larger scale, with ATM switching, which demands the use of fast packet switches to move the little 53-byte ATM cells along their respective virtual paths. ATM switches route incoming cells arriving on a particular input link to an output link (also called the "output port"), associated with the cell's route. Three techniques have been proposed to handle the routing (switching) function: Shared-memory, shared-medium, and space-division.

Shared-memory (SM) switches consists of a single dual-ported memory shared by all input and output lines. Packets arriving on all input lines are multiplexed into a single stream that is sent to a common memory cache for storage where the packets are grouped into separate output queues for each output line. Simultaneously, an output stream of packets is formed by retrieving packets from the output queues sequentially, one per queue; the output stream is then demultiplexed, and packets are transmitted on the output lines.

On the other hand, shared-medium switches are based on a common high-speed bus, much like present day computing systems. Cells are launched from input links onto the bus in round-robin order, and each output link accepts cells that are designated for it. Buses don't scale very well, however.

It's the third scheme, however, that interests us: The space-division switch, which is a crossbar switch, a kind of technology that has served circuit-switched telephony networks for many decades. A crossbar switch's inputs and outputs are connected at switching points called crosspoints, which results in a matrix type of structure that's very scaleable.

PLX claims that their ring topology has some advantages in reliability over crossbar switch approaches, since to make a crossbar architecture redundant and provide high availability you generally need to have two ports on every line card, one for each fabric card. With the PLX system, however, there are two rings built into the architecture. So if failure does occur on one path, the traffic is immediately routed onto an alternate path.

Some critics charge that ring schemes have limitations in terms of scalability and performance, suggesting that after eight or so bus segments the aggregate bandwidth of the ring is maximized and that every node you add after that actually decreases the average bandwidth available to any given node.

For some reason, lots of MIS guys (such as Yours Truly) have avoided ring-based concepts over the years, whether topologically real or logical. IBM's Token-Ring never even came close to displacing Ethernet. Also recall that, in the late 1990s, National Semiconductor and Apple Computer got together to create QuickRing, a high speed point-to-point data transfer architecture (really a bus-like low latency network). QuickRing was going to be Apple's fabulous local bus technology.

The QuickRing architecture used National Semiconductor's Low Voltage Differential Signaling (LVDS) CMOS technology, which could provide very high-speed, low voltage swing differential signals that minimize common mode noise interference. Each QuickRing Data Stream Controller was capable of moving high bandwidth data streams of up to 350 million samples per second per line. Very high bandwidth systems could thus be constructed, with peak theoretical rates of 1.7GBps for a 16 node ring, according to National Semiconductor.

Apple ultimately decided to pull out of QuickRing and go to USB, partly because it's a "hub" architecture rather than a "ring" architecture. Besides that, QuickRing was based on a non-redundant single ring, so if a node on the ring got into trouble, the whole ring died. QuickRing could only be made redundant by using two rings and doing some electrical gymnastics to hook them together.

Perhaps PLX will finally make ring architectures commonplace. PLX has released its GigaBridge GBP64, the first 64-bit controller based on PLX's adaptive switch fabric architecture. Bustronic, Elma Electronic (Fremont, CA – 510-656-3400, www.elma.com), and PLX have a partnership to develop backplane solutions based on the PLX GigaBridge PCI Switching Solutions product line. Bustronic will begin integrating the PLX GigaBridge technology into their extensive line of switch fabric backplanes and Bustronic's parent company, Elma Electronic, will also implement the GigaBridge technology into their line of NEBS-compliant chassis with "five-nines" High Availability characteristics.

Different Strokes

Any way you look at it, there's going to be a plethora of new computer architectures in the 2002 to 2005 time frame. At Pete Holmes of Intel says: "Our outlook is that we see different strokes for different folks. We see different architectures making sense for different applications."

Or, as Joe Pavlat's father likes to say: "the number of things I know nothing about is increasing at an alarming rate."

Maintainability

With such tremendous scalability now possible for fault resilient systems, technicians now face the daunting task of actually maintaining huge communications systems consisting of hundreds or thousands of computers.

The tasks associated with controlling today's computing environment can be divided into four parts (thanks to Don K. Harrison and James Honey of RadiSys for these):

- **Asset Management:** Suitably equipped or "instrumented" components can identify themselves to management software, and provide data for inventory control and asset management. Hardware considerations include components modified to do this as well as data acquisition hardware components such as temperature sensors and monitors. Such "instrumentation" enables management applications to understand and change the state of a PC and to be notified of state changes. This assumes a common methodology and syntax for defining and reporting the management features and capabilities of all hardware, software and attached peripherals of the PC.
- **Off-hours and remote maintenance:** Remote boot and remote control capabilities can automatically "wake up" a PC and install or upgrade software, without disrupting the user and without a technician's direct involvement.
- **Initial system configuration:** A service boot feature allows unattended installation of new systems, including installing the operating system and user software.
- **Remote problem resolution:** Support technicians can take over a system remotely, increasing first-call resolution rates and minimizing the need for personal visits to the equipment. Hardware

monitoring, prediction of impending failures, and notification of problems before they bring down a system are an important component of problem resolution. Also, the modern vision of this idea has been extended to include access to the World Wide Web so that systems can now be managed on a global basis as with WBEM (Web-based Enterprise Management).

Hot Swappable Components

This challenge of maintainability at first led to the innovation of hot swappable components. Normally, systems are powered down for repairs and/or reconfiguration. However some systems (such as those for telecom and datacom) are configured so that it's not acceptable to power down all or part of the system. If a power supply, disk drive, or even a single board computer or communications resource card has failed, hot swappability (also called "live insertion") allows a technician to remove the defunct device and "swap in" a new one within minutes or even seconds, all while the system continues to run.

CompactPCI's Hot Swappable Boards

The hallmark and big selling point of CompactPCI systems is the hot swappable peripheral "card" or "board." The ability to replace a defective board while the system is running is totally beyond what's available in the desktop computing world. With cPCI, the hot swap sequence is generally initiated by the operation of a button or micro-switch in the ejector handle. This asserts what's called an ENUM signal on the CompactPCI backplane causing the hot swap software to unload the device drivers for the board (as well as software for any bridge chips if the component happens to be an add-on "daughtercard" or PMC module). Then a front panel blue LED lights up indicating to the user that it is safe to physically withdraw the module from the backplane without having to shut down the rest of the system. When the replacement module is inserted, "first make-last break" contacts fitted to the power and ground pins apply power to the board in the correct order and the blue LED again lights up. The board or PMC module then powers up. By returning the component's ejector handle and/or micro-switch to the proper position, the operation is completed and the hot swap software extinguishes the blue LED.

The spectrum of board hot swappability ranges from basic hot swap systems where the backend power circuits and bus interface circuits require only that power remains on, to more complex "full hot swap" systems where software applications must run continuously, reallocating accessible hardware resources in real time as the systems go on with their business.

Keeping hot swap systems up and running thus means occasionally pulling and replacing boards in subracks that are running 24x7. One can discuss the abstract nature of the electronics and various levels of hot swappability but these aren't the only issues – also vitally important are the design of handles, sockets, supporting structures, and how you can wreck a system with them if you're not careful.

The issues involved relate to the sheer physical act of inserting and removing a board, and what kind of trouble that can get you into: The force of insertion / extraction needed, the effect on board handles, avoidance of bowing / deforming the backplane, inserting the wrong board in the wrong slot, the danger of one board gouging another or getting so close that electrical discharges occur between them, etc.

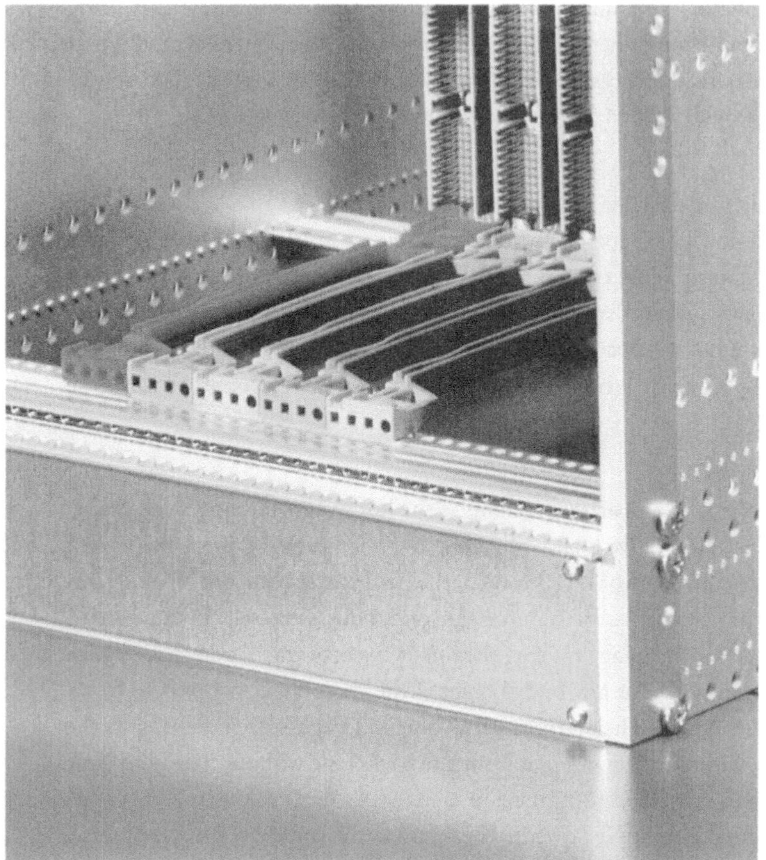

To help keep technicians from plugging the wrong board into the wrong slot, color coding is often used. Here are examples of Rittal's color-coded board guide rails.

Goof-Proofing Hot Swap

When you're going about the process of integrating ruggedized CompactPCI systems into your business, you soon discover that, unlike a conventional PC with PCI slots, you can't plug any CompactPCI board into just any old slot on a passive backplane. When backplane manufacturers lay out their backplane for their customers, they design the backplane with the intent of having certain boards plugged into certain slots. This is because some of the board's jumpers and connectors (such as J4/P4 and J5/P5) carry user-defined signaling and voltages (I/O voltages can be either 3.3 volts or 5 volts), and there may be timing issues (a board with superfast circuity can't be too far along the bus from another one). Also, some slots are designated for CPU boards.

Similarly, with many VME64 and VME64x systems, you must be careful to insert certain boards into specific cage slots to match I/O cables attached to the backplane.

Moreover, in-rack CompactPCI power supplies are designed to fit into the card cage too and resemble fat boards, so it's important that you don't do something really catastrophic such as damaging an AC power supply by inserting it into a backplane wired for DC and vice versa.

It comes down to this: If you plug the wrong board into the wrong slot your system may become inoperable, or worse, damage may occur to the plug-in board and/or backplane.

One easy, though not completely effective way, of inserting the correct board into the correct slot is color coding. The color red is now associated with CPU boards, for example, so you can order a subrack with a red card guide as well as a red front panel overlay or marking for the board. The PICMG 2.11 specification says that a pluggable power supply's guide rails shall be green in color, but later in Appendix B it says that the existing practice of using any color guide rail is also acceptable for a 38 pin in-rack power supply connector!

Slot Keying CompactPCI Boards

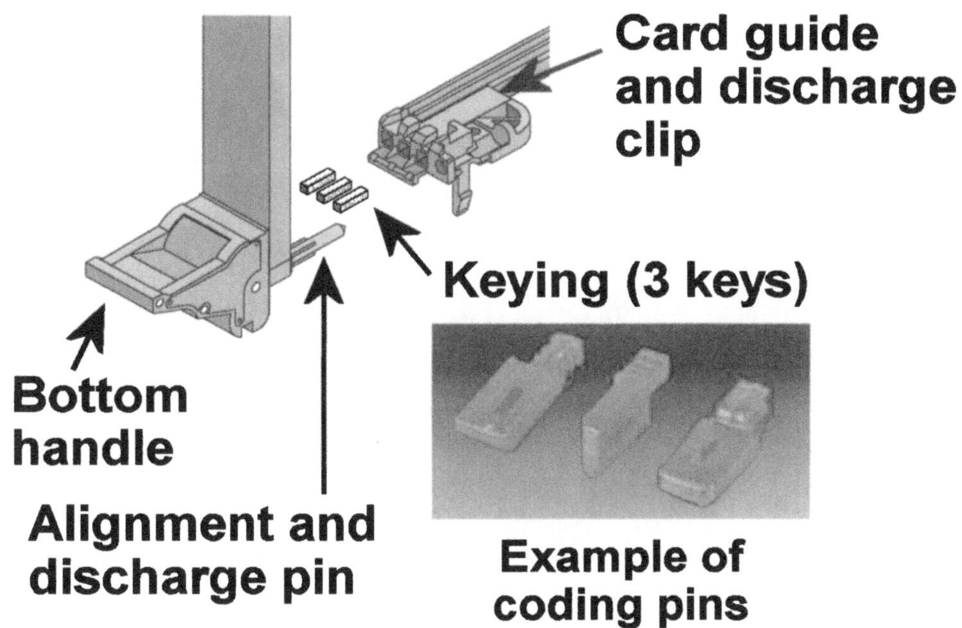

Rittal offers color-coded guide rails. Another company that offers a huge range of card guides is Bivar, Inc. (Irvine, CA – 949-951-8808, www.bivar.com). One can't resist listing Bivar's wonderful naming scheme for their various guides: Circ-o-Gide, Comp-o-Gide, Econ-o-Gide, Grip-o-Gide, Narr-o-Gide, Stat-o-Gide, Temp-o-Gide, and Vert-o-Gide. They also offer conductive and horizontal card guides (which I guess should have been called "Horiz-o-Gide").

Even when you employ color-coding, you'll be surprised how many fumbling technicians suddenly claim to be colorblind after they've pounded the wrong board into the wrong slot.

Fortunately, for customized applications where a dedicated slot for a certain board must be designated, CompactPCI and VME64x allow for what's called "mechanical keying" or "slot keying" of the boards, so you can't make a mistake.

At the bottom and back of a board's faceplate (or "front panel" as it's called), there's an alignment pin, and to the left of that are three other rectangular plastic pin-like structures in certain configurations. These are the "keys" which sit in a keying chamber. The keys snap into the card guide on the subrack, fitting into corresponding "mating keys." When a plug-in board and a slot are properly keyed, the card can only be plugged in the one correct slot. If the board's keys and the slot's mating keys don't match each other, the board won't fit.

The keys can be positioned in various ways (rotated, for example), and the Eurocard / IEEE 1101.10 standard allows for up to 4,096 unique programming "key" possibilities per slot for cPCI boards. It's said that VME64x allows up to 15,625 mechanical keying combinations to prevent board insertion into the wrong slot.

Rittal's "programmable" keying tool prevents manufacturers, integrators or end-users modifying their systems from messing up slot coding key installation.

However, I/O board and system assembler / integrators have difficulties with mechanical keying. When it comes to installing the special keys in a system, it's very cumbersome to install three keys behind the board's bottom handle and perhaps three keys behind the top handle too. And you have to set up a corresponding configuration on the mating card guide too.

To solve this problem, Rittal developed a "programmable" keying tool which enables easy insertion of keys with one simple move for both the front and rear of the subrack. This tool is about four inches long. One end of the tool is designed for inserting in one motion up to three keys into the front end of the subrack mounted guide rail keying chambers. An integrated alignment pin orients the tool during the insertion process. On the other end, the tool is designed for inserting in one move up to three keys into the rear of the cPCI module front panel keying chambers. Here, an integrated alignment hole orients the insertion process.

The procedure is quite simple:

1. First you pre-set the tool by using the existing keys.
2. Load the keys for the subrack or module into the pre-keyed tool backwards.
3. Push the loaded keying tool against the subrack or module keying chambers by first aligning the keying tool using the pin / hole orientation feature.
4. The keys will simply snap into the designated positions.

Many boards and systems have to be keyed before they're shipped, and Rittal's tool has made things a lot easier for integrators, printed circuit board manufacturers and ultimately, end-user technicians.

Of course, it would be great if one day CompactPCI's physical limitations could be overcome and we could just plug a cPCI card in any cPCI slot. One step in that direction came in 2000 with the SENTINEL bridging chip from Force Computers (San Jose, CA – 408-369-6000, www.forcecomputers.com). The SENTINEL enables a cPCI board to work in every slot of a CompactPCI system.

For example, you can now put Force's SENTINEL-equipped universal cPCI single board computers into any cPCI I/O slot, not just a System slot (SENTINEL is only available on Force products). SENTINEL was also the industry's first hardware application of industry standard Hot Swap for all cPCI I/O boards, including intelligent and non-intelligent I/O boards. Indeed, SENTINEL is optimized for cPCI Hot Swap and asymmetrical multiprocessing.

Getting a Handle on Things

Once you can feel confident that you've purchased a system allowing you to insert or hot swap a board in its correct slot without looking like a bumbling idiot, the actual physical act of doing so generates all kinds of mechanical forces that could wear out or even damage parts of the computer system. Don't worry though, a whole series of products is ready and waiting to counteract any harmful activity, even if you or your technician tends to handle components like King Kong.

Take something as simple as handles, for instance. Rittal's Type IVs ("s" for superior) handles are a new kind of CompactPCI handle. Regular CompactPCI handles for 3U and 6U high boards used to be completely made out of plastic and wore out quickly. Below the bottom handle there's a little claw actuator that grabs an aluminum extrusion (called the extended lip extrusion) and it helps insert the board into the subrack. It also assists you when trying to eject a board. Because there are so many pins in the connector on your plug-in board, the mechanical force required to install and remove each board is about 48 pounds (215 newtons) for a 308 pin 3U high board, and 118 pounds (525 newtons) for a 749 pin 6U high board. Under these conditions, plastic handles have a limited life expectancy (anything from 50 to 200 insertion / extraction cycles).

If you insert or remove these boards too many times, the handles (especially the little claw) will start to wear or even fracture or break off entirely. To address this problem, Rittal developed a metal handle. The handle itself is a really a tough plastic, but the claw actuator that interacts with the extrusion is made out of metal. So basically they're nearly indestructible, and you can subject them to a nearly unlimited number of insertion and ejection cycles.

As CompactPCI designs move to even larger 9U cards, Rittal recommends that customers use their Type VII handle. These "low profiles" are a little taller than the Type IVs and fold up flat against the panel instead of sticking out at an angle like conventional cPCI handles. The black plastic Type VII is rated at the same 118 pound insertion and ejection force of the standard cPCI handles. For a 1,190 pin, 9U high board, you need about 180 pounds (815 newtons) of force to move an I/O board in and out, so board makers and others should go with the metal version of the Type VII handle (these handles are already a hit with swappable power supply makers).

Rittal's Type IVe handle is new too. Recall that the metal Type VII handle folds and sticks up next to the faceplate panel, instead of sticking out like an ordinary cPCI handle. When the Type VII handle folds up, it occupies about two inches of space along the panel. That's okay if you don't need your full front panel spacing for I/O. But typically, from top to bottom, you need every inch of real

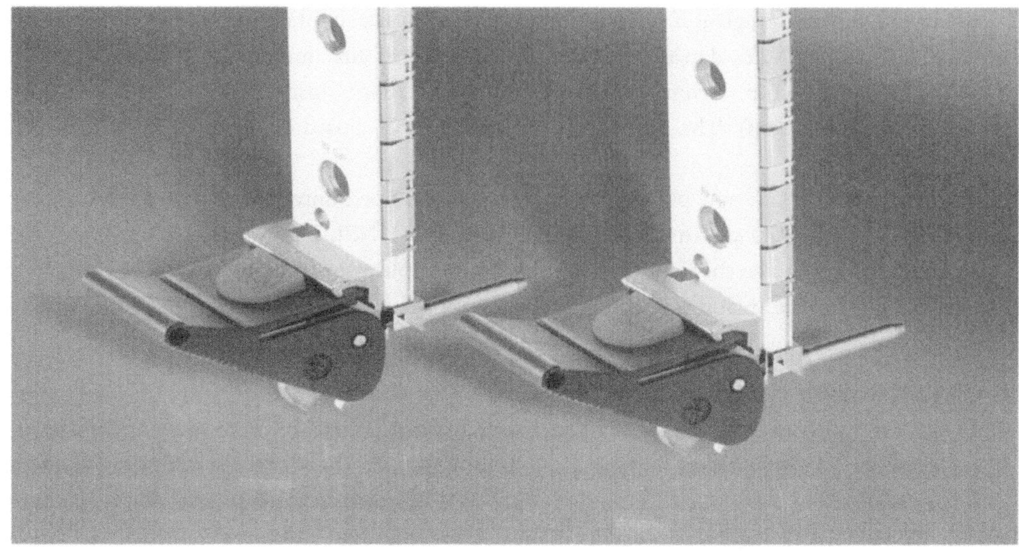

Rittal Type IVs Handles

Example of cPCI Hot Swap and VME64 handles

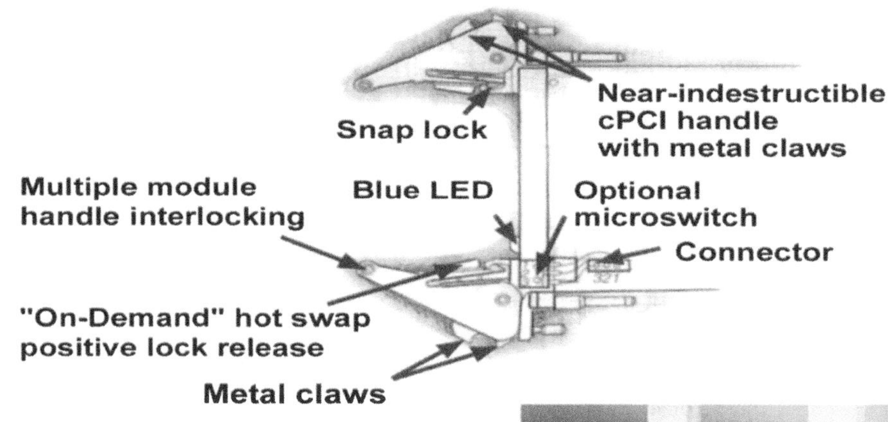

Low-profile Telco Type VII metal handles will extract 9U boards demanding up to 180 lbs. of force and are also available in plastic (up to 118 lbs.)

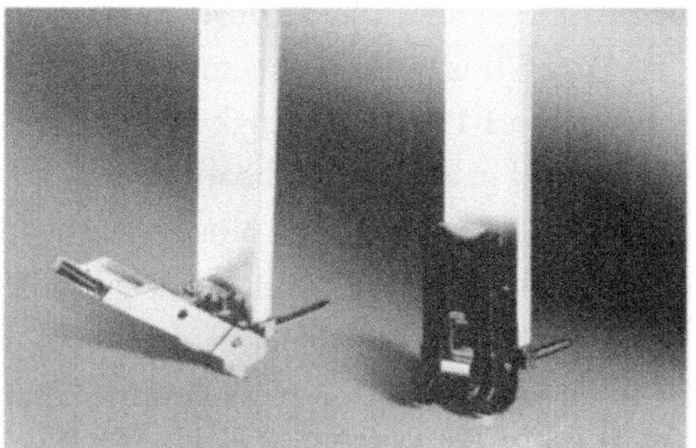

estate for connectors, the "hot swap" LEDS and anything else you can cram onto the panel.

To reclaim the panel space, Rittal developed a Type IVe handle which looks like a standard Type IVs handle, except that it sticks outward a generous 2.5 inches. It's longer than a regular handle and has a bit of a curve to it, providing greater leverage. It's made with a casting process that leaves it nearly indestructible. It can withstand the same 180 pound insertion or ejection force of the Type VII handle.

Rittal's handles can also be ordered so that the board is offset to the right by 0.10".

Not to be outdone, Pentair Electronic Packaging (Warwick, RI – 401-732-3770, www.pentair-ep.com) also offers an advanced CompactPCI insertion/ejection handle as part of its Schroff product family, that incorporates several unique design features that solve industry-wide handle problems.

Offsetting Your Problems

It may come as a surprise that boards inserted in passive backplane slots are offset from the slot's center. Offset? Why displace the board slightly? Well, if you look at a front panel of a board module, it's usually about 4HP wide, which is 0.8 inch, or 20.3 mm. (each "HP" or Horizontal Pitch is a unit of measurement for width – 1HP is 0.2" or 5.08 mm wide). When a circuit board is bolted onto a handle, the soldered side of the circuit board has one tenth of an inch space before it would make contact to the board to the left of it. Companies are now putting more and more components on what's called the "Side 2," "backside" or "solder side" of the board, and there's a danger that while inserting or removing a board the components will bump against or electrically interact in some way, damaging the circuitry.

To lower the chances of that happening, manufacturers engineer the handles and card guides to move the board to the right, centering it in the front panel. This is called an offset. Rittal's Type IVs and Type VII handles can be ordered with an offset giving the end user more space on the solder side for mounting components such as surface mount resistors or capacitors. This shift in the plug-in card location nearly doubles the space for all sorts of components, allows for increased airflow in some of the card slots and improves electromagnetic safety and emissions (EMI/RFI) control.

Both the VITA 30-20mm Eurocard Mechanical Packaging standard and the PICMG 2.11 cPCI Power Interface specifications (for modular power supplies using the P47 power connector) refer to this 0.1-inch shift to the right, so it's now mainstream technology. For cPCI modular plug-in power supplies, the card guide for the supply also must be offset 0.1 inch in order for the connectors to mate. At least two different offset card guides have emerged. This is not too surprising, since even "standard" card guides for cPCI come in at least two different designs depending on the enclosure makers' front rail configuration, or how the card guide snaps into place in the enclosure.

When Offsetting Isn't Enough – Board Covers

Of course, if a customer modifying a plug-in board decides to put too many components on the solder side of the board, a simple offset isn't enough to protect the board from interacting with its neighbors, particularly if you're going to be periodically inserting and extracting them. To prevent any bumping, scraping, arcing or other damage to board components on cPCI front mounted modules and/or cPCI rear mounted I/O modules, you install a ***solder side cover*** which is a protective sheet of material that's mounted over the component side of the printed circuit board.

Rittal's solder side covers come in both solid and perforated versions. In a situation where there are heat-producing active components such as resistors and capacitors on the solder side of the

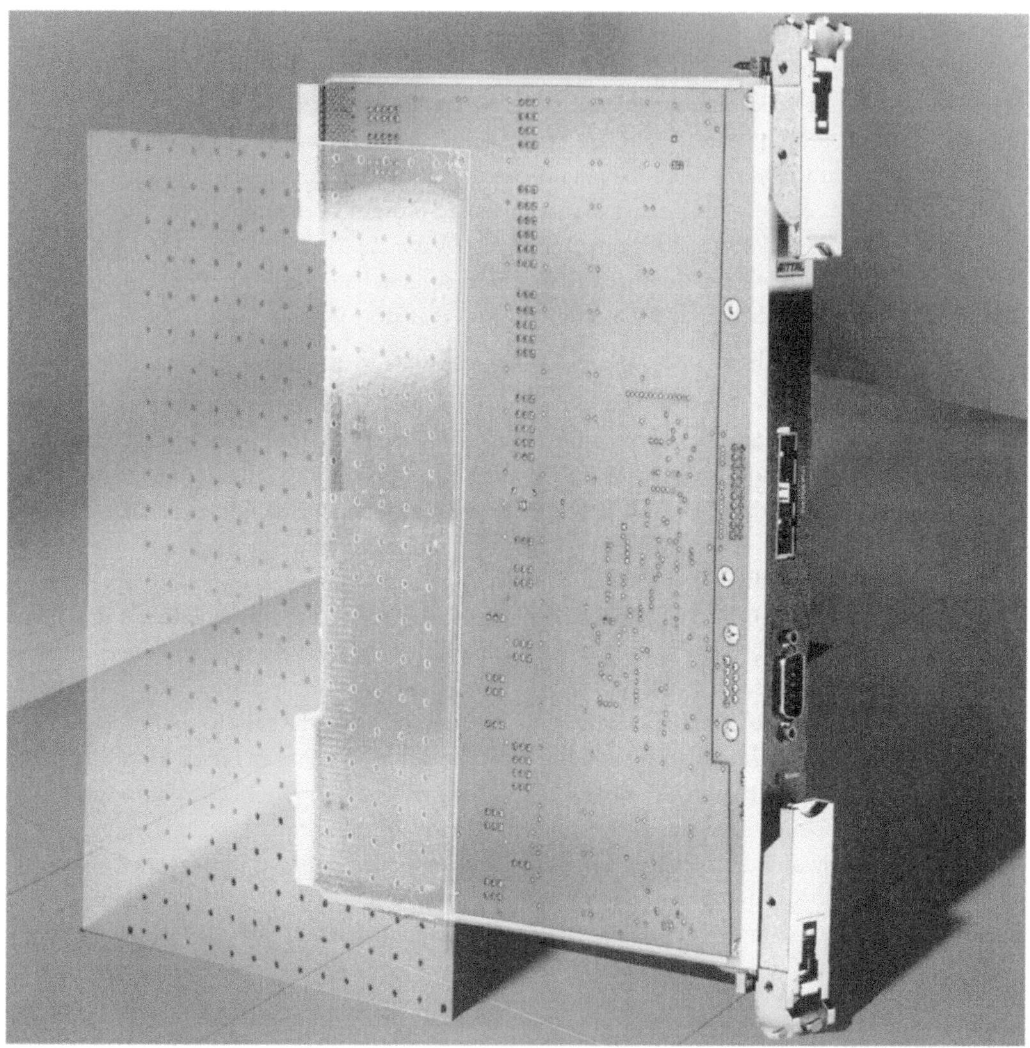

Rittal's perforated solder side cover protects hot swappable electrical components while allowing for a cooling airflow to pass through.

board, you want that side of the board to be ventilated for airflow, so you'd use Rittal's perforated solder side cover. The cover material is a UL-recognized nonconductive, nonflammable plastic that's vacuum formed and has 2mm diameter round holes punched through it.

While the solder side covers are optional per the PICMG 2.0 specification, the PICMG 2.1 hot swap specification demands that a means shall be provided for mounting a protective cover (solder side cover) on a board anyway, so even if you don't have any highly active components on the solder side of your board, you should still use at least a cover.

Solder side covers can also be obtained from companies such as APW Electronic Solutions and Bivar, whose SSC series made of proprietary AntiStat material easily mounts to existing boards via four

Rittal's non-bowing and non twistable support rails. At top is the standard version, at bottom is the new double thick cPCI COMPACT-I version.

small screws at the corners of the board. Elma Electronic (Fremont, CA – 510-656-3400, www.elma.com) sells a 0.2 mm thick polyester film mat protective cover for a 6U board. Some interesting 6U high Black Neoprene covers for cPCI boards can also be had from One Stop Systems (Escondido, CA – 760-745-9883, www.onestopsystems.com). They cost $20 apiece, $19 if you buy ten.

Support Rails

As we've said, as hot swappable CompactPCI boards grow in size, the sheer physical force necessary to insert or remove them approaches 200 pounds. Aside from the handles, there are several other issues in the hardware area influenced by the card size and the amount of insertion/extraction force used.

Imagine if you will, inserting a 6U board into the very center of a subrack. The handles' little claws grab the "extrusion" (what are part of the support rails that run left to right across the subrack) and pull the board inward. The forces of this action alone magnifies the natural tendency for the top and bottom extrusions to bow outward, placing distorting forces on the subrack and the other boards.

Even if you don't touch the board, the combined weight of cPCI modules, drives and power supplies that are plugged into a cPCI system may be so great that the bottom module support rails may deflect (or sag), causing modules to misalign or even to "fall out" of their guidance features. To prevent this, rugged horizontal support members are needed.

Originally, Rittal tackled the problem by putting two extrusions on top of each other so that there was double support. This solved the bowing problem but testing showed that the "double rail" design led to electromagnetic interference (EMI) and electromagnetic compatibility (EMC) emissions problems, so they had to install a gasket between the two layers to seal it up.

Rittal's aluminum backplane stiffener cuts the kind of backplane bowing that occurs when you insert a 6U CompactPCI card.

In 2000, however, Rittal introduced a new double support rail design specifically for CompactPCI subracks. Their new extrusion, which they call a horizontal extrusion, is essentially two extrusions in one. Instead of looking like two separate extrusions sitting one on top of the other, it's layered and appears to be a double thick rail. This eliminates the need for a gasket, and the layered double thick metal gives enough structural support so that you can insert or remove even a 6U or a 9U board without the rails bowing. Indeed, the double rail extrusions were found to eliminate bowing even in the really large 23-inch wide telco subracks.

Also, because of the double retention, the cPCI injector/extractor handles are unable to "twist" these double support rails, thereby improving the injection / extraction action. These rails are therefore considered ideal for NEBS applications.

Rittal's double horizontal support rail can be positioned near the outside of a cPCI system at the bottom only or at both the top and bottom of the system. There is no double rail required inside the system, where the rear support rails are tightened together with the backplane. However, just as the leg bone is connected to the ankle bone, the forces that occur at the point where the board connects to the backplane causes its own set of problems. . .

Backplane Stiffeners

When you press a 6U or larger cPCI board with 100+ pounds of force into a subrack, there's friction on all of those hundreds of pins entering a backplane slot. While larger slot count cPCI 6U backplanes may be rigid enough to counter the 100+ pounds of force necessary to inject / extract a cPCI module from the system, certainly the lesser slot count 6U backplanes don't accommodate such forces easily – they resist the board's insertion by bowing outward and away from the board. This results in poor pin connections.

To address the bowing, various companies such as Rittal have developed a backplane stiffener. This is a piece of extruded aluminum that attaches to the rear side of the backplane using the existing mounting screws that hold the backplane to the subrack assembly. Rittal's version reduces the backplane flex on a three slot 6U backplane by as much as 70 percent. You can install from one to 20 or more stiffeners, at any location along the backplane. Backplane stiffeners can be installed in the factory before the unit ships, or later out in the field.

VME64x and cPCI backplane stiffeners are also offered by companies such as Dawn VME Products (Fremont, CA – 510-657-4444, www.dawnvme.com) and Hybricon (Ayer, MA – 978-772-5422, www.hybricon.com).

Hot Swappable Power Supplies
Redundant hot-swappable power supplies can also come in load sharing, (also known as "load balancing" or "current sharing") and N+1 (need+1) configurations.

Two load sharing power supplies could each be supplying 50% of the total system power plus or minus 20%. So you might have two 500 watt power supplies running in a system designed to consume only 500 watts. Each power supply is therefore subject to only a 50% load, so both hot-swappable power supplies run at cooler temperatures and their life is extended. When one of them finally does fail, the other power supply takes up the full load while you pull out the failed power supply and replace it with a new one.

N+1 architecture generally means "more than two." In an N+1 configuration load sharing still exists but the load is now balanced among more than two power modules.

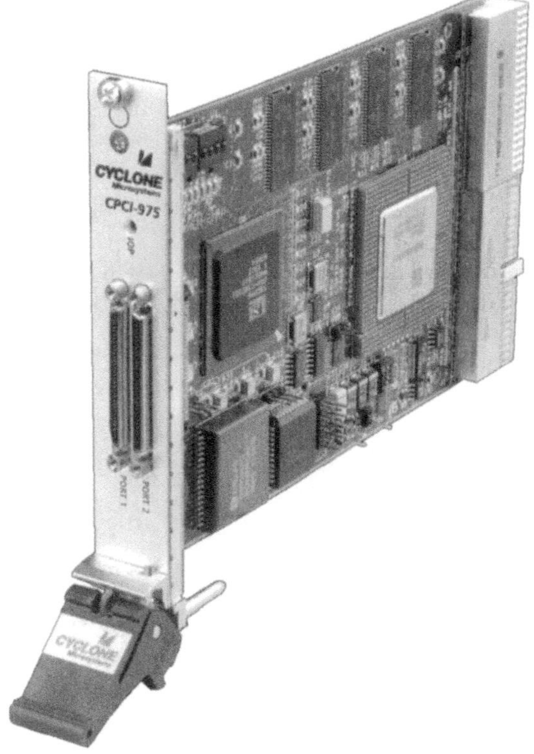

For high-end CompactPCI systems, Cyclone Microsystems (New Haven, CT — 203-786-5536, www.cyclone.com) offers hardware RAID controllers such as this CPCI-975 UltraSCSI controller, which supports various RAID levels as well as disk caching incorporating up to 128 MB of SDRAM.

RAID and Hot-Swappable Drives
Dealing with drive failure and lost data also became a lot easier with the rise of hot-swappable, redundant data storage. This generally comes in the form of a multiple disk drive RAID subsystem. RAID means Redundant Array of Independent Disks. Although any set of disk drives put into a common enclosure could be called an array, in terms of RAID technology those drives are subject to a hardware or software-based controller that makes them appear to be a single drive to the host CPU or operating system. A RAID disk subsystem architecture will in most cases write (or "stripe") data across multiple hard disks to achieve fault tolerance, if not continuous availability. One drive in the array is the "parity" drive, which contains computed data that can be used to recreate data lost on any one of the other drives, assuming, of course, that the other drives remain operational.

For example, you might have a five-drive array where one drive is designated as the parity drive. If a data drive fails and is then replaced, the drive array controller will rebuild the data on that drive using the parity drive and the other functioning drives.

An example of a RAID subsystem is the ProARRAY 2 RAID storage subsystem from CSS Labs

An example of a RAID subsystem: Advantech's Rackmount RAID-500U2 Data Storage Subsystem hold a quarter of a Terabyte and can be expanded to handle 56 SCSI drives via six expansion chassis.

(Irvine, CA – 949-852-8161, www.csslabs.com). It's a 4U (7-inch) high, 19-inch wide rackmount that can hold 15 1-inch high hot-swappable SCA2 SCSI drives (each having a transfer rate of up to 160 MBps) that are configurable as one array or they can be segregated into three individual 5-bay array modules, with each module having a single connecting backplane consisting of five 80-pin SCA2 SCSI connectors. Total storage is in excess of an 1.1 Terabytes (TB). Three internal hot-swap chassis fans (one fan per each drive module) keep the ProARRAY 2 cool. The system is powered by dual redundant, hot-swappable, 250 watt, auto-sensing power supplies in an ATX PS/2 form factor.

CT Division of Advantech also offers the RAID-500U2 Data Storage Subsystem. By using eight top-notch, 80MBps, half-height drives, Advantech has managed to fit a 0.25 Terabyte RAID array into a rackmount subsystem only 5U (8.75") high. For even greater expansion of storage, the RAID-500U2 can support six expansion chassis via an Ultra2 or Ultra Wide SCSI bus for a total of 56 hard drives. All drives, fans and the internal temperature are monitored by an alarm board. There's also a 128MB DRAM data cache.

Many PC customers often don't realize that there are four levels of hot swappability regarding SCSI drive arrays. Level 1 hot-swap is really a "cold swap" in that the power is not being applied to the hard drive or device. Level 2 is a "hot swap while reset" where power is applied to the device and the bus is held in a reset state. Level 3 is "hot swap while bus is idle," which means that no ongoing I/O processes are occurring on the bus during insertion/removal of the device. Finally, Level 4 is the Holy Grail of "Hot

Swap on an active bus," where the bus may have live I/O activity on the bus during the removal / insertion procedure, but the device being inserted or removed (such as a SCSI drive) must be idle.

Thus, the main difference between Level 4 and other levels is that the bus is allowed to move data or operate in any legal SCSI bus phase.

Adding Hot-Swap to an Operating System

One fly in the ointment is that while the latest, most advanced CompactPCI hardware supports hot-swap, all operating systems do not, at least not yet. To solve this problem, Jungo Ltd. (Natanya, Israel – 877-514-0537, www.jungo.com) has released an operating system extension which adds hot-swap support capability to all major OSs. Called the "GO Hot-Swap" OS Extension, it enables cPCI hot-swap capabilities under Windows 2000, NT, CE, NT-Embedded, 9x, Linux, Solaris and VxWorks – where these capabilities are not natively available.

➤ Types of RAID

There are various types of RAID, each indicated by a number or "level." The most simple level is RAID 1, which is where one or more disk controllers writes data to two drives simultaneously, which means that each drive is an exact duplicate, or "mirrors" another.

RAID-0 – Uses disk striping without parity information. RAID-0 is the fastest and most efficient array type but offers no data protection and is actually subject to the ***opposite*** of fault tolerance, since the failure of ***any*** disk will bring down the system. It's the one RAID level that's not really "RAID" at all.

RAID-1 – Stands for Redundant Array of Inexpensive Disks. Uses disk mirroring or duplexing. This is the array of choice for performance-critical, fault-tolerant environments if you don't want to buy more than two drives. One drive crashes. The other keeps the system up.

RAID-2 – Uses disk striping at the bit level with parity. This is seldom used since error-correction codes are embedded in sectors of almost all disk drives. Some implementations exist for supercomputer storage.

RAID-3 – Uses disk striping at the byte level with one disk per set dedicated to parity information. This can be used in single-user environments and performs best when accessing long sequential records. However, RAID-3 does not allow multiple I/O operations to be overlapped and needs synchronized drives in order to prevent performance degradation involving short records.

RAID-4 – Same as RAID 3 but stripes data in larger chunks (sectors or records). This allows multiple reads to be overlapped but not multiple writes. Like RAID-3, there is a dedicated disk to store parity information.

RAID-5 – Same as RAID 4 but parity data is also striped across the disks along with the data. Supports both overlapping reads and writes, but write performance is slightly degraded because of the need to update parity data. RAID-5 is the most popular RAID. It writes data across all its drives. If one crashes, it will continue to work. If two crash, the system will crash. The key with RAID-5 is to know when one has crashed (sometimes you get error messages) and replace it quickly. You can hot-swap a RAID-5 hard disk. And when you replace a bad one with a good one, the system will recognize the good one and rebuild its data.

RAID-6 – Same as RAID 5, plus additional striping so two disks can fail simultaneously, redundant controllers, fans, power supplies, etc.

Maintaining Call Center PCs with "Concentrated Computing"

Of course, even a single fault resilient PC-based phone system can serve a huge call center consisting of hundreds of computers distributed among several floors of a facility. During the course of each PCs working life, a technician will have to visit it several times, resulting in a lot of exhausting legwork.

Earlier we spoke of how the "shoe-box" computer has made possible what the author calls "concentrated computing." This uses what's called "extender" technology to allow hundreds of workers, students or call center agents to use PCs that are in fact sitting together not on their desks, but in a seven-foot rack perhaps hundreds of feet away, locked securely in a single rack that sits in an air-conditioned, controlled room (or an armor-plated bunker, if you like) environment.

Perhaps the best known of this technology appeared in 2000 when the call center world was shaken up by ClearCube with their C3 architecture, consisting of the ClearCube CPU Blade (an incredibly thin, 5.5" high Pentium-powered PC), the Cage (a metal housing that provides Ethernet connections, airflow management and power leads for up to eight CPU Blades per rack shelf), and the

➤ Fault Tolerant Software Tips

Although your hardware may be fault resilient or even completely fault tolerant, your software (operating system and / or applications) could be a real klunker. Here are some tips from QNX Software Systems Ltd (Kanata, Ontario, Canada – 613-591-0931, www.qnx.com), which makes a real-time version of UNIX called QNX.

1. High availability, fault-tolerant systems are only as good as the software that runs on them. The OS technology you use for your solution must not only allow for hardware issues like hot-swapping, but it's equally important that the OS allows you to recover rapidly from any potential software fault – preferably without a system shutdown.

2. Develop a mission-critical mindset. Even if you're an experienced programmer and strive to write error-free applications code for your business, you should implement your software in a way that assumes errors can occur. Common programming errors such as stray pointers and out-of-bounds array indices can evade detection during development. An OS that supports the Memory Management Unit (MMU) can help protect against these errors. If a memory-access violation occurs, the MMU will notify the OS, which in turn can abort the rogue process at the offending instruction. The OS should not only detect these errors, but also have the ability to inform a "software watchdog" of the violation – the watchdog can then decide how best to recover.

3. What are the benefits of a software watchdog over a hardware watchdog? A software watchdog can make an ***intelligent*** decision on how best to recover from the software fault. Instead of forcing a full reset, the software watchdog could:

- restart the failed process,
- abort any related processes, initialize the hardware to a "safe" state and restart the related processes in a coordinated manner,
- if the failure is critical, it can perform a coordinated shutdown of the entire system, perhaps sounding an alarm to control or maintenance staff. By employing the partial restart approach, your system can survive intermittent software failures without experiencing system down time, and perhaps without users or operators being aware of these quick-recovery software failures.

The software watchdog can also record logs of problems and perhaps process dump files for post-

C/Port (a small box located at the user's desktop that connects standard peripherals including monitor, keyboard, PDA or other serial device via a serial port, mouse and speakers) which connects to the CPU Blade via Category 5 (CAT 5) wire.

ClearCube's technology uses standard CAT 5 wire to carry a mixture of digital and analog signals. Video, audio, keystrokes, USB and mouse click signals can travel from the desktop to a ClearCube Blade, and video and audio signals can be sent from the ClearCube Blade to the desktop, with no degradation even after traveling 200 meters (656 feet).

In November 2001 ClearCube introduced their C3-RM platform which delivered the highest density of Pentium 4 blade computers on the market by employing advanced thermal designs and cooling methods to squeeze Intel's desktop processor into a blade computer format. With eight blades per shelf, ClearCube's solution allows you to take 112 PCs strewn around a huge call center, telecenter or other business and fit them snugly into a single 19-inch rack, where a single technician can manage and service them, allowing for dramatically increased control, security, mission-critical

mortem analysis. Compare this against the mysterious resets and interruptions in service that come from a hardware watchdog approach

4. Reducing kernel faults is critical. Programming errors don't occur only in application code. You may be developing device drivers or other system-level services. In traditional OS architectures these components run as part of the kernel – in kernel mode. Code running in kernel mode runs without MMU protection. As a result, errant pointers and such can cause kernel faults which only a hardware reboot can remedy. The more code built into the kernel, the higher the likelihood of kernel faults.

In a microkernel OS like QNX, only the kernel (32K of code) and interrupt service routines (ISRs) run in kernel mode, drastically reducing the possibility of kernel faults.

5. Redundant online resources – hot backups. With a distributed OS such as QNX, a network of computers functions as a single logical machine. A process running on any node can access all the resources of the network – your redundant resources can be located on any node on the network. If a device fails you can automatically restart a process to use a device, or even a file system, on another machine. QNX also supports multiple network links for further fault-tolerance.

6. Hot-swap hardware. The OS must support live insertion and removal of hardware, even cards. The card going in may have no resemblance to the card coming out – it may not even go into the same slot. Thus the ability of the OS to start / stop drivers dynamically is particularly important.

7. Quality control through OS architecture. By dividing software into a team of cooperating, memory-protected processes, the application developer (and the OS vendor for that matter) improve their ability to reuse these processes "as is" in new projects. The exact binary image of the process is being reused – no recompiling, no relinking, no relocation. By not modifying the binary image and having MMU protection provides a confidence that processes proven in the field will perform likewise in new applications. From a development perspective you can be assured that processes cannot clobber each other's memory as is the case for systems where all modules are linked into a single target executable.

8. Conclusion: Fault-tolerance is just as important in software as in hardware. A well-designed OS contributes to the level of fault-tolerance that can be achieved. Having the ability to intelligently recover from software faults is particularly important.

reliability, and significant uptime improvements.

ClearCube also has a blade-switching backpack, which is a network device that plugs into the back of each rack Cage. This "switchblade" allows you to switch a user to another PC on the fly to eliminate downtime. And once you've swapped them to a new PC, the IT staff can take control of the defective PC and troubleshoot it from anywhere.

So if, say, an agent at a help desk has a problem, he or she calls the IT staff, but instead of relying on a verbal ability to describe the problem over the phone, the IT staff can actually look at the PC screen from their end too and watch the agent generate an error. It's resembles a hardware version of pcAnywhere software, but the technician can see what's going on even if the PC is frozen.

ClearCube is thus bringing computers back into one central location (a "one rack, one office" solution" as they call it) and then providing either a internal or external help desk ability to control those PCs from anywhere in the world.

ClearCube's technology also allows USB I/O to be brought across CAT 5 to the desktop, yet it maintains the same level of security associated with the previous part of their architecture. ClearCube supervisors can block out the usage of unauthorized equipment on a class type basis, so you can specify that nobody can plug in a floppy, a CD-ROM, a scanner or printer. There are two ways of doing this: Taking advantage of the software preferences in user profiles that's built into Windows 2000, or you can use ClearCube's hardware version that's actually on the blade in the rack. This consists of a jumper setting that restricts the use of, for example, all mass storage devices, so you can't plug in a SCSI CD burner or whatever.

Because ClearCube's technology is a natural for use in call center environments, they have a strategic marketing and bundling agreement with call center software solution vendor Altitude Software (Milpitas, CA – 877-367-3279, www.altitudesoftware.com). Their UCI 2000 integrated software suite enables companies to manage self-service and assisted interactions across all customer touchpoints – including the Web, telephone, e-mail and WAP – and is built on a single Web-centric platform that consolidates, manages and reports on all customer-driven business transactions. ClearCube and Altitude do cooperative marketing and joint sales calls to both new and existing customers who are either building new call centers or revamping old ones. Both companies also offer a seminar series, which includes input from several architectural firms who specialize in call centers.

Aside from the safety and tight management capabilities offered by ClearCube's concentrated computing concept, an added benefit is that it can actually save your business money.

In interviews with ClearCube Technology's customers done by IDC for a study in 2001, it was found that ClearCube's technology can enable customers to radically maximize the efficiencies of managing their PC infrastructures while at the same time significantly improving worker productivity. This can enhance both top-line and bottom-line performance while reducing operating costs. Overall, the 17 customers examined for the study realized significant benefits from leveraging ClearCube technology. IDC found that a customer with a 100-seat ClearCube implementation could save the equivalent of $35,120 annually in IT time. Based on this data, IDC calculated that a firm of 5,720 employees – the average used for this study – could realize potential savings of $2 million annually in desktop management costs as a result of implementing ClearCube. Moreover, the potential increase in employee productivity for this size firm could amount to the equivalent of 25 employees at essentially no additional cost. The study concluded that while ClearCube may not be appropriate for all organizations, it does allow certain types of companies to do more with less.

ClearCube has its competitors. Although they haven't focused on the wide range of extender technology to the extent that ClearCube has, Crystal Group (Hiawatha, IA – 319-378-1636, www.crystalpc.com) does offer a KVM (keyboard, video and mouse) extender system called DataReach.

Crystal instigated the DataReach product in an effort to help MCI, which comes to Crystal periodically when they have a unique need. MCI systems engineers complained to Crystal that they were experiencing a specific set of problems in their call centers.

The first problem was employee theft of PC RAM. The second problem centered on employees reconfiguring the machines for one reason or other. For example, agents would bring in a bitmap photo of their child or a vacation spot and use it as a screensaver. The agents would bring in the photos on diskettes which also happened to be harboring PC viruses. MCI thus wanted better site security. Third, MCI didn't want their internal LAN made available to agents on the floor, because anybody who brought in a laptop or like device could log into their corporate net and cause mischief. Fourth, MCI had a big problem with foot-raised dust in carpeted areas, such as call center cubes.

All of these problems spurred MCI engineers to ask Crystal whether the PCs could be moved away from the operators and locked up in a room of their own.

Crystal looked around and found an interesting KVM extender technology from a company called Cybex Computer Products Corporation, which merged with Apex Inc. in 2000 to form Avocent (Huntsville, AL – 256-430-4000, www.avocent.com). Avocent has locations in Redmond, Washington; Shannon Ireland; Boston, London; Steinhagen and Munich, Germany; Tokyo and Toronto.

Crystal liked Avocent's LongView technology, which involves connecting a transmitter unit into the keyboard, video monitor and mouse ports of the PC, then plugging in a CAT 5 cable directly (not via the network), then plugging your local keyboard, video monitor and mouse into a receiver unit.

LongView offers both PS/2 and Serial connection, if you are using a PS/2 mouse you can even install a second serial device, such as a touch screen or graphics tablet. All major mouse types are supported, Wheelmouse, MS Intellimouse, etc. The user and the receiver unit can be up to 150 meters (492 feet) away from the PC and the transmitter unit. LongView supports video resolutions up to 1600 x 1200 at 85Hz. Audio support is also available.

Crystal noted that Avocent was using external transmitter boxes at the PC end, so they partnered with Avocent, repackaging the transmitter as a card that can fit in either a slot in a PC. The card and its accompanying cables and receiver unit is called DataReach. The DataReach card is interesting because it has ISA connectors on the bottom and PCI connectors on top, so if the card doesn't fit in a PCI slot, you just flip it over and it will now fit in an ISA slot! Crystal also reoptimized Avocent's technology and re-certified it with the regulatory agencies.

The DataReach card has a NIC jack on the back that makes it look like a NIC card. The KVM data travels through this port and out across the CAT 5 cable to the agent's unit into which is plugged the actual keyboard, video monitor and mouse.

Unlike ClearCube's Cage device, DataReach is one of the few extender products that actually can fit inside a PC, be it one of Crystal's PCs or even an old legacy machine with an ISA bus.

And oh yes, MCI did use Crystal's DataReach card to transform its call center. The centralization of computers in one rack and the stringing of DataReach signaling out to each agent desk has brought kudos from the IT staff, which can now fix or alter a computer by just walking into one room, just one room – they don't have to figure out where somebody's particular working "cube" happens to be.

DataReach, like many extension products, has a maximum range dependent on the frequency of the video signals. Since all of the first-generation extension products send video frequencies directly over CAT 5 and don't actually turn a monitor's video signals into actual TCP/IP data packets, the distance they can be sent is dependent on the resolution of the image. If you run a high resolution image at a high refresh rate, you must send a higher frequency along the wire, which causes it to act like an antenna and radiate the signal away before it reaches its destination.

ClearCube can send their signals 600 feet. With the DataReach card, Crystal finds that an 800 x 600 pixel screen refreshing at 70 Hz will allow you to extend about 500 feet of CAT 5 between the PC and the agent.

There are clever ways of extending this distance, however, to allow construction of mammoth, multistory call centers or offices. For example, using flat LCD screens instead of ordinary, cheaper CRTs in call centers is now a trend, since agents are finding themselves in smaller and smaller cubicle spaces. It just so happens that if a call center uses LCDs as displays, the images on LCD screens persist longer than on phosphor coated CRTs, and so you can bring the image refresh rate way down. So far down in fact, that video with screen resolutions of 800 x 600 or 1024 x 768 can be sent extremely long distances over CAT 5.

Still, companies such as Crystal continue to operate on the classic paradigm of the CRT, and they farthest distance Crystal recommends for DataReach is 800 x 600 at 70 Hz.

There's also an Israel-based company called Minicom (Linden, NJ – 908-486-7788, www.minicom.com), that just makes extender products using CAT 5 cable.

Minicom's PC arrangement is a little different from the other vendors in that the other systems are based on a one-to-one correspondence between computers and agent stations, while Minicom's systems are built up around a one-to-many classroom paradigm. Minicom thus sells a lot of equipment to schools or training centers, where one can find classrooms having one teacher supervising 20 students on computers, and the teacher wants to be able to view individually, one at a time, what each student's desktop looks like.

Minicom's approach is to daisy chain everything, using CAT 5 between the cards installed in the PCs. The last card in the chain also has a CAT 5 connection, one that goes up to a box on the teacher's desk, which he or she sits behind with a keyboard and a mouse and a monitor. The teacher can take control of the computer keyboard and mouse on a student's station and demonstrate something to him or her. For example: "No Fred, bring your cursor over here and click that just like this."

Minicom does offer what it calls a Duet Extender component, which allows you to place a monitor-keyboard-mouse-unit at a location up to 120 meters (393 feet) away from a computer. Another version of Duet, the Duet PC-Splitter, allows you to connect one local and two remote KVM workstations to your CPU. Yes, you can buy one PC and have three users work with it without needing an expensive network and with no signal degradation!

Indeed, yet another version of Duet, called Duet Twisted Pair (Duet TWP) can connect one local and one optional remote KVM-equipped workstation to your CPU. Duet TWP is available in two versions: "Duet TWP Regular" with a KVM to CPU distance up to 110 meters (360 feet) and "Duet TWP Long," which has a KVM to CPU distance up to 250 meters (820 feet). Duet Twisted Pair allows you to access a CPU from its local and a remote KVM workstation at distances of up to 250 meters (820 feet) using a single, low cost, CAT 5 UTP/STP twisted pair cable. Duet Twisted Pair doesn't need any software installation.

Interestingly, Minicom first conceived the Duet UTP-based products as being ideal for dangerous and sensitive work environments (such as the aftermath of a bioattack), since they allow personnel to control and monitor sensitive equipment from a remote station. They can also be used for secure data retrieval – one can avoid the tampering and manipulation of sensitive systems by installing the CPU in a secure environment and remotely accessing and viewing it from various locations.

Since Minicom has focused on classroom technology rather than call centers, they have all sorts of interesting teacher-student interactive networking products, some involving grouping PCs together, others allowing them to be scattered about a classroom and connected in a big daisy chain, such as their Classnet Twist, an interactive, single cabled, multiworkstation management solution, using CAT 5 cables for high resolution VGA transfer to 31 other PCs.

In a similar vein is Minicom's Supervisor Phantom, a KVM switch that's based on PCI switching cards for each server interlinked in a daisy chain bus pattern by CAT 5 cables, which allows a supervisor to access several CPUs that are connected by only one CAT 5 cable, be they in a rack or many feet apart from each other. Their Supervisor Phantom Specter, ideal for rack environments, is a matchbox sized, remote unit (measuring 5 cm across) with dedicated KVM cables and no external power supply needed.

Also, the Minicom Video Splitter allows a teacher to simultaneously broadcast a VGA screen to an unlimited number of remote monitors over CAT 5, each of which can be located at various locations within the facility at distances of up to 50 meters (165 feet) from the broadcasting CPU.

An amalgam of all Minicom's ideas can be found in their high-end Classnet 3.15, an interactive computer classroom system which puts at the teacher's control every monitor, mouse, keyboard and multimedia resource in a computerized classroom of up to 99 computers.

Finally, Minicom recently announced AristoClass, an innovative breakthrough for managing software programs in their CAT 5 linked computer equipped classrooms, which they now call Computer Equipped Learning Environments (CELEs). AristoClass, resembling a sort of Intranet-ASP, is a time saving "umbrella" solution that enables synchronized, interactive teaching, integrating hardware, software, Internet utilities and additional applications, including distance learning. It is possible for AristoClass components to be deployed as individual units with limited features, but they really shine when integrated with a LAN. When combined with Classnet, AristoClass can merge and manage all of the technological ingredients that formerly required non-centralized purchases and installation. A teacher or supervisor can have complete control and monitoring options over student workstations along with access to real time views of what's happening on multiple student workstations.

Minicom's products work with Windows 9x, NT, 2000, Novell, DOS, Linux, and Unix.

As impressive as all this sounds, a new generation of extender technology is appearing. Up until now we've been discussing the first-generation "extenders" from ClearCube, Crystal, Cybex / Avocent, and Minicom. All of these are limited in that although they use CAT 5 cable, they don't actually communicate their peripheral signaling data using TCP/IP packets. They just use CAT 5 cable, standard connectors and some interesting signal modifications and wiring techniques.

There is, however, another type of extender technology that's starting to appear, one more in tune with the ASP model. Avocent has such a device, called the DS1800, as well as Cybex's earlier

Cybex KeyView II and Apex's Emerge. These boxes have powerful microprocessors built-in that are capable of actually taking the video, keyboard and mouse data, compressing it, then sending it over a packetized network such as the Internet, which means that you can safely view and work with PCs that are at opposite ends of the earth from their users and IT staff.

The pluses of this technology include the fact that you actually get to see the Windows desktop, or the actual display output from the computer you're monitoring. However, these systems are very processor intensive, they're a bit expensive, and they can chew up a lot of network bandwidth. They also exhibit a little bit of "lag" – if you continually move your mouse there's always going to be a slight discrepancy between where the computer that's being monitored "thinks" its mouse pointer is positioned and where the person who's doing the monitoring actually sees it when moving it directly. When you stop moving the pointer, of course, the other system will catch up to you. The greater the distance or the more congested the network, the longer the delay.

Still, these products are quite attractive, are now available, and they're being miniaturized to the point where you can buy a KVM switch that has one of these "super extender" devices built in. They're very expensive, because the processing that's necessary to do the compression and video capture is not trivial.

Faultless Software

Even on a standalone system, software can "break" in various ways. One major category entails "failure to recognize and adapt to hardware failure." An easy way to fix this is to start out by developing on a fault tolerant OS.

"No carrier or business based on absolute real-time software uses Windows NT," says Eli Borodow, CEO of Telephony@Work, Inc. (La Jolla, CA – 858-410-1600, Telephony@work.com) which offers CallCenter@nywhere, a comprehensive call center solution that can be deployed at the carrier level as a sort of super Centrex.

Eli says that "Greenfield ASPs and enterprises use NT and we support that. But carriers use Sun Solaris or Compaq 364, and we sell to those markets too. Most developers are working in venture funded environments that demand they get to market in two years. This forces them to use NT COM objects and Microsoft shortcuts, or else their investors will institute a change in management. If they just spent a little longer writing code, the same code could be made OS independent. But if you're under that kind of financial pressure, odds are you're creating an enterprise product because you won't be able to sell a non-Unix/Solaris product to the carriers – UNIX is a veritable religion there."

Grover Righter, VP of Technical Strategy for Kabira (San Rafael, CA – 415-446-5000, www.kabira.com), concurs: "To be perfectly blunt, we don't have any customers deploying commercial grade services and service level agreement managers on top of PC architectures. We have an Intel product which runs on NT that people use primarily for development. People tend to deploy the important core of their Operations Support System (OSS) software on Sun Microsystems machines running Solaris and high-end HP-UX servers running UNIX. Still, in Windows 2000 we're starting to see more commercial designs incorporating higher-end applications"

Indeed, as Charley Pitcher, the solution unit manager, Web Services and Application Hosting within Microsoft's (Redmond, WA – 425-882-8080, www.microsoft.com) Network Service Providers (NSP) Group says: "I would be the first to admit that if I were going to build the very high-end, high reliability convergence applications on NT 4.0 and SQL 6.5, I would have to think twice about it. But on Windows 2000 and SQL 2000 it's a very different story. When you spend five billion dollars on R&D the products do change."

"If you're writing stateless middle-tier applications," says Pitcher, "then the key issue of availability is the SQL database. We introduced clustering in NT 4.0, significantly enhanced it in Windows 2000, and now another set of enhancements will appear with .NET Server."

"If your SQL server fails, clustering automatically does failover, preserving the system state and the user continues on," says Pitcher. "We did a specialized version of that called the Windows Datacenter Server. It's only sold by OEMs who have been through a rigorous certification and test process, which tests every kernel driver. Our TerraServer on the Web is a huge (10 TB) SQL database, and since they moved it to Windows Datacenter Server its had 100% availability."

Windows 2000 Datacenter Server is considered the most powerful (and perhaps the most expensive) server operating system ever offered by Microsoft. It supports up to 32-way symmetric multiprocessing (SMP) and up to 64 GB of physical memory. It provides both 4-node clustering and load balancing services. It also provides Internet and network operating system (NOS) services to all the versions of Windows 2000 Server. It's optimized for large data warehouses, econometric analysis, large-scale simulations in science and engineering, online transaction processing (OLTP), and server consolidation.

"Windows Datacenter produces very 'locked down' versions of the kernel which then lead to much higher levels of availability," says Pitcher. "We do this because in our studies we've found that about 80% of the causes of outages on our boxes didn't have to do with our software. Third party software can be the problem if it's not well behaved."

Windows Datacenter requires that you write cluster-aware applications, which means that you must pay some attention to your architecture. "It's not a big deal," says Pitcher, "any SQL programmer can figure out how to do it. It's not rocket science."

Moreover, Microsoft has taken the SQL clusters and literally split them so that they can run in separate datacenters. One whole cluster mirrors another whole cluster. You take two nodes of a cluster and the Storage Area Network (SAN) and run it in one datacenter, and take the other two nodes of the cluster and the SAN and run it at another datacenter. The two datacenters are connected with fiber and both are running datacenter level load balancing software so you can do real-time load balancing across two datacenters. If either datacenter fails, you'll get real-time disaster recovery in the neighborhood of five to 10 seconds with no loss of the user sessions.

"It's all a combination of Microsoft software, EMC SANs and Nortel Networking gear to connect the centers," says Pitcher.

"That brings us to carrier grade kinds of solutions," says Pitcher. "AT&T has signed up to trial our solution, working with a number of system integrators who target high-end enterprise and hosted kinds of services. There is a very small list of OEMs who have actually certified their hardware with Windows Datacenter and people who have SQL database applications: Honeywell Bull, Compaq, Dell, Fujitsu, Fujitsu/Siemens, Hitachi, HP, IBM, ICM, Next, Stratus and Unisys."

Pitcher says that XP is even better. XP has more sophisticated clustering functions available in what's called the .NET Server family. XP is the client, while the server family is .NET Server, .NET Advanced Server and .NET Windows Datacenter Server. Pitcher says that the underlying OS is more robust than Windows 2000 and that Microsoft is seeing "quite extraordinary performance gains in .NET Server compared to the previous version."

Software and Hardware Must Not Fight Each Other

Part of the software reliability problem has to do with it's interaction with hardware. It's taken longer than expected for "ruggedized" standardized hardware such as CompactPCI to achieve the level of reliability demanded by carriers and large businesses, partly because it took quite a while to formulate all of the cPCI specifications and partly to assure carriers that the Windows operating system could be made as real-time and fault tolerant as UNIX.

This is one reason why the Motorola Computer Group has introduced a Windows 2000 high-availability (HA) hardware and software platform that makes it possible for Windows 2000 to be deployed within mission critical "five nines" environments. Their program includes an "HA Aware" certification program and "HA Aware" documentation, training and support services. Motorola is also working with several specialized hardware and software technology firms to bring complete HA Windows 2000 solutions to the marketplace, such as Audiocodes, Brooktrout, Dialogic and Natural Microsystems.

Motorola also offers two starter kits to help telecom OEMs evaluate the their high availability architecture. Both starter kits are based on Motorola's CPX8216 NEBS compliant cPCI platform, which has an 8+8 slot, dual-segment cPCI chassis with front and rear I/O access.

But some telcos and other companies think that even if CompactPCI and Windows can be made fault tolerant, it's still "bloatware" and they can get more "bang for the buck" (greater scalability and more usable ports given a limited number of CPU cycles) from the more streamlined UNIX-like operating system, Linux.

This kind of talk has spurred Motorola to deploy their High Availability Linux (HA Linux) offering. HA Linux – with hot swap capability and support for system platforms based on both Intel and PowerPC architectures – is said to be the first Linux for carrier-grade networking, wireless and Internet applications demanding five nines availability. NMS Communications (Framingham, MA – 508) 271-1000, www.nmss.com) has already integrated its hot swappable AG2000 and AG4000 family of DSP and network interface modules with HA Linux.

The Add-On

Of course, if your operating system of choice doesn't have fault tolerant features or superb maintenance capabilities, you can add them.

As mentioned earlier, Jungo (Netanya, Israel – +972-9-885-8611, www.jungo.com) sells add-on software that imbues hot-swap CompactPCI card capability to such operating systems as Windows NT /2000, Linux, Solaris (both x86 and SPARC processors), and VxWorks 5.4. It's a multi-operating system solution. Jungo also offers tools so you can add hot swap capabilities to existing drivers or develop your own hot swap drivers. Jungo's driver development tools support other non-PCI buses such as USB and ISA.

If you're too lazy to alter your code in any way, Marathon Technologies (Boxborough, MA – 978-266-9999, www.marathontechnologies.com) offers a hardware-software combination that can take your debugged application and run it at fault tolerant reliability for years on end, even on Microsoft Windows NT.

With Marathon's technology, you have the ability to choose Windows-based, off-the shelf software (which of course has been debugged and proven), and then just run it on the Marathon Assured Availability hardware to achieve high availability. You can use standard off-the-shelf servers from IBM, Dell, Compaq or HP.

"Take an ASP like Scantron (www.scantron.com) which does survey solutions," says Linda Mentzer, Marathon's Vice President of Marketing. "They have a Web hosting service product that allows their clients to poll people for information and data. Scantron was able to put together a solution using off-the-shelf software and in particular Windows NT, SQL Server and Dell and HP hardware. They put together a fault tolerant system without having to sit down and write a lot of fault tolerant-specific software."

"Our solutions consists of four servers," says Mentzer. "There are two compute elements (CEs) which run in what we call 'instruction lock step' so the application and the OS are really the only things running in those boxes. If anything should happen in one of the servers, the other server continues running."

"We've also disaggregated the I/O from the server; we have all of the I/O being done by two Input-Output Processor (IOP) boxes that are connected to the two CEs," says Mentzer. "Normally, you'd run a single or mirrored Windows server loaded with your application, cluster services, OS, your device drivers, and your hardware interrupts. By disaggregating in this way, your application isn't running into any kind of contention with I/O which is one of the major issues in causing blue screens on NT servers in clustered environments. So in essence we've separated out the application and the I/O."

Any I/O that's asked for by the app or the OS is redirected to Marathon's hardware and software to an I/O processor. The two IOPs run in parallel but they're asynchronous. A app request for a disk write, for example, goes simultaneously from the app running on both compute elements to the two IOPs, each of which then reports back through the Marathon hardware and software when they've committed the disk write. That info is reported back up to the application just as the OS normally would.

Four PCI cards that are called Marathon Interface Card (MICs) connect the two CEs to the two IOPs. All four units are interconnected via a fiber cable.

Each "half" of the system (one CE and IOP on each half) can be separated up to 10 kilomenters (over six miles). In California, systems using Marathon Software and hardware have been split across power grids to enhance disaster tolerance.

Platform Management Software for Fault Tolerance

A common requirement of any system that includes redundant hardware and software is some level of manageability. The minimum goal of platform management in a high availability system is to provide monitoring and control required for the detection, diagnosis, isolation, recovery, and repair of faulty hardware and software. Beyond that, additional platform management capabilities aimed at predicting and preventing failures may be included.

GoAhead! Software (Bellevue, WA – 425-453-1900, www.goahead.com) sells GoAhead Service Availability technology allowing communications providers (using equipment and software from Ziatech, Force Computers, Motorola, I-Bus/Phoenix, WindRiver and many more) to achieve extreme high availability. Service Availability provides a collection of tools that allow a network to take care of itself – monitoring and managing all system resources (hardware and software) in real time to maintain customer connections despite failures – enabling on-demand, uninterrupted service and "5 nines" uptime for the communications infrastructure.

Michael O'Brien, CEO and Founder, says "GoAhead answers the question 'How do I create a fault resilient system?' GoAhead provides the middleware which manages the hardware, the operating system, and all the components in the system and provides APIs that application developers can write to so as to achieve five nines availability."

"For some hardware platforms such as a Sun Netra, we just do the whole lot ourselves, we provide a layer of middleware that sits above the operating system," says O'Brien. "On CompactPCI systems, however, we may work with the various vendors such as RadiSys and Force Computers to interface our software to the specifics of their hardware platform."

"Either way, application vendors don't want to worry about the specifics of the particular hardware platform, but you have to write your application to use this redundancy type of model," says O'Brien. "The missing piece here is what you could call availability management. It's middleware which manages all of the components of the system, orchestrates switchover or failover to redundant components, and provides the APIs for the application to do checkpointing [the process of copying data from a unit to memory or to the unit's standby to allow for switchover while maintaining state], heartbeating [sending a periodic signal from one component to another to show that the sending unit is still functioning correctly], and persistent preservation of state."

"Also, the application needs to participate in what we call a 'system model'," says O'Brien. "the availability management function is all about understanding what are the components in the system. How do they depend upon one another and when something fails, what else fails? Maybe the network stack fails or a hardware network interface card fails. But does that necessarily mean that you're application has failed or not? It may or may not. That's what our Availability Manager does, it's like the conductor of the orchestra. It's the 'conductor' of the whole system, so the applications can just be applications without having to be the conductor and the orchestra at the same time."

Another "monitoring-management" package is from Concord (Marlboro, MA – 508-460-4646, www.concord.com). Concord's eHealth Suite manages the Internet infrastructure – networks, systems, and applications – to detect and resolve faults and potential outages, maximize availability, and optimize performance, in real-time.

Rather than just reporting real-time status, Concord backs its real-time information with comprehensive historical data so you get both the real-time events, and a historical context to help you manage your IT infrastructure. This combination of real-time status and historical context enables eHealth to not just detect faults, but to also detect subtle variations that signify potential outages before they happen.

Products in Concord's eHealth Suite are organized into three solution sets that focus on network, system, and application management. Thus you can choose products pertaining to your project – whether it's just applications, or the whole underlying systems and network infrastructure on which your applications depend. Concord's other Suite-Wide products provide management capabilities that work across all three technology areas – apps, systems, infrastructure. eHealth Suite is tied together by a common console and customizable interface.

Brian Burba, Concord's VP of Product Marketing says "Our software is management software that actually looks at the behavior of an application, and we'll tell you if its behavior is degrading, so you can go do something about it before your end users are impacted."

Burba's advice to developers: "As you're developing your application, consider from the beginning how you're going to manage it. Look at different ways of what are the key metrics that will describe the behavior of this application. Make sure that you have a way of monitoring them. Concord's product eHealth Suite is one way to do that."

"We have an Application Response Agent that sits on the end user desktop," says Burba. "We have a Service Response Agent that does transaction testing. We have a SystemEDGE Agent that sits on the box and manages the app and looks at the system. All of these products are part of the eHealth suite, which is an end-to-end fault and performance management product."

Concord sells software to service providers, telcos, and enterprises in the service provider space.

Just Plain Stupid Errors

For all our concern over sophisticated conundrums, the biggest group of software problems can be racked up to stupid stuff occurring on your application host computer: There's always a bug or two in the OS or in your development tool libraries. Has the world's greatest programmer divided by zero in the code today? Perhaps someone has inserted some accidental memory leaks, interrupt handling errors, or have failed to write code that can handle certain error codes returned by the OS. Perhaps there's a failure of the software (or the programmer?) to deal with the eventual non-availability of certain peripherals (hard drives, CDs etc.)

Sometimes software just doesn't scale up very well and becomes highly inefficient.

Bad "legacy" software can be temporarily fixed by "wrapping" it in another piece of software, or carefully monitoring it using tools such as Concord's. But companies such as Marathon Technologies practically twist your arm to get you to test your app to perfection long before you ever get the privilege to run it on their platform.

If you're building your own software you'd better be familiar with the traditional techniques for constructing usable software: Structure programming, version control, internal documentation (comments), good team discipline, pre-certification of purchased components, purchase of components (ActiveX, etc), rather than writing these things internally, use of LINT, software checkers, etc. The latest innovation is open source – can 40,000 programmers be wrong?

If you're not a hot shot programmer you could use some kind of 4GL or object-oriented development environment (formerly called "app gens") that obviates the need for you to concern yourself with hardware limitations.

For example, Apex Voice Communications (Sherman Oaks, CA – 818-379-8400, www.apexvoice.com) makes a sophisticated development environment as well as prepaid and billing systems.

Osvaldo Gold, VP of engineering of Apex Communications, tells the author that "Within the Apex runtime environment we developed a framework which consists of a monitoring process, an area of memory shared by the monitoring process and all other processes, and a directory where certain files are saved."

"So let's say that we develop a new process to take care of some signaling issues," says Gold. "So this new process registers it's name and process ID (PID) in this area of shared memory. And the monitoring process can see it. It records in this area of the directory something like a simple script that can later be used to restart itself if necessary. Thus, every process knows how it comes into existence. The monitoring process then monitors everything via shared memory, it knows what to look for – is this process running? If a process fails, then it knows what script to execute to restart the process. And it can send a notification."

Apex's products work with Windows 2000, SCO Unix, Solaris, and Linux.

Elix (Montreal, CA – 514-768-1000, www.elixonline.com) offers the Maestro product (picked up from its merger with Prima).

Robert Townsend, systems architect at Elix, says: "Many of our products must be fault tolerant, because we're in the telco space and also in the high end call center space (2,000 agents or more). These systems are never deployed without some kind of fault tolerant sensibility at all levels: at the hardware level, at the LAN /router /WAN level, and at the application level."

"Elix doesn't use the Microsoft Clustering in NT *per se*. What they've done is to write processes that are autonomous and run on multiple boxes. These processes will keep track of one another and if one ultimately times out, a failover process will take over. One must be careful using the word 'cluster' because that means so many different things to different people. There's easy things you can do with hardware and you basically do the risk analysis and that determines what investment you should make in hardware."

"With our IVR family of products you install client software on more than one machine," says Townsend. "Now you have what we call an N+1 architecture, which is more than two machines. There's automatic load balancing on the various services, and you could provision your environment to have additional resources by at least one extra of everything."

"You can push the fault tolerance closer and closer to the client, but at some point you need to recognize that risk analysis needs to be done and you need to balance the trade-offs of the risk of a failure," says Townsend. "What's most likely to fail? What's the cost of the failure? Recognize that in the public-switched network, a certain small percentage of failures are expected to occur and are, in fact, part of the architecture of the network."

"The only area, really, in our typical solutions that we don't control is the trunk," says Townsend. "The trunk is a given. Even if we had a system where we could flip the trunk over to another box, that's hardly worth the investment, because the chances of the trunk breaking an inch beyond the flipover switch is higher than the T1 card actually failing. So it rarely makes sense to do that."

According to Henrik Thome CEO and president of Envox (Westborough, MA – 508-898-2600, www.envox.com), by the time you read this, his company will have released the powerful Envox 4.0 Communications Development Platform which can be used for building custom communications solutions for the enterprise or public network.

"In Envox 4.0 we've added a multinode awareness functionality that is good for large scale solutions," says Thome. "First of all, we've added a function working with the nodes so that applications that are run across multiple nodes will believe that they are in the same node. Calling it a virtual machine is probably inflating it too much, but the applications can act that way."

"We call it a group function," says Thome. "It's simply another name for multinode solutions. But for large switching we've also included the support for ATM, using boards from the former IML, which is now part of NMS. These ATM boards enable you to use this multinode awareness – or rather multinode 'unawareness' because the application doesn't really care if it's in the same node or different nodes."

One could create a huge switch with 10,000 ports, spread across as many nodes as you wish.

One of the oldest GUI communications development environments is Pronexus (Kanata, Ontario, Canada – 613-271-8989, www.pronexus.com). David Lee, director of professional services says that they pretty much rely upon the fault tolerance provided by Microsoft platforms.

"Customers who use our product will do fault tolerant hardware solutions too so they can have redundant T1 sites – in case one site goes down, calls are routed to the other site," says Lee. "Our customers have separate database servers that back each other up, and they have multiple IVR systems or CT systems with hot swap power supplies and RAID hard drive systems. They'll often use clusters, such as Windows 2000 clustering."

"If a customer has four T1s and wants to have failover capability, what they can do is have two T1s in two boxes," says Lee. "If one fails a technician can take those T1s out and plug them into the other box where each machine is configured for 96 lines. And our software will immediately recognize the new T1s when they're plugged in."

"In Quebec there's a company called CML Versatel," says Lee. "We did an integration with their product and basically their product provides automatic failover on T1 lines, so it's kind of a unique product."

One ASP using Pronexus to develop services is Mitercom (Ottawa, Canada – 866-238-4566, www.mitercom.com). Mitercom, delivers advanced telephone based voice commerce solutions that integrate natural language speech recognition, text to speech and Internet technologies.

Top of the World, Ma

If you're looking for a tremendously advanced fault tolerant development platform to automatically program fault-free mission-critical applications, particularly in the world of services delivered over networks, look no further than Kabira (San Rafael, CA – 415-446-5000, www.kabria.com)

Kabira's ObjectSwitch software technology enables eBusinesses to create new services directly from business rules defined in high-level Unified Modeling Language (UML), Interface Definition Language++ (IDL++) and XML. These business rules are converted to "Kabira Service Engines" that can go online within a few minutes of final testing. The new Kabira-based service can then be changed on-the-fly by simply editing the business rules and replacing the running service with the updated Service Engine – all without shutting the system down or taking users offline.

Kabira's Grover Righter tells the author: "We've actually built an adapter factory that automatically generates the interfaces that talk to OSS software systems such as billing systems, and databases and switching systems. It's more complex than a Wizard, it's really more of a compiler."

"Take a switch, for example," says Righter. "An intelligent switching device may have hundreds and hundreds of APIs in it. In the old days you learned the APIs and wrote everything yourself, which took forever. Now we can create a model of the system and automate the function of programming against the APIs. We turn the model into an executable engine, which is a state machine. Underneath that we have this adaptive real-time infrastructure that actually handles the events in real time. Our system can automatically generate a complete interface to a switch or billing system or database. What would take us two to three months by hand a year ago is now taking us about 12 minutes."

Use a premier modeling tool such as Rational Rose from Rational Software (Cupertino, CA – 408-863-9900, www.rational.com) and you can build a model that defines a system. Rational Software is a partner of Kabira's. Kabira's software interfaces with Rational Software as well as those from other partners, though Rational is the 800 pound gorilla in this area.

"Every one of the big telcos and most of the small telcos use Rational Rose as their UML modeling tool," says Righter. "You can formulate your logic, push a button and you generate the service engine, which is half a million lines of C++ code. That thing you generate with that engine is in fact the application that then gets plugged into the system.

"We want to give out fewer APIs because they get outdated fast," says Righter. "Instead of having programmers specifically write to a CORBA interface, they define what they want to have done, not how it's implemented. The generator even allows you to change from a CORBA interface to an Java or an XML interface. The model will still be good, as long as your business rules haven't changed."

The initial description file can be written in IDL, or it can be Java or CORBA information or XML. It can be anything that has a deterministic combination of both defined subsystem behaviors and data. If you've got it, then you can generate code against it.

"We also provide customers an applications services environment we call an Adaptive Realtime Infrastructure," says Righter, "which is becoming a way to describe this very flexible, high speed backbone as opposed to an older style applications server architecture. What they're buying from us is the ability to go online and change their running service without shutting stuff down, and the ability to tolerate faults and then scale up to millions of users."

As for pricing, Kabira's customers tend to deploy pretty big operating networks – a small project would be $500,000. A typical project is a couple million dollars.

"But consider that they're potentially replacing hundreds of software people," says Righter. "Every ten software people cost you more than $1 million a year, and $2.5 million replaces at least 20 programmers. Productivity is much greater with an automated system."

The Service Availability Forum's Middleware

With the continuing shift to packet-based, converged multi-service networks, every manufacturer has recognized the need for dependable, interoperable and scalable platforms and components.

That said, most vendors have in fact shunted these noble ideals aside, concocting different technological schemes willy-nilly. Fault tolerant hardware, for example, can range from automatic failover between processors in the same box to clustering different boxes. As for software, forget it. Different middleware abounds, each picky about what operating system or hardware it'll deal with. Fault tolerance can be implemented in the application, in the OS, or all over the system.

The result? High development costs, a lengthy time to market and confused customers.

Now, however, 20+ companies (including heavy-hitters such as Intel, IBM, HP, Compaq and Sun) have gotten together to form The Service Availability Forum (www.saforum.org), which should bring some order to the chaos. The SAF encompasses hardware and software vendors, with a particular

➤ Faultless Software Developer Summary

For maximum bang-for-your-development-dollar:

1. Use a Fault Resilient OS on appropriate hardware.
2. Use some kind of abstraction platform that handles some aspects of fault resilience. (e.g., Envox).
3. See what you can do about securing distributed sub-functions of your app: e.g., the e-mail part, the mass storage part, etc., by using ASPs, VPN services, and other forms of outsourcing.
4. Make a commitment to engineering on a comprehensive application platform that spews forth automatically programmed code, like Kabira's.

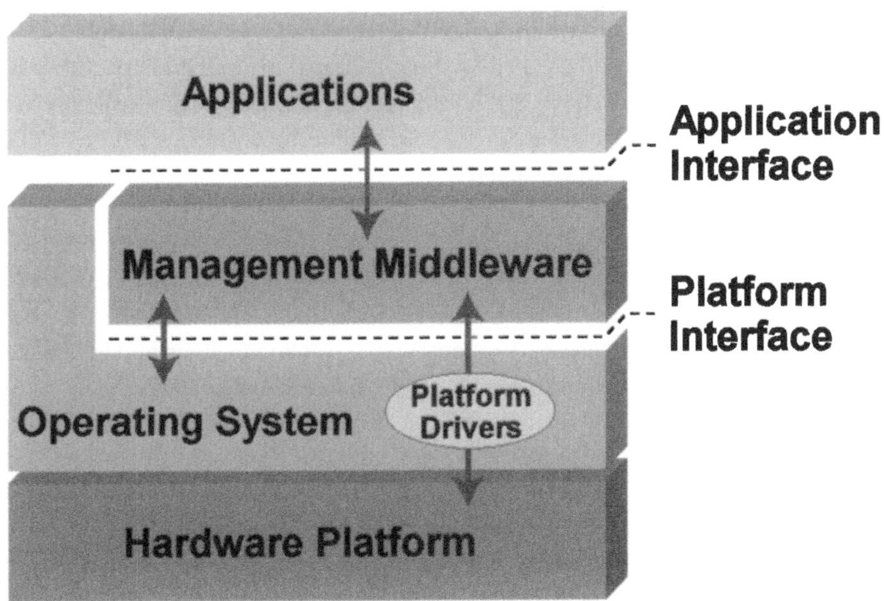

Open standards at last! The SAF interface specifications allow the vertical integration of applications and systems to Service Availability middleware without access to or modifying any source code.

interest in devising standard middleware and clearly defined and well-specified standard interfaces between the middleware, the OS and the applications.

The SAF definition of an official "Service Availability Solution" involves two underlying requirements:

- **High Availability** – 99.999% system uptime (or better) for on-demand services, and
- **Service Continuity** – This means that customer sessions and session state data are maintained for uninterrupted services without disruption across switchover or other fault-recovery scenarios.

To achieve this, the SAF envisions its members contributing to an "ecosystem" of standard building block network elements: open, standards-based systems and components designed and managed for ultra-dependability. To this end, the SAF is promoting open, standard programming interface specifications that in turn shall expedite development of fault tolerant infrastructure equipment and applications.

The SAF initially will develop two interface specifications: an Application interface specification and a Platform interface specification. The Application interface will be a programming interface between applications and Service Availability (SA) middleware. This deals with such questions as: How do I detect heartbeat signals in my app? How do I re-start? How do I switch from active to a standby state? How do I detect what the "level" (active, unassigned, standby) should be as specified by the availability manager?

On the other hand, the Platform interface deals with the interface between the OS or hardware platform and the SA middleware. For example, say you're a resource board vendor and you want your board to be used in a redundancy configuration. How do you ensure that the board will work as a good citizen in the overall system? That's the provenance of the Platform interface.

The SAF interface specifications will provide the ability to vertically integrate applications and systems to SA middleware without requiring access or modification to source code.

Michael O'Brien, Chairman and CEO of GoAhead Software, Inc. and president of the SAF, says: "We have a big, long-term vision. To get leading companies committed to that vision, is very important, which is why we've got not just the company names but the key individuals, the Number One architect from each firm, to make this happen."

"The SA middleware will just be a basic building block developers will need to incorporate into their products," says O'Brien. "Just as vendors don't write their own OS, they shouldn't have to write their own middleware. They should be confident they can write to a real middleware standard."

"SAF-defined middleware is the next logical layer above the OS and will ultimately become commercial off-the-shelf stuff," says O'Brien. "Vendors should be able to focus on building their solution and get to market quicker with far lower risk by using standard SA middleware."

"So, if you're trying to create a media server, media gateway or a computer telephony system," says O'Brien, "you'll select your hardware components, use commercial off-the-shelf SAF-compliant middleware, then spend the bulk of your investment in your application and system design, not in reinventing the wheel by writing your own OS or your own high availability or SA layer."

"Let's say you're selling a media server and it's being OEMed by a large telecom equipment manufacturer," says O'Brien, "and they insist on it running under a specific OS. Well, you can now freely port your application to that system because you're not locked into a particular middleware. My company, GoAhead and some other vendors already offer development kits and libraries to do this. The model is proven, it works. It just hasn't been standardized and ratified by the SAF yet. But we are taking the next step to make the market understand that this is now a formalized layer and that we're going to drive the standards above and below it too."

The SAF decided in 2002 to no longer be merely a collection of interested parties. Instead, it'll be incorporated as a freestanding legal entity just like other standards bodies.

Can the SAF produce the kind of open standards that will at last ensure fault tolerant systems and uninterrupted landline and mobile network services? If it does, your worries over high availability will be greatly alleviated.

BUYING FAULT RESILIENT COMPUTING

The popular image of a fault resilient computer maker is one of a giant warehouse like facility where sheet metal is bent, and backplanes, power supplies, disk drives and other components are assembled into 50 to 80 pound, 19-inch wide rackmountable boxes that are shipped to distributors, integrators and VARs.

This classic hardware manufacturing model served the industry well during its formative years, but many PC manufacturers now are diversifying in an effort to escape the small price margins inherent in the manufacture and sale of hardware. PC makers suddenly realized that all-encompassing, PC-based communications systems had become easier to install and modify thanks in part to telephony, media and networking standards such as TAPI, TCP/IP, CT Media, S.100, H.323, SIP, H.100 and H.110.

Also, it was found that customers prefer to deal with a single source as they search for ways to make telco, enterprise and call center communications systems as easy as possible to install, operate and maintain. While preparing to buy today's incredibly complex telecom hardware and software, they increasingly ask the originating PC manufacturers if they can deliver the systems in final form.

Also, customers purchasing high customized and unproven equipment may require additional services such as certifications, and NEBS testing.

In short, customers want turnkey solutions (or at least they want someone else to worry about getting a new system up and running), and this has spurred hardware companies into becoming solutions companies.

Not Just Boxes – The Changing Role of the Fault-Resilient Manufacturer

PC makers such as Alliance Systems (Plano, TX – 972-633-3400, www.alliancesystems.com) are being drawn further into the "back end" piece of the business, expanding their offerings far beyond mere hardware customization to now include complete system integration, "application tuning" and other "hand-holding" services that include field support, logistics, and some new ASP-related delivery models.

Another example is Crystal Group's Manufacturing Integration (Mi) services that can provide manufacturing and integration expertise to help you design complete end-to-end systems or improve an existing configuration. Also, Force Computers can deal with the resolution of problems and faults in your system caused by unauthorized third parties or user error, and will even consult on problems with non-Force products if they're part of your overall solution.

As Jean-Michel Briere, vice president of marketing for Teknor Applicom, a Kontron company which began doing business on the Kontron name on July 11, 2001 (Boisbriand, QC Canada – 450-437-5682, www.kontron.com), says: "Most of Teknor/Kontron's big OEM customers such as Nortel and Lucent are looking for turnkey solutions providers who can deliver one stop solutions. That includes not only single board computers and chassis but also software platforms that can reliably work with the application software for which Teknor/Kontron's customers are responsible."

"There's two major reasons for this," says Briere. "The first is time to market, which is a critical issue for our customers in the telecom market. The second reason is that our customers can no longer afford to buy expertise for in-house development of SBCs or platforms. Therefore, they're looking for partners such as Teknor/Kontron who can do that for them."

"I don't think we have any real competition with VARs because the OEMs' financial power allows them to attract reliable partners with whom they can deal directly," says Briere. "So if you look at this market trend, there will probably always be two levels: Large orders will be outsourced to companies such as Teknor/Kontron or RadiSys, while the VARs and integrators will work on the smaller volume purchases."

Aside from changing their relationship in the channel, Briere says that companies like Teknor/Kontron will also have to transform themselves and acquire additional expertise as the communications convergence trend continues.

"Teknor recently acquired Memotec Communications of Montreal, Canada, which brings us a step further to a new convergence strategy that goes beyond SBCs and chassis," Briere says.

Memotec is a networking company that provides data, voice and video convergence solutions for carriers, Internet service providers and corporate customers. Its product line of access devices and edge switches provides consolidation of multiservice traffic over packet-based and cell-based networks such as Frame Relay, IP, and ATM.

"Packet protocols such as IPSec and Quality of Service considerations are becoming a major consideration," says Briere, "And our customers now require companies like ours to be able to speak their language, which is why we acquired Memotec's expertise in these areas."

"Above all, of course, we must define where the bottom line is," says Briere. "What applications are customer comfortable with? Where is the real added-value?"

Briere gives an example: "The Memotec products, as they exist now, are really competing with companies such as Cisco and 3COM. What we want to do is to unbundle their technology and integrate them with our PCs so that large OEMs such as Cisco, Nortel, and Lucent can port their voice over IP applications software to our PC-based platforms."

"Don't get me wrong, we're not getting into the router business," says Briere. "We very much believe that PCs are the most efficient platform in the largest telecom / datacom industry segment, and the fault tolerant PC will be the basis of for platforms that enable companies to bring their products as fast as possible to the market. Remember that one of Intel's major customers is Cisco!"

Online Purchasing and "Virtual Configuration"

Some disasters give you some warning: Hurricane arrivals can be predicted several days in advance, as could the Mississippi River floods of the 1990s. Impending disasters (or strong rumors of them) are sometimes the thing that "swings the ball" as it were; finally convincing some fence-sitting companies to invest in fault tolerant computing equipment. And, of course, the disasters themselves can result in new equipment being ordered. Problem is, the fault tolerant computing industry has been traditionally steeped in the ability to do customization, and to do it slowly. Only recently have vendors applied rapid production techniques to their products.

To make quotes and purchasing as quick and easy as possible for customers, many computer companies are dabbling in e-commerce.

Industrialcomputer.com, a division of Comprehensive Computer Solutions, is a custom integrator of fault resilient computing solutions that follows an eBusiness model. Industrialcomputer.com allows a prospective customer to surf to their Web site, view up-to-date product documentation and information, use their 24 hour "Easy Quote" configuration and quoting screen to custom configure 1U to 4U high rackmount systems and Panel PCs, then execute the order in real-time. Later on, the customer can view the product's shipping status online.

Over at Kontron America, a Kontron acquisition company formerly known as ICS/Advent (San Diego, CA – 858-677-0877, www.icsadvent.com), they have a product family called Omnix, with individual units delivered via a combination of Web-based virtual configuration and an advanced, modular, real-time manufacturing process. A customer can go online and use a Web-based configurator tool to pick a chassis, power supply, backplane, drives, SBCs/CPUs, etc. Your initial conception of your system can be sent directly to the manufacturing floor, or further refined through communication with an Kontron America sales agent operating the same, Web-based client.

Since fault resilient computers have always been nearly infinitely customizable – more like nuclear reactors than mass-produced products – one wonders how a customer can get on a Web site, specify a few characteristics out of thousands possible, and then expect to have the resulting virtual configuration be assembled in near real-time.

The answer lies in the fact that if computers are broken down into small enough standardized, compatible and cheaply mass-produced components, then the mathematical laws of combination and permutation allow systems of nearly any possible configuration to be quickly and inexpensively assembled on a suitably equipped assembly line.

In the case of Kontron America, even the backplane can be assembled from modules. Instead of buying and keeping on hand 25 to 30 backplanes with differing numbers of slots and bus structures (ISA, PCI, etc.), custom backplanes can now be assembled from combinations of just four possible "building block" segments: A CPU segment (1 PICMG slot, 2 ISA slots, 2 PCI slots and 2 ISA/PCI shared slots); an ISA segment (with 3 standard ISA slots); a PCI 32 segment (with 3 PCI 32-bit slots); and a PCI 64 segment (with 3 PCI 64-bit slots).

The CPU segment acts as a starting point from which expanded backplane configurations can be incrementally created by adding the appropriate mix of three slot ISA and PCI segments (or you can take the CPU section and use just that as your backplane and leave the rest of the chassis empty, if you so desire). For maximum flexibility, the CPU segment can be installed at any position within the 20-slot chassis, with ISA segments added to the left and PCI segments added to the right, up to a total of six segments in the chassis.

All of this gives the higher-end members of the Omnix family incredible configurability while using a relatively small number of core system components.

The PCI segments themselves are bridged using the Intel PCI bridge chip technology. Kontron America is in very tight with Intel and their Applied Computing Platform Provider program. Indeed, the actual engineering work that was done on the modular backplane concept was in lock-step with Intel.

Thus, a PC maker's expensive low-volume purchase of many varied components now gives way to cost effective, high volume purchases of fewer, less expensive, manageable and more reliable sub-components. Quality goes up, cost comes down, and flexibility to the end customer increases dramatically.

But some components produce more heat than others, while others restrict chassis airflow in different ways. Allowing customers to configure any configuration they fancy might lead to a Frankenstein's monster device that may overheat. Fortunately, manufacturers have devised some incredibly sophisticated thermal modeling and simulation techniques to eliminate internal hot spots and optimize cooling efficiency.

Kontron America, for example, has combined the thermal databases of each Omnix chassis, so their software can quickly simulate the thermal characteristics of any customer-defined set of components. Indeed, their simulations have led to system designs that actually exceed the company's own established cooling specifications.

The Omnix modular manufacturing process is very much an organic process. Everything is linked. This has several tremendously beneficial side effects. First, whereas a conventional manufacturer of a fault resilient PC product has anywhere from a 20 to 44 day lead time cycle to get the product to the customer, over 50 percent of ICS Advent's Omnix machines can be produced within 24 hours. The end customer may see about a five day cycle because there's an up-front sales event which involves time taken for the classic financial approval process, but the Omnix manufacturing line and the inventory management models are put together in such as way so that, for a majority of Omnix products, the actual manufacturing cycle within the factory is just 24 hours. Customers should love this since traditional customization time lags forced companies to do elaborate forecasting and even pay money to hold a "buffer" of a large inventory of PCs to try to decrease the cycle and during a crisis situation a 24-hour turnaround is a godsend.

Amazingly, a typical Omnix unit only takes about 35 minutes for the core assembly and then about 35 minutes for basic configuration and testing. The manufacturing line was designed to deal with all manufacturing and testing variables, particularly the variables that Kontron America calls "A" items that can be handled rapidly.

Kontron also recognized that they would occasionally be confronted with more complicated integrations that could not be satisfied in the standard 24 hour cycle. Anything above that has to be set up on what's called the "B" item line that may take three to five days to deal with any unique configuration or setup issues – still less than the usual 30 to 40 days found in other assembly environments, again thanks to the very tight links with the vendor base, the manufacturing process, the inventory management process, all the way through the logistic process.

All machines have a "burn-in" testing period and the customer has the right to specify "enhanced burn in" if necessary, which is part of Kontron America's automated testing procedure.

Yet another major benefit of the Omnix manufacturing process is that the end user receives a fully integrated solution. With some Asian-based competitors, the end-customer, the Value Added Reseller (VAR) or the Independent Software Vendor (ISV) often find themselves buying incomplete units, so at their factory site or location they still have to do the software integration and the product testing. In Kontron's case, the Omnix process does that for you online. They fully load the OS and all applications, they configure the system and they validate it so that what you get is almost plug and play, unless you have proprietary pieces that must be inserted into the configuration at your end. Even the custom paint job and logo is treated as a "standard build event" which is highly unusual. It takes about two weeks for the sales engineers to go through the qualification process for a customer's paint and logo. After that the customer can for future orders go to the Web site and have their own custom paint and logo treated as a standard high availability build item.

The whole Omnix manufacturing process bears more than a passing resemblance to a Just In Time (JIT) manufacturing approach, but one that can amazingly produce about 2,800 configurations of a fault resilient PC.

Fault Tolerant Manufacturing of Fault Tolerant Computers – Assembly, Certification, Software Pre-installation, and Testing

Producing customer-defined, virtual configurations on a Web site and running thermal simulations on them sounds wonderful, but how does an actual assembly line put together the resulting system in real-time without a multitude of errors appearing? Isn't the idea of a built-to-order system instantly rolling off of an assembly line a contradiction in terms?

In the old days workers would have tried their best, made many mistakes, and would finally just throw up their hands and blame it all on Murphy's Law ("anything that can go wrong will go wrong").

Today, however, highly customized fault tolerant PCs can be made with new and equally fault tolerant assembly techniques. The most advanced assembly and quality assurance techniques are found

Kontron America's customers can configure a fault resilient Omnix computer on their Web site, which relays the specifications directly to a Kanban-based assembly line.

at companies such as Kontron America.

Many of these techniques are borrowed and modified from Japanese manufacturing methods that in the 1980s helped Japan capture a large share of the global market in automobiles, copiers, and personal electronics. The new approaches include the Kanban materials management system, Poka-Yoke techniques, Kaizen continuous improvement programs implemented through Hoshin-Kanri goal-setting and timelines processes, and of course the more conventional ISO 9000 series validated quality control inspection system.

Kanban

Kanban is a Japanese word meaning "card" or "label." The Kanban system establishes inventory windows and a reordering procedure that's both simple and automatic. In Kanban, the depletion of stock triggers reordering. This is sometimes referred to as a "pull system." Many of the complications inherent to traditional Materials Requirements Planning (MRP) are bypassed in Kanban. Kanban has been compared to driving a car using the cruise control, freeing the driver from some of the routine of driving the car while maintaining a steady speed even while traveling over arduous terrain.

The Kanban process keeps the active inventory of components in a very small bucket right at the workstation where it's used.

At Kontron America, for example, after a customer fills out his order on the Web, each of the workstations on the Omnix product line are adjacent to inventory stations, each of which is called a Kanban. Each workstation corresponds to one complete step in the manufacturing process. For example, the first workstation ("Station One") is assigned the first step of assembly, which is chassis selection: the whole chassis inventory – 1U through 6U – is stored near that station.

The next station on the assembly line is the backplane and power supply station, where a worker can select from among all possible versions of these components sitting right there in a bin.

When a Kanban bin empties out, someone immediately replenishes it. Thus, production doesn't stop because no lines ever form in the stock room.

As you walk through the process, the backplane is done at the next station. The next major sequence of events is either the motherboard or the single board computer being assembled with the appropriate CPU, fan, heat sink, memory and jumper combinations. That then gets installed in the unit, and the unit then goes to the next station, which handles drive cables. Again, based on the customer order, the appropriate drive cables are inserted.

The next set of stations are where disk drives and drive cages are put together, installed in the unit and cabled in.

The final station is the final "cosmetic and configuration" station where the front panel rack mount "ears" and exterior items are put in place.

There are many inspection points in the process as well as a mid-point test and a validation point, reflecting a strong obsession on quality and control.

Perhaps the most startling impact of the Kanban-influenced Omnix manufacturing process is that it enjoys nearly a 70% decrease in direct labor to build these units compared to the traditional PC manufacturing model. This is partly because the Omnix manufacturing process doesn't use production control or warehousing clerks or work orders to move material. The Web interface delivers instructions straight to the manufacturing line where inventory is maintained directly on the line in a Kanban fashion, to satisfy each of the workstations, so there is no production control.

From an industry engineering standpoint, the Omnix Kanban line is called a "continuous flow, lot size one, manufacturing line." It's not a traditional model where there might be, say, 10 or 12 people involved in building up backplanes. In the case of Kontron's Omnix there's just one work area where that's done, so management can always in real-time alter, correct, enhance or troubleshoot a particular element of the process.

Poka-yoke

Poka-yoke (pronounced Poh'-kah Yoh'-kay, or the Americanized "poke-yoke") is a Japanese term meaning "mistake proofing." The phrase was coined by Japanese industrial engineer Shigeo Shingo while working at Toyota in the 1960s.

Encompassing poka-yoke is a systems philosophy called The Zero Quality Control System (ZQC). ZQC accepts that "To Err is Human," and instead of blaming or terrorizing workers, it finds ways to keep errors from becoming defects. It does this by monitoring manufacturing conditions at the source and correcting errors that cause defects via mistake proofing devices called poka-yoke that are used to check and give feedback about each operation in the manufacturing process, not just the inspection of a sample product. A poka-yoke device can be any mechanism that either prevents a mistake from being made or makes the mistake obvious at a glance. Poka-Yoke is thus a tool for improving flow, and can reduce or eliminate human error.

An early example of poka-yoke actually occurred in the US during World War II, in a plant manufacturing hand grenades. If a worker dropped a grenade while the safety pin was missing, it would explode. The manufacturer tried to prevent this error from happening by assigning two inspectors to observe and ensure installation of the safety pins. Even so, a grenade managed to slip down the line without a safety pin and then accidentally fell to floor, detonating and killing a worker.

The poka-yoke solution to this problem was the development of a special rack used to transport the grenades down the assembly line. Under the new design, the grenade hung on the rack by the safety pin. If the safety pin hadn't yet been inserted into grenade, the worker couldn't absent-mindedly hang the grenade on the rack and so the grenade couldn't continue traveling down the production line to wreak havoc. Thus, both the problem and the need for redundant inspectors were eliminated.

In the case of fault resilient PCs, suppose a worker must assemble a front-panel device that has two push-buttons. A spring must be put under each button. Occasionally, a worker will forget to put the spring under the button and a defect occurs. A simple poka-yoke device to eliminate this problem is for the worker to count out two springs from a bin and place them in a small dish. After assembly is complete, if a spring remains in the dish, then this indicates that an omission error has occurred. Upon seeing this, the operator can quickly add the other spring, with minimal cost of rework.

Testing

Another key point to the fault tolerant manufacturing processes of high availability computers such as the Omnix is the way they do the final configuration of the unit and the system testing. Traditionally, testing is done by multiple operators occupying multiple work cells. Kontron America instead uses an automated system test and configuration rack. Each one is called an Automated Test Equipment (ATE) unit. ATEs are based on a central server that maintains all of the possible software that can be loaded in a unit, be it operating systems such as Windows 98, NT, 2000, or applications.

Kontron maintains released versions under document control of all software on the central test server. Also residing on that server is the prime test software that they use, PC Doctor.

When a unit goes into a test station, a set of queries is sent to the server defining what that unit is and what it's made of. The server then builds an image of the hard disk for the system, including the operating system, the drivers, any configuration issues, and the .INI files. Customer specific application software can also be installed. Then via "Blaster" technology the hard disk gets reformatted and the new image is placed on the disk drive.

The unit then goes into a test sequence to validate that in fact it's loaded with software appropriate to its associated worksheet and then it kicks off PC Doctor, which is the core testing and validation program. If the PC is a generic type product it will be finished in a single pass. If the system is something sophisticated it may be subjected to multiple test passes or if the customer has requested a specific burn-in period, Kontron will continue to burn everything in during an extremely "hard exercise" under PC Doctor.

The key point to the Omnix configuration and test process is that it's completely automated, not operator dependent. Therefore, from a quality control standpoint, if there is ever an issue Kontron has a single point on a workflow diagram (and a single location on the factory flow) to look at and resolve.

Other companies could adapt these manufacturing and testing techniques to computers with different kinds of processors and operating systems.

Kaizen and Gemba

Embracing all of these manufacturing concepts is Kaizen, the Japanese word for gradual and orderly continuous improvement. Literally translated it means change ("Kai") and good ("Zen"). The basic supposition behind Kaizen is that lives deserve to be constantly improved, both in the home and at work.

Kaizen first appeared in the US in Toyota plants during the 1980s. Kaizen involves prioritizing, standardizing, and improving processes in a sustained series of incremental adjustments (even a change that reduces manufacturing time by as little as half a second is acceptable). Kaizen is decentralized in nature; all workers at all company levels can make changes that increase quality and production. The process dissuades the expenditure of capital to solve problems – instead the worker is told to use creativity instead. Generally, Kaizen worker teams learn how to use statistical tools and become familiar with the art of problem-solving in the gemba, or the "real place" where value is added, which is generally right on the manufacturing floor.

CHAPTER 4

Data Management and Storage Technology

THE QUANDARY OF DATA MANAGEMENT IS that business (or just about any organization for that matter) needs instant and constant access to data, yet there must be zero downtime and the files must be protected at all costs, which usually involves the seemingly conflicting activities of real-time backup and replication of data at various locations.

The good news is that, with the rise in popularity of data warehousing and mania over data mining, data is more important that ever. There is a plethora of storage options: ASPs, managed systems, hardware and software tools with high performance, fault tolerant storage systems architectures, and backup systems and services. SAN/NAS (Storage Area Network/Network-attached storage) solutions, and disaster-recovery procedures are the flavor of the moment.

EVERYTHING IS DATA, EVEN MONEY

Communications networks are a conduit for data. Data from e-commerce sites, distant data warehouses, customer orders, financial transactions, call centers – an awesome real-time mass of information. Data may travel at the speed of light over networks, but it's not ephermeral. Data is sent from somewhere to a destination, and at some terminal endpoint it must be stored, analyzed and acted upon. Not only transactions but voice conversations recorded between customers and call center agents can be digitized, transmitted as data and stored for later analysis. Data has become more mission-critical. People work with computers all the time, so everything from bank account transactions and hospital patient records to personal income tax documentation and product engineering designs are "data" that is stored somewhere. Indeed, there are those who say that, next to diligent employees and paying customers, data is an organization's most important asset. Losing it can be catastrophic.

In the old days before the PC could handle global enterprises, a 19" rackmount was a novelty, seeming to be rather rugged with its "server grade" SCSI drives that were faster and more durable than the "dinky" IDE drives found in mass-produced desktop PCs – now, of course, IDE systems with the ATA-133 (Advanced Technology Attachment#133) specification provide up to 133-Mbps transfer rates in bursts, and during 2002 the Serial ATA specification appeared, further pushing transfer rates to 150 Mbps.

And back in the ancient days of yore, databases hadn't yet become the globe-straddling behemoths they are today, and used to fit quite nicely on 100 MB of a single 512 MB SCSI drive from a respected manufacturer (I used Conner), the contents of which would be backed up on tape whenever the network administrator got back from the gym and yelled at somebody to do it.

RAID Revisited

If more protection was needed, one would set up a then-exotic RAID storage subsystem, like the kind mentioned in Chapter 3. Thanks to the ability of the SCSI bus to cable together seven (later 15) devices such as multiple disk drives and a special drive controller, real data redundancy and recoverability was possible, at least on a small scale.

Readers may be amused to discover that RAID was originally invented to boost data I/O performance and not protect data. After some monitoring of disk drive I/O workloads in various business and scientific computing environments, it became evident that in nearly all cases disk drive activity tended to be concentrated in certain areas of disks. That 100 MB database and the database program that accessed it were "hot" files. The now archetypal "80/20 Rule" was found to apply to storage: About 20% of disk storage capacity gets 80% of the read and write requests issued by end users, or by some software application.

Since the incredibly popular 20% of storage capacity (our 100 MB database) happened to be located on a single drive in a large system, I/O performance was degraded. The increasing CPU power and disk drive capacity of newer computers and the increase in the numbers of network users and their concomitant data requests stymied conventional disk drives, creating real I/O bottlenecks.

The early researchers suspected that if data (and therefore the "hot spots") could be distributed across several disk drives – a "drive array" – then performance could be improved.

In the mid-1980s, IBM co-sponsored a study at the University of California at Berkeley for ways to efficiently harness CPU power by providing it with the most disk storage data in the least amount of time – to transform the 80/20 Rule into a 50/50 Rule. This resulted in a 1987 paper entitled "A Case for Redundant Arrays of Inexpensive Disks (RAID)" by David Patterson, Garth Gibson and Randy Katz at Berkeley. The RAID concept was to get around the access time barriers of what was called the Single Large Expensive Drive (SLED) found in all computers of those days by replacing it with multiple, lower-capacity, inexpensive disk drives. These drives would be wired into an array that would appear to the host computer to be a single big logical drive that had unusually good I/O performance, which would be the combined performance of all the drives.

They realized, however, that by adding drives together in this way the Mean Time Between Failure (MTBF) rate of the aggregate system would decline. The failure of any drive would cause the whole array to fail, and the MTBF of the array is equal to the MTBF of an individual drive divided by the number of drives. Therefore, an array of 100 disk drives would be sure to fail in just a month or so (remember now, we're talking about circa 1987 disk drives).

To solve the MTBF problem, the Berkeley RAID paper proposed using ***redundant*** disk storage to protect data and ensure its continuous availability. Thus was born the modern concept of RAID,

User's view of storage is one large disk

A | B | C | D | E | F | G | H | I

RAID-0 Data is divided and striped across multiple drives. No data protection, but has high performance.

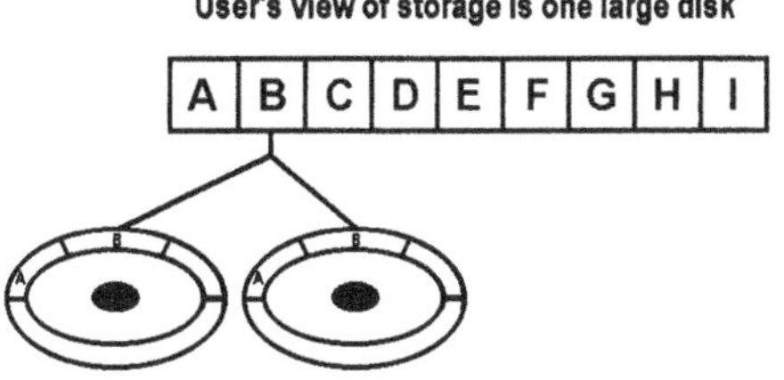

RAID-1 Data is completely copied or "mirrored" onto second disk.

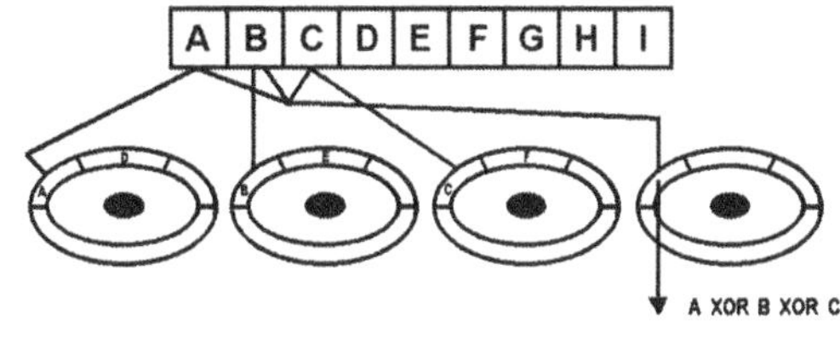

RAID-3 Data is striped across multiple disks one byte at a time. Parity is also calculated byte-by-byte and stored on an extra "parity drive." All drives have synchronized rotation.

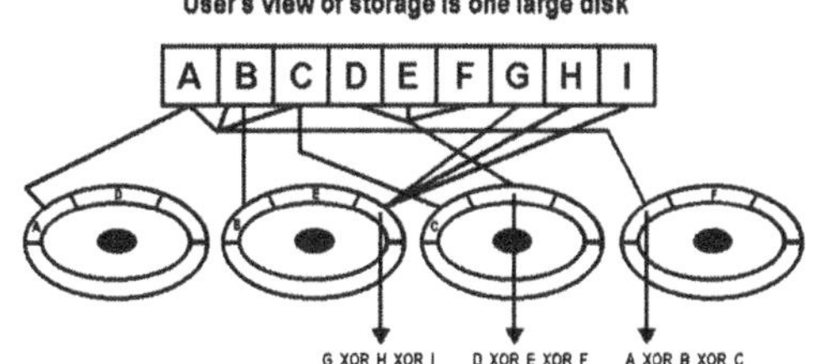

RAID-5 Data is striped across multiple drives in large, sector-sized blocks. Drives spin independently. Parity information is striped along with the data.

along with its three principal attributes:

1. It is a set of physical disk drives appearing as a single logical device.
2. Data is distributed across the physical set of drives in a defined manner.
3. Redundant disk capacity is set aside so that data can be recovered even if a drive crashes.

Modern definitions of RAID are pretty much the same: Essentially, a RAID subsystem needs to have data redundancy, functional redundancy, the capacity for regeneration, the ability to reconstruct data online, and an ability to handle I/O bottlenecks.

Two features often implemented but not defined in the original RAID paper are the ability to hot swap drives (replacing failed drives while the whole system is online and continuing to function) and "spareset" or "hot spare" drives – spare drives kept on-line and waiting to instantly take over when one fails. These are discussed in Chapter 3.

The Berkeley paper described various RAID research and development that had been done up until that time, and described five possible RAID methods that embodied the above attributes, which were labeled as RAID levels 1 through 5.

Each RAID architecture uses a different approach to achieve fault tolerance and increased performance. The method by which data is distributed on the disk array and the manner in which the protective redundant capacity is engineered are the distinguishing characteristics among the five RAID levels, all of which have become industry standards. Higher RAID levels are not necessarily better or faster than lower ones, although the higher numbered-levels do tend to be more sophisticated.

While the RAID levels 1 through 5 (as described on page 149) are considered to be the true industry standards, most RAID products usually don't fit nicely into any particular category, with each manufacturer adding to its product its own idiosyncratic technology. In recent years a sixth level, RAID 0, has also become generally accepted, as has RAID 6, RAID 7 and several other "hybrid" levels.

One of the original hopes for RAID back in 1987 was that RAID would be less expensive than big SLED disks. That's why the "I" in the RAID acronym stands for "Inexpensive." Ironically, by the time the first real RAID products reached the market in 1991, improvements in disk technology had enhanced the reliability and performance of small disks (their MTBF has skyrocketed from 40,000 hours up to over a million). Indeed, these days the cost of the hardware and software to tie the array together now exceeds the price difference between large and small disk drives! This is why the term RAID now actually stands for Redundant Arrays of ***Independent*** Disks.

RAID as a State of Mind

RAID 1 is considered to be the most secure, reliable (and most expensive) form of RAID. With RAID 1, also called "mirroring," one drive is a perfect copy of another. The redundancy is achieved in the simplest manner imaginable, called ***mirroring***, "dual copy," or "shadowing:" Whenever data is written to a file, it's duplicated and written simultaneously to both disks. Thus, the data on one disk is a copy or "mirror image" of the other. In case of the failure of a disk drive, the mirrored drive with its preserved copy of data continues to function and can be accessed immediately. You then just replace the failed disk and copy onto it ("reconstruct") all of the data from the functioning disk, which restores the system to full redundancy. This is in contrast to the popular RAID 5, where the lost data must be reconstructed using the error-correcting information held on what's called a parity disk.

But what if your server goes down? With conventional direct attached storage (DAS), the drives are trapped in the server, unable to dispense data to clients. Moreover, what if a disaster great enough to destroy a whole system occurs? Then there are no drives and no data, except perhaps tape backups. Obviously, both drives should be at different locations on different systems. In the old days this would be impossible, as one could not string a SCSI cable across town (SCSI is both a protocol and a physical transport for sending blocks of data between disk drives and a file server, but because the physical-layer transport is a parallel cable with 32 data lines and several control lines the electrical phenomenon of skew (see Chapter 3) keeps the cable length to about 15 to 25 meters

Today however, RAID has become more of a generic concept, and the interconnect between drives is no longer just SCSI but high bandwidth networking protocols. First came Fibre Channel, then Gigabit Ethernet and now IP and iSCSI (Internet SCSI).

The Birth of NAS

Putting drives on large networks initially became popular without any considerations involving RAID or any sort of data protection. Initially, such network-attached storage (NAS) served as mere solutions to storage problems. For example, let's say you need more disk space for a major undertaking, but there's no time to power the server down, add more hard disks and reformat the whole volume. Indeed, the server's drive bays are already bursting full of drives, and the budget doesn't allow for a new server. The quick, neat solution is to buy a standalone NAS device, generally a 19" wide rackmount which is filled with Just a Bunch Of Disks (JBOD) on a common backplane and which has its own operating system, gigabytes of storage and most importantly an Ethernet connection – Ethernet network interface circuitry can rely on a unique 6-byte MAC (media access control) address when data is exchanged between devices on the same LAN segment, but to communicate with a different device on a different segment, a network protocol such as TCP/IP or Novell IPX/SPX must be used along with a "real" unique network address.

A NAS device is essentially a server that's dedicated to sharing files among multiple hosts or client workstations simultaneously. NAS gets as much mileage out of the existing IP network as possible by incorporating well-known file-serving protocols such as NFS for UNIX and CIFS for Windows NT that enable servers to communicate efficiently with a file server. Since the NAS is assigned its own IP address it can be accessed by clients either via a server that acts as a gateway to the data or directly without an intermediary. You can plop down NAS devices anywhere on your network and in just seconds data can be shared with multiple platforms, including UNIX-, Microsoft Windows-, and Apple Computer Mac OS-based clients.

Some of the first NAS devices were inexpensive, relatively low-capacity (a few gigabytes at most), and small enough to fit under a desk. Cobalt, Hewlett-Packard and Quantum were among the early popular vendors of lower-end NAS systems. Now the NAS phenomenon has become huge, and the capacity, size and performance of these devices have increased tremendously. Even vendors such as IBM (White Plains, NY – 800-426-4968, www.ibm.com), EMC (Hopkinton, MA – 508-435-1000, www.emc.com), and Network Storage Solutions (Chantilly, VA – 703-834-2222, www.nssolutions) offer huge cabinet-size products that can run faster than many servers and store terabytes of data.

Although NAS was not originally conceived of as a fault tolerant storage system (and unless you run some kind of RAID software logic on the system the drives will just be vulnerable drives), using NAS does allow data on the network to be accessible by clients (all clients) even if servers go down.

Midrange NAS devices are often the most cost effective kind to buy, if only because that market has felt the most pressure from both the top as high-end equipment prices drop, as well as from the bottom as low-end NAS equipment offer more features such as data snapshots, high capacity IDE disk drives and external SCSI connectors allowing you to attach a local tape drive. To stay in the game, midrange vendors continually lower prices while increasing density and augmenting features.

NAS-device management continually improves as faster and more user friendly interfaces appear, both Java-based and Web-based. You can set up and run most NAS devices in a few minutes, but customers are demanding tighter integration with storage management packages, so make sure your NAS device is software upgradable.

Perhaps the fastest and least expensive NAS approach of all is Novell's (Provo, UT – 1-888-321-4727, www.novell.com) NetDevice NAS product, which is just software on a CD-ROM that you load in your legacy PC. Over the course of a 20 minute installation your PC gets transformed into a dedicated NAS. The PC must have an Intel or AMD chipset, run at 600 MHz or better, and house at least 384 MB of memory, 9 GB of storage, a CD-ROM drive, and a Network Interface Card (NIC). Configuration and administration can be conducted from any workstation with access to the same subnet as the appliance, with Internet Explorer 5.x or better having the Java Virtual Machine enabled. You can also use a console-based interface if the PC has a monitor and keyboard, or you can just Telnet into the system and use a Command Line Interface.

The Rise of the SAN

Simple network-attached storage devices are limited in that, aside from some of the more elaborate, expensive solutions, they don't generally provide effective backup, causing traffic congestions in the

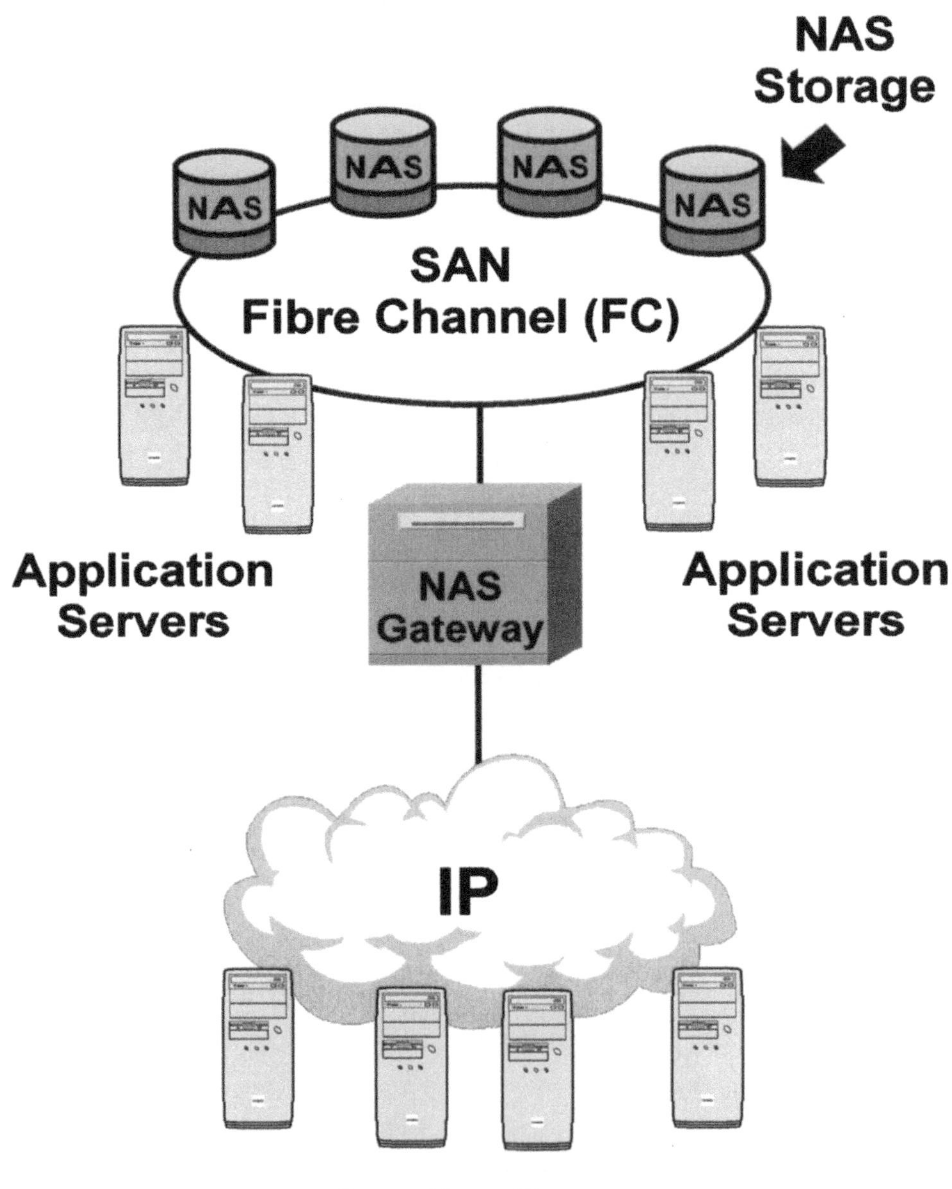
NAS and SANs
NAS Storage
NAS
NAS
NAS
NAS
SAN
Fibre Channel (FC)
Application Servers
NAS Gateway
Application Servers
IP
Application Servers

network. Storage-area networks (SANs) were introduced to eliminate these deficiencies, though NAS devices are often integral components of SANs.

The term storage area network suggests some kind of network independent of the LAN. The most primitive SAN can be just an architectural modification to the network, a subnetwork of storage devices connected to each other and to a server, or cluster of servers, which act as an access point to the SAN. In other, more abstract configurations, a SAN is a totally separate entity that connects at points to the regular network. Indeed, most SANs tend to be separate computer networks. But whereas NAS devices leveraged existing IP networks, SANs appeared in the mid 1990s almost solely because of the development of Fibre Channel (yes, "Fibre" spelled with an "re") interconnection technology. All early and many present-day SANs are based on a "fabric" of Fibre Channel, which uses special switches and hubs that connect storage devices to a heterogeneous set of servers on a many-to-many basis. These switches, which resemble Ethernet networking switches, are the connectivity point for SANs.

SANs can provide increased performance, data protection, scalability, resiliency, availability and manageability. Performance is heightened because servers talk directly to storage and other servers on a separate 1 Gbps network (soon 10 Gpbs SANs will proliferate). The storage traffic on a SAN doesn't have to compete for bandwidth with the data traffic normally found on an organization's LAN, while the LAN no longer has to contend with the storage traffic generated by doing backups to tape.

SANs are also highly scalable, since they use advanced networking technology such as Fibre Channel, which, unlike SCSI, can handle dozens of devices stretched across six miles (10 kilometers). You can configure such a network for maximum availability by creating a single storage area to which all servers and users can point.

All of these reasons are why SANs are becoming immensely popular, and why the research company IDC has forecast that by 2004 SANs will account for 70% of all network storage.

The SAN — NAS Love-Hate Relationship

It used to be that if many users needed low-volume access to a large amount of storage, NAS was the solution, but if you needed tera- and petabytes (a petabyte is 1,048,576 gigabytes) of storage or multiple, simultaneous access to files, such as streaming audio/video, SAN was the answer. These days the lines are becoming blurred between the two technologies, and while there is a SAN-versus-NAS debate going on, the fact is that both technologies actually complement each another. A NAS transfers data in the form of entire files, while a SAN transfers data in blocks and depends on a file system to organize the blocks into files. If you look at NAS devices as simply appliance file servers, you soon realize that you can use a NAS appliance as an Ethernet head-in for your SAN infrastructure instead of a traditional server. NAS boxes are a natural for use as storage modules for SANs since they can easily hold many more drives than an ordinary PC-based file server can. These NAS boxes can easily be backed up, since you can affix tape devices on a Fibre Channel network too.

All of this, however, means that you must forego the traditional NAS connectivity of IP and instead connect your NAS devices together with a Fibre Channel SAN topology.

"Fibre" is something of a misnomer, as to most people it implies an optical fiber. Fibre Channel did in fact start out as a system-to-system or system-to-subsystem ("box-to-box") interconnect standard that could handle both High Performance Parallel Interface (HIPPI) and the Internet Protocol over optical cable, but the FC-AL subset of the standard first defined in 1991 allows for such non-optical media connections as coaxial and twisted-pair ("twinaxial") copper wire.

When the first Fibre Channel SANs were installed, they had to be compatible with all of the legacy systems, where were based on SCSI. Fibre Channel iteself isn't really a bus like SCSI, it's actually a simultaneous two-channel bidirectional ring or loop architecture consisting of "channels." Indeed, Fibre Channel attempts to combine the benefits of both channel and network technologies.

A channel is a closed, direct, structured, and predictable mechanism for transmitting data between a few devices. Once a channel is set up, there is very little decision making needed, thus allowing for a high speed, hardware intensive environment. A "channel," therefore, simply transports any kind of data at the highest possible rate from one point to another independently of the overlaying logical protocol. Channels are often used to connect peripheral devices such as a disk drive, printer, tape drive, etc. to a workstation. SCSI and HIPPI are common channel protocols.

The "channels" of Fibre Channel have a data transfer rate of 100 MBps (signalling at 1.062 GHz using an 8B/10B code) in single-port configurations and up to 200 MBps in dual-port configurations (this is theoretical – in practice a dual-port, duplex configuration is about half that).

Networks, however, are unstructured and unpredictable. Examples of common networks are Ethernet, Token Ring, and Fiber Distributed Data Interface (FDDI). Networks are able to automatically adjust to changing environments and can support a large number of connected nodes. To be able to do this, much more decision making and processing must take place so as to successfully route data from one point to another. Much of this decision making is done in software, making networks inherently slower than channels. This "software intensive" nature of ordinary networks is exacerbated by the fact that the transport layers are bound to a specific logical protocol that also handles data integrity and communications handshaking. Fibre Channel, on the other hand, places the burden of communication handshaking and simple error correction on the hardware – the channels – without any need to waste overhead by involving the logical protocol. By using such a hardware-intensive technique along with a token-passing method for loop access, FC-AL thus assures the full bandwidth for each connection.

Fibre Channel manages to combine some characteristics of both channels and networks through a clever technique: Since all channel and network protocols send and receive data via buffers, at the core of Fibre Channel's design is the concept of simply moving data between one buffer at a source device (such as a computer or RAID subsystem) and another buffer at a destination device. These devices are connected by a "fabric," which is whatever hardware is doing the switching between the source and destination devices. Theoretically, devices aren't concerned about the fabric's internal workings since they only transfer data from buffer to buffer. Any kind of data can be in the buffer, and various protocols can still manipulate data before or after it's in the buffer.

Perhaps Fibre Channel's biggest early advantage over other serial interfaces is that developers could leverage their existing investment – all supported protocols can be used on the same facility at the same time. You can create "virtual" logical layers to handle anything over Fibre Channel such as ATM, HIPPI, IPI, SCSI, TCP/IP or whatever. Since data doesn't have to be converted between protocols, a sophisticated controller isn't needed. Since SCSI technology was extremely important when Fibre Channel was being developed, the first version of Fibre Channel was Fibre Channel/Arbitrated Loop (FC/AL) the SCSI variant of the Fibre Channel standard that enables it to be used as a direct disk attachment high-speed interface. The actual disk interface protocol used is SCSI-3, technically referred to as the SCSI-FCP (Fibre Channel Protocol) for FC-AL. With it you can connect PCI, SCSI host, and RAID adapters from companies such as Adaptec and Symbios Logic to PCs, printers, disk drives, servers, minicomputers or even mainframes without a switch or a hub. For additional redundancy, computers can be attached to more than one loop.

Manufacturers have adapted to the new Fibre Channel 2 Gbps bandwidth standard, but one wonders whether Fibre Channel will ever have a bandwidth of 10Gbps, such as competing fabric and networking technologies.

But are SANs all really "married" to using a Fibre Channel fabric? After all, one could turn this argument on its head: the key driver to NAS is the fact that it is connected to the open IP/Ethernet network and, consequently, its ease of integration. In contrast, the switches within current SANs are not interoperable with each other and conflict over standards prevails. SANs introduce the problem of an additional network to worry about. Like LANs or WANs, a SAN has its own management software, maintenance, security, operation system, support, utilities and training. Even in terms of hardware, it often requires higher-bandwidth cabling. All of these things increase the network staff's workload.

One solution is to use the same protocols for both LAN and storage networks, thus essentially making them one network. The main influence here is the ubiquity of IP. Favored alternatives include storage over Gigabit Ethernet, the Internet Small Computer System Interface (iSCSI), and Fibre Channel over IP. The Gigabit Ethernet alternative at first appears the most attractive since it obviates duplication of software, training, and test equipment, which would all make life easier for smaller businesses. iSCSI, however, would conserve existing investments in storage devices, software, and management techniques. It, in short, is a relatively new Internet Protocol-based storage networking standard for linking data storage facilities. Developed by the Internet Engineering Task Force (IETF) iSCSI can transmit data and iSCSI commands not just over a regular SCSI cable, but over LAN / Intranets, WANs and the Internet, allowing you to manage location-independent storage over any distance by carrying SCSI commands over such IP networks. The promise of iSCSI is that storage management software which was orginally written for the well established SCSI standard, can now be used to make a remote disk or tape drive on a network operate just like a local disk. Ironically, the aged, venerable SCSI channel protocol wasn't designed for storage networks, but iSCSI promises to give both legacy and cheaper hardware tremendous longevity. In most iSCSI implementations, one or more Fibre Channel SAN "islands" working even in different locations, can be bridged to a single IP network with devices like Cisco Systems' SN 5420 Storage Router.

The latest challenger to Fibre Channel includes the use of the more exotic switch fabrics mentioned in Chapter 3, particularly the emerging Infiniband standard, which also promises to provide some sort of upward migration path.

In any case, the success of NAS will not only lead it to be absorbed by the SAN world, but perhaps drive SANs towards more open storage networking environments such as Gigabit Ethernet or IP / iSCSI.

The Future of SANs

While tape or optical disks may still be used for long-term backup storage of a SAN, the SAN itself will usurp these technologies to become the principal fault tolerant, quickly accessible backup sys-

➤ From Fibre Channel to iSCSI

STAGE 1

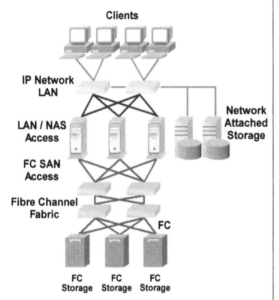

Clients

IP Network LAN

LAN / NAS Access

Network Attached Storage

FC SAN Access

Fibre Channel Fabric

FC

FC Storage FC Storage FC Storage

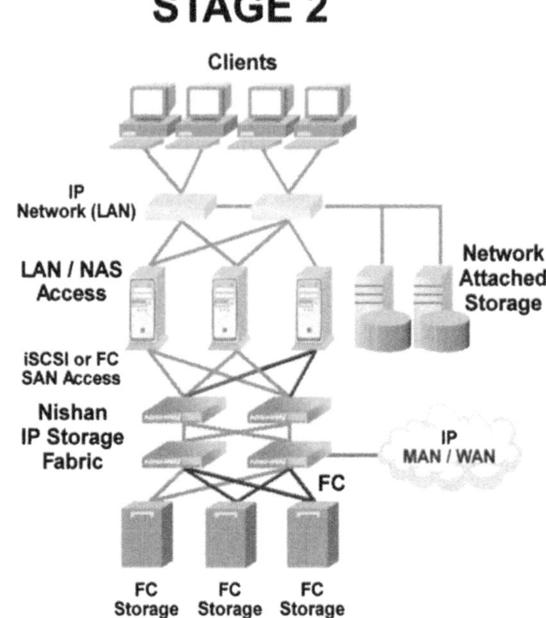

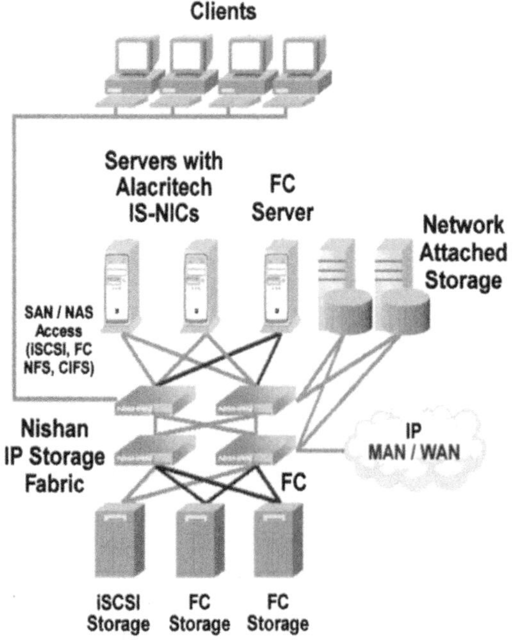

The iSCSI vendors Nishan (San Jose, CA — 408-519-3700, www.nishansystems.com) and Alacritech (San Jose, CA — 408-287-9997, www.alacritech.com) have outlined a three-stage migration path for customers wanting to move from a Fibre Channel network connecting disparate NAS and SANs to a newer iSCSI network. Customers can mix and match parts of any of these three phases of development:

Stage 1: The enterprise has separate NAS and SAN networks, each with different managers and different management software.

Stage 2: Network administrators still have to deal with a separate NAS and SAN, but have replaced the Fibre Channel switches with an IP fabric. The old Fibre Channel storage devices are still in place, but when new storage devices are added to the system they are iSCSI-compliant products.

Stage 3: Network administrators use Alacritech accelerators to attach the NAS to the IP storage fabric, and Nishan's switches enable wire-speed non-blocking access to varioius devices on a single storage network that is easier and less expensive to manage.

tem for enterprise computing. First of all, the time interval traditionally set aside to do a backup (late at night) is no longer available because of around-the-clock e-commerce, Internet, and data warehousing applications. And data itself is increasing at a geometrical rate, with comical results: While tape devices can still produce nice, compact cartridges that can be transported into a vault or offsite, Michael Ruettgers, EMC's CEO, says that it takes six months to restore a petabyte from tape. Moreover, as we've seen on countless occasions, backups are not infallible, as tapes and cartridges don't always work properly, backup software can become corrupted, and users can accidentally back up corrupted or incorrect information.

On the other hand, a single SAN can effortlessly store the data for an entire WAN as long as the links have a high enough bandwidth to handle the load. As even medium-sized organizations start to deal with systems in the terabyte range, the best solution will be disk-to-disk backups, where the enterprise disk subsystem is continually copied or replicated to another enterprise disk subsystem. A LAN-independent SAN can do this, as it can be made to encompass the entire storage hierarchy with NAS, disks, tapes, tape libraries, and other archiving facilities. Achieving short recovery times even in the face of catastrophic disaster of the September 11, 2001 variety demands that enterprises incorporate the best data replication architectures into their disaster recovery plans to replicate from a primary processing site to an alternate site, which can then be used in the event that the primary site becomes unavailable.

This is the logical extrapolation of the concept of "data vaulting" but realizing this idea brings considerable costs related to networking interconnects between production storage and the remote vault/mirror. If networks lack sufficient bandwidth to perform data transfers in real time without impacting production application performance, intermediary storage platforms such as caching appliances may be required. Services have appeared (and disappeared) over the years to promote and facilitate this approach. Some existing electronic vaulting companies, which handle remote backups for businesses, offer the option of receiving the data to be recovered on a NAS device. For example, E-Vault (Walnut Creek, CA – 925.944.2422, www.evault.com) returns their customers' data back to them on a portable version of E-Vault's servers. This can be faster and more convenient than shipping and receiving a box of tapes, particularly if coupled with a NAS device.

Adopting internally the most superlative, fault tolerant approach means that the SAN must provide mirroring and fault-tolerant devices to provide better resiliency and highly available data. This brings us back to applying RAID principles to NAS and SAN and so that RAID-protected data can move back and forth between servers. Any RAID subsystem links an array of conventional hard disk drives into a coherent unit. The subsystem must have some kind of "controller" to coordinate all of the disk activity when a CPU sends a request to read or write data from the array. RAID subsystems can be controlled either via user-installed software alone or a combination of hardware (the chassis with multiple drives) and software (now found in UNIX, NetWare and Windows NT, Windows 2000, etc.).

For a series of disk drives in a single box, RAID is easy. One would use a hardware-based solution, generally a host bus controller-based subsystem where the RAID microprocessor and other microchips are on a controller card plugged into an expansion slot of the host PC. Also, in the old days, "SCSI-to-SCSI," also called "SCSI Hub" or "Subsystem Controller-based systems" were the cat's meow. These put the RAID intelligence into a separate cabinet containing the drive array and

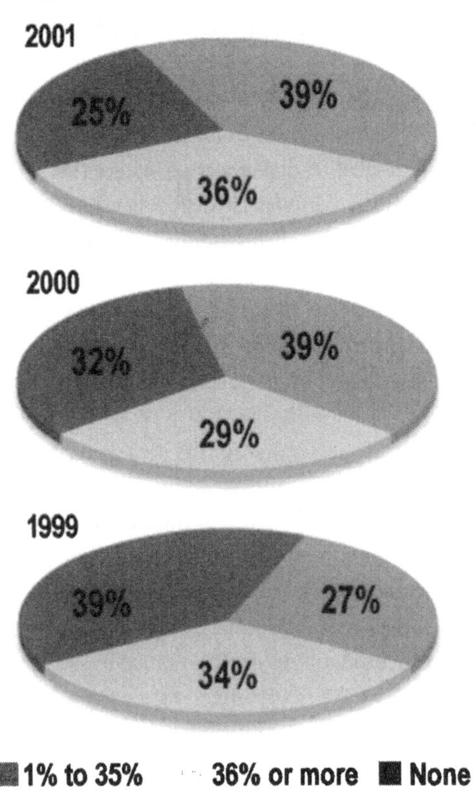

Source: InformationWeek *research survey of* InformationWeek *500 executives.*

attached it to the host PC via a standard SCSI controller. Unlike a host-controller, no special drivers are needed, just one for the standard SCSI controller. The SCSI-to-SCSI configuration is therefore operating system independent and can be used even on such "exotic" machines as the Macintosh. SCSI-to-SCSI allows multiple RAID subsystems to use a standard host adapter occupying just one PC slot. They place the drive array processor inside an external cabinet (or tower, rack, or what-not) and present the array to the host server as a single SCSI device, allowing other redundant devices (tape, CD-ROMs) to share the bus, making it the most scalable of array implementations. Since the subsystem itself is independent of the host server bus (EISA, MCA, PCI, etc.) there is little extra cost involved when upgrading the server. Also, in dire emergencies the SCSI cables can easily be daisy chained to standby servers.

But for bringing the wonders of raid to a bunch of JBOD boxes scattered around a SAN, things become more complicated, since software must subsitute for hardware, and over a network too. How the separate disk drives within a JBOD are used for data storage is settled by the host server or workstation, or by RAID intelligence on a host-based adapter.

Windows Disk Administrator, for example, can be used to create individual volumes from individual JBOD disks, or can assign groups of JBOD disks as a volume composed of a striped software RAID set. Software RAID will increase JBOD I/O performance but will also restrict access to the JBOD's striped set to a single server. Only volume-sharing middleware can enable multiple servers to simultaneously manage striped data on a JBOD without corrupting the data, otherwise unsanctioned sharing will cause an endless series of check disk sequences as a host struggles with unanticipated reorganization of disk data. Generally, software RAID on JBODs offers higher performance and redundancy for dedicated server-to-storage relationships, but does not lend itself to server clustering or serverless tape backup across the SAN.

Storage Virtualization

One exciting way of using JBODs for high availability shared storage is by using storage virtualization appliances and/or software that are positioned between the host systems and the JBODs. Storage virtualization was a hot topic that burst onto the scene in 2002. These products will let you collect all your discrete storage into one logical pool, an idea that goes back to the mainframe era but hasn't appeared on PCs until now.

Research companies such as Gartner Dataquest (Stamford, CT – 20-316-111, www.gartner.com) report that most companies have an excess of storage capacity, with an average utilization of between 30% and 60%. Virtualization gathers up all discrete and SAN storage, making it one logical pool. Yes, literally any storage connected to the network can be virtualized. Once in this pool, the software can dole out logical "slices" of this storage whenever an application server is about to run out of space, all without having to re-cable or rezone the SAN. The ideal is for the entire SAN's utilization to be around 85%. Visualization products can automatically allocate space on the network where needed and provide accompanying e-mail or other notification.

More importantly, virtualization manages the positioning of data to multiple JBODs or RAID arrays, while tricking each host into seeing a single logical storage resource. Software RAID can thus be eliminated on the host because this same function is now performed by the appliance or software. Storage virtualization thus imitates an intelligent RAID controller, except that the virtualization appliance and storage arrays now sit in separate enclosures across the storage network.

Veritas (Mountain View, CA – 650-527-8000, www.veritas.com) has done much work developing products in the virtualization marketspace. You can do host-based virtualization via their Volume Manager product. They also market a line of products called storage appliances, these are actually software applications that you can install in a storage device you buy separately.

Other competing products include HP's (Palo Alto, CA – 650-857-1501, www.hp.com) Storage Allocator, DataCore Software Corporation's (Ft. Lauderdale, FL – 877-780-5111, www.datacorp.com) SANsymphony and storage-virtualization features already embedded in Sun Microsystems' (Sun Microsystems (Palo Alto, CA – 800-681-8845, www.sun.com) Solaris 8 operating system, as well as products appearing from FalconStor (Melville, NY – 631-777-5188, www.falconstor.com) and LeftHand Networks (Boulder, CO – 303-449-4100, www.lefthandnetworks.com).

At last count, five differing approaches to sharing virtual disk capacity have emerged in the SAN market, spanning about 10 discrete implementations. Ranging broadly in price, performance, and utility, these virtualization solutions can be categorized by the methods they use to translate the physical reality to the host's logical view. The effectiveness of each technique is essentially determined by where in the SAN the mapping takes place and what platform is used to deliver the services:

1. Multi-host storage arrays.
2. Host-based LUN masking filters.
3. File system redirectors via outboard metadata controllers.
4. Specialized virtualization engines.
5. Dedicated storage domain servers.

Storage Virtualization Selection Criteria

While virtualization suppliers' claims are often indistinguishable, the DataCore Software Corporation, lists seven criteria that one should use to determine the viability of each approach:

1. The degree of independence that these products provide from a host's operating system and file system.

2. The broadness of support for a mixture of storage hardware.
3. The ability to protect investments in legacy storage assets.
4. The ability of the security policy to share virtual resources while adequately excluding uninvited guests.
5. The effectiveness of the technology at minimizing losses due to planned and unplanned downtime.
6. The breadth of devices consolidated into a centralized management view.
7. The ability to leverage commodity hardware and storage devices for improved performance and functionality at reasonable cost.

Ultimately, the best choice for virtualized SANs must provide unprecedented levels of reliability, availability and scalability, while serving as the basis for advanced storage services and management. Once all of one's storage is pooled, when your system tells you new storage is needed, you'll definitely need more storage quick and so you'd better have hardware that can scale up, such as the MetaStor family of storage systems from LSI Logic Storage Systems (Milpitas, CA, 408-433-8686, www.lsilogicstorage.com) that scales a thousand fold, from 36 GB to more than 39 terabytes (TB) of on-line data storage, and features advanced software for centralized administration, replication, virtualization and disaster recovery. In direct attach storage SAN or NAS configurations, the MetaStor system can be connected to computers running Windows NT, Windows 2000, NetWare, Solaris, Linux, HP-UX, AIX, and/or IRIX operating systems.

Virtualization will also spur increased acceptance of iSCSI or proprietary SCSI-over-Ethernet technologies.

SAN Management is Improving

SANs are powerful and incredibly flexible. They also can be incredibly complicated for unskilled personnel to handle. SANs should be easy to manage, because they're all part of the same network and can be located in the same area, but that isn't yet the case. There's some help in the form of SAN-management software, which simplifies administration while preserving the storage hardware's underlying flexibility and capability. Packages such as Computer Associates International's (Islandia, NY – 631-342-5224, www.ca.com) Unicenter, Hewlett-Packard Co.'s HP OpenView and IBM's Tivoli are beginning to fully integrate SAN management. One interesting package is from SANavigator (San Jose, – CA 408-232-1000, www.sanavigator.com).

SAN management will improve as open standards slowly begin to appear. Some vendors consider SAN management to be device management, and tend to emphasize their support for the Storage Networking Industry Association's (www.snia.org) HBA API version 2. This API provides a consistent, uniform standard interface to access information. But by itself, device management leads to inefficient use of storage resources. SNMP will always be with us, and as we've seen, the latest buzzword has been "virtualization" which joins "replication" and "backup" in the marketing/PR lexicon, and which particularizes SAN management too much ("see forest from the trees" kind of thing). All of this will keep your staff running around in circles.

Ideally, SAN managers must be able to view, configure and troubleshoot components and devices from end to end.

Data Hosting and Remote Backup Services

Smaller and even many medium-sized businesses don't maintain extra storage networks on the other

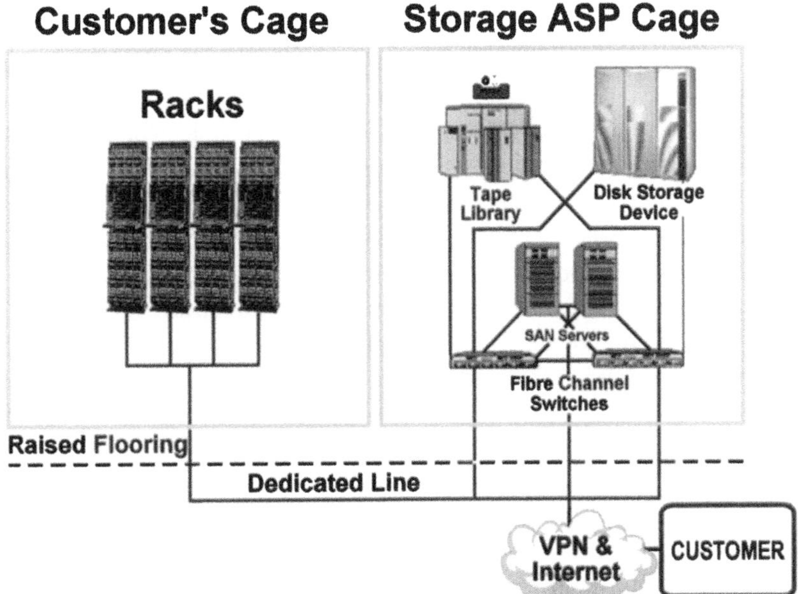

side of the country that perform real-time remote vaulting of data. Some small businesses and SOHOites have trouble keeping their lone tape drive in working order, and their Zip drive as of late has been sounding off "the click of death." What do these guys do for reliable data storage that won't occupy all of their time and money?

The answer may be a data hosting service, which is essentially a secure building filled with disk drives are equipped with high bandwidth connections to the Internet and other kinds of networks, and which are eager to reliably take care of your data storage needs. Such hosting services come in all sizes, from huge ones comprised of vast, secure datacenters that can serve a multinational corporation, to smaller, more flexible hosts that serve smaller companies and even individuals working in SOHO or teleworking situations.

There are both manage and unmanaged data hosting services. A business large enough to have its own in-house expertise could simply use a relatively unmanaged co-location service from companies such as Exodus, a Cable & Wireless Service company (Santa Clara, CA – 888-302-8855, www.exodus.net) to house the company's servers and work with them from afar. In such cases the host only provides data-center rack space, power, and a network connection. Managed hosting services, however, can combine maintenance, professional services, and other options in integrated bundles at predetermined prices. Naturally, hosting vendors want to move customers from low-margin basic services to more expensive and profitable managed services packages that include a variety of hardware, software, security, monitoring, and network options.

Digex (Laurel, MD – 240-264-2000, www.digex.com), a WorldCom company whose customers include J.P. Morgan Chase, Nestlé, and United Airlines, now offers about 20 managed service packages through its own hosting centers and from some run by UUnet (San Jose, CA – 800-967-5326, www.uu.net), another WorldCom company. The leader in the hosting is Exodus, a Cable & Wireless Service, with over 50 centers and thousands of customers, including British Airways, General Electric, and Microsoft.

AT&T, with 13 centers in the United States, the United Kingdom, and Japan serving companies that include CBS.MarketWatch.com and FAO Schwartz, also has joined in the remote data hosting craze. Qwest Communications International Inc., (Denver, CO – 303-992-1400, www.qwest.com) has spent billions of dollars to extend its hosting presence worldwide, and has introduced managed services, created in conjunction with Hewlett-Packard.

Managed services go beyond the mere leasing of data-center space to include a wide variety of services and capabilities that are managed and monitored by the hosting company. The packages – which comprise both new services and some already available from hosters 'a la carte – range from basic offerings that include leased servers, software, and monitoring to integrated setups with Internet infrastructure staples such as security firewalls, caching systems, load balancers, Web switches, storage, and content-distribution systems. Professional services and overall site management is also be available.

The value to hosting vendors is clear: Lower costs, thanks to automated processes that support packaged, rather than custom-designed, services, and higher revenue, since the more services they sell, the more money they make.

Hosting executives say customers can add – for a price – any type or number of servers or services, including sophisticated capabilities such as burstable network bandwidth to handle peak demands. They offer products from a variety of leading vendors, including Cisco Systems, Compaq, Microsoft, and Sun Microsystems.

Vendors say packaged services will save businesses money, especially when compared with the costs of setting up the systems and integrating them internally.

Certainly, hosters will have to do something if they want to meet growth projections. Overall sales of hosting services will climb to $19.77 billion in 2004, according to Forrester Research (Cambridge, MA – 617-613-6000, www.forrester.com), thanks in large part to managed hosting services, which are expected to jump from around $2 billion in 2002 to $10.97 billion in 2004. Co-location, however, will show modest growth, rising from $730 million in 2001 to $1.03 billion in 2004.

Managed services are one way to recover some of that revenue. Another is to automate the mostly manual process of setting up and configuring systems for each customer, which results in sky-high staffing and operational costs. Digex says its automated processes let it set up and deliver systems in 10 days, even though the number of servers it deploys has jumped from 100 per quarter in 1997 to nearly 700 per quarter in 2000.

Even individuals and SOHOs can benefit from the right kind of hosting service. At backupmystuff.com (www.backupmystuff.com) a division of Connected Corporation (Framingham, MA – www.connected.com, 508-808-7300) you first select a backup plan that suits your personal needs: The "Critical Data Backup" plan protects up to 100MB of PC data files to ensure your data is safe and off site (it's only $6.95 per month), while the "Full Data Backup" plan similarly protects up to

4GB of PC data files ($14.95 per month).

Next, a small software agent will download to your PC's hard drive. This will start your 30-day free trial of backup service. After the 30-day trial is complete, simply input your billing information.

Under the backup plans, your data will be compressed, encrypted, and sent over your Internet connection via SSL (secure sockets layer technology), automatically, every day. The process is optimized for Internet access with line speeds as low as 28.8 Kbps. The information is ultimately secured from your PC and stored at an off site data center in its secure form. You can recover data backed up in earlier versions for up to 30 days; deleted files are available for 90 days. Since your privacy is important, the data is never accessed or used for any purpose at the site where it is stored fully encrypted. Only you have the rights to your data.

There are also corporate plans available from backupmystuff.com's parent company, Connected Corporation. IT workers who use these services will recognize this as an outsourced version of products such as Veritas' Backup Exec or Computer Associates' ARCServeIT. Like those products, this is a server-based backup system where the software works from the client PCs, automatically protecting them.

Smaller businesses have a choice of many such companies; for example @Backup (San Diego, CA – 858-720-4500, www.@backup.com) and Novastor Corporation (Simi Valley, CT – 805-579-6700, www.norastor.com).

The proliferation of remote data hosting services indicate how distributed computing is turning to the time-sharing model to recapture some of the virtues of old-time centralized-computing. That's not to say that individuals have more power than ever before, but that valuable assets are best warehoused, as the song goes, "somewhere out there." It should be a relief to have industrious little elf-like processes spirit your data away while you sleep, without any effort on your part, and stored safely off-site.

Other advantages of using a backup service is that access to your archive can be granted to co-workers as a simple way of doing remote file sharing. Travellers in foreign countries can continue working on files last touched in the office.

And when a hosting service falls victim to economic troubles, others tempt customers with migration paths to their own systems. In 2001, when hosting leader Exodus Communications Inc. lost hundreds of millions of dollars and its CEO Ellen Hancock departed, Web-hosting provider Conxion Corp. (Chicago, IL – 800-266-9466, www.conxion.com) "coincidentally" jumped in and offered a new option for businesses wanting to transfer Web-site operations from co-location providers facing financial difficulties. It lets customers deploy their own Compaq, IBM, or Sun Microsystems servers within Conxion's managed hosting operation. Conxion joined other Web-hosting companies in offering migration programs that target particular kinds of customers. Conxion provides the kind of Internet connectivity and rack space that its co-location rivals supply. It also offers hardware, software, performance monitoring, system administration, and data storage and recovery.

As for Exodus Communications (Santa Clara, CA), after its bankruptcy it was acquired in February 2002 by Cable and Wireless plc and its wholly owned subsidiary, Digital Island, Inc. and the combined hosting services of all three companies is currently called Exodus, a Cable & Wireless Service, as mentioned previously.

TOP 10 DATA PROTECTION TIPS

Most businesses are not global enterprises and will not be implementing some of the more grandiose ideas outlined in this chapter. Still, the average owner or IT person would probably like to have some general guidelines regarding data loss and its prevention. To this end, in 1997, CBL Data Recovery Technologies Inc. (Armonk, NY – 914-765-0373, www.cbltech.com) published a wonderfully concise report entitled "Industry Sources – Data Recovery Report" which concluded with these time-honored data protection tips:

1. Back up data and test restore capabilities on a regular basis. Verify that the correct data is backed up.
2. Keep your computer in a dry, controlled environment that is clean and dust-free. Set up your computer in an area with very little traffic to ensure that it does not get bumped.
3. Only entrust your data to someone who has the training and expertise to properly maintain and repair it.
4. Use diagnostic and repair utilities with caution. Never use file recovery software if you suspect an electrical or mechanical drive failure.
5. Use anti-virus software and update it at least four times per year.
6. Check all incoming disks for viruses. This includes packaged software, software carried on-site by users and software downloaded via e-mail, modem, bulletin board services or the Internet.
7. Never attempt to operate a visibly damaged hard drive. Do not use any storage device that has been exposed to heat, moisture or soot.
8. Do not shake or remove the covers on hard drives or tapes.
9. Use a UPS (Uninterruptible Power Supply) for proper power protection.
10. Immediately turn off your computer if it begins making an unusual noise. Further operation may damage it beyond repair.

Truer words were never spoken.

CHAPTER 5

Power Protection

Obviously, modern telecommunications and computing systems run on electricity. Not so obviously, electricity is remarkably unpredictable. It even can be spectacularly unpredictable, as when Mother Nature, Zeus, or whoever is up there hurls a bolt of lightning that follows the path of any sort of conductor – be it a power line, a modem line or a clothes line – straight to your computer, frying it in an instant. Or it can simply be annoyingly unpredictable, as in the case of a blackout or a brownout. It can even be subtly unpredictable, with brief surges in voltage graciously provided by your friendly local power company, slowly damaging your sensitive electronic equipment. Or it can be downright mysterious, as when tiny static discharges from your fingers shortens the life of semiconductor chips by leaps and bounds.

The origins of these anomalies are as varied as their characteristics: Some types of power anomalies you can expect from the power company include blackouts (no power), dropouts (very short blackout), brownouts (lower voltage than normal), surges (higher voltage than normal) and phase shifts. Some power problems are caused by the customer (you or your neighbor). These can be blackouts (overloading local substation), brownouts (starting a large motor), phase shifts (using low power factor loads that require the power company to switch in power factor correcting capacitors), transients, and high frequency noise (such as from arc welding). Mother Nature can occasionally be blamed for transients (lightning hitting a power line) and blackouts (wind blowing down a power line); so can corporate greed (California's rolling backouts).

The Best Power National Power Laboratory's five-year Power Quality Study, conducted over a five year period in the 1990s, estimated that the average computer is subjected to 289 potentially damaging power disturbances per year, which is a little more than one "event" per business day. Since little has changed since then, you definitely need some kind of protection.

MORE POWER TO YOU

As telecom and routing equipment becomes more sophisticated, it requires cleaner electricity with a nice smooth "waveform" to run. Delicate integrated circuits are not only friable, but expensive, and a good surge or frequent transients can knock them out or slowly burn them out, forcing expensive repairs or replacements: In a word, downtime. It can be especially frustrating if you don't realize that the cause of your switch locking up, data becoming irretrievable, or memory/data getting corrupted is electrical. It may seem like faulty manufacturing or an incompetent telecom or IT manager. Unless an obvious blackout, brownout or surge hits, you may not realize your power is your problem.

Calming the Waves

Indeed, although you think you're buying your power protection system to simply supply power when the utility fails, the power conditioning aspects of your UPS or power system are just as important. Research by IBM indicates that up to 45% of all computer and electronics malfunctions leading to data loss are attributable to the effects of power disturbances.

Many UPSs use "ferroresonant" or constant voltage transformers (CVT) that allow primary input voltage to vary while the output voltage is held constant. Although "ferros" normally produce nasty square AC waves, high-end ferros use a harmonically compensated secondary winding, producing a waveform that's nearly sinusoidal. Ferroresonant inverter transformers thus protect the UPS battery from transitory high currents as it recharges.

Electrical problems are thus a real threat, and things can get rather ugly if the proper measures aren't taken to monitor, detect and suppress the myriad surges, sags, lightning strikes, etc. Electrical problems cause millions of dollars of damage each year and thousands of hours of down-time. Most everyone has heard or experienced at least one business horror story related to power failure. Fear tactics alone are usually sufficient to convince most IT and telecom managers to take basic steps toward protecting critical equipment such as phone systems and back office servers. Dealers and VARs help,

➤ A Menagerie of Power Line Disturbances

Category 1 Voltage Spikes – from lightning, power network switching and equipment operations such as elevators, spot welders, etc. Individual voltage spikes are normally brief (10 to 100 microseconds), but bursts of such pulses can sometimes last for several milliseconds. Also, nanosecond (billionth of a second) spikes can occur. Spikes take place around 50.7 times a month and make up 39.5% of total power line disturbances.

Category 2 Decaying Oscillatory Transients – from switching power factor correction capacitors and other network or load switching equipment. Oscillatory transients are sudden, out-of-sync waveforms that span the frequency range from 400 Hertz to over 5,000 Hertz. Their initial amplitudes are often 100% of peak voltage and usually decay to zero within one cycle. These occur 62.6 times a month and make up 49% of disturbances.

Category 3 Under-Voltage and Over-Voltage – caused by line faults and the action of fault clearing devices, power cutbacks, and sharp load changes. Power utilities' equipment normally responds relatively slowly to large changes in their networks, certainly not as fast as delicate computer equipment. These events occur 14.4% of the time and make up 11% of disturbances.

Category 4 Outages – total voltage loss for longer than one half cycle. Although outages are probably the reason why you've bought a UPS in the first place, these events occur only 0.6 times per month and make up just 0.5% of all disturbances.

—*from a study conducted by IBM*

too, by including a UPS (with built-in surge suppression and power conditioning) in the installed price of a new telephony or data server, messaging or phone system. But such basic systems tended to be installed and kept out of sight and on the back burner, as it were.

But in the wake of California blackouts and terrorist attacks, power protection, once a ho-hum affair, is now at the forefront of business strategy. Now, when one thinks of buying a UPS, visions of disastrous outages come to mind. For example, when the World Trade Center disaster disrupted the 80 megawatts of power normally flowing into the Center and its environs, UPSs in surrounding offices kicked in, as well as the occasional diesel generator to keep freight elevators and minimal services running.

But such colossal disasters are (hopefully) few and far between. Instead, UPSs are called upon more often during the kind of power alerts common in California nearly a year ago. A Stage Two Alert, for example, happens when the power supply gets to within 5% of the projected demand. Voltage drops, brownouts occur, damaging computer power supplies and delicate electronics.

Fortunately, many UPSs have what's called a ferroresonant transformer that corrects brownouts. And when things really get tough in the guise of a Stage Three Alert (power reserves have fallen below 1.5% and rolling blackouts are imminent) many UPSs can now be controlled remotely via software.

Power technology itself has been rapidly evolving, as we shall see. But you also must integrate this technology into a realistic disaster plan so you can maintain service through short failures; fall back, offload, and shut down cleanly in longer blackouts (or cut in the generators); and get your organization back up and running in one piece.

THE WEIRD WORLD OF STATIC ELECTRICITY

Before we get to the rundown on the standard equipment designed to protect your precious phone, data and other electronic equipment, here's what always gets ignored:

Always wear anti-static protection when working on internal server components. This is one of the easiest things to overlook involving platform maintenance, since few people realize that humans don't feel an Electro-Static Discharge (ESD) shock unless the voltage exceeds 3,000 volts, even though ESD events of only a few volts can cause serious permanent damage to semiconductor devices. The damage usually shows up as punctured MOSFET gates or PN junctions.

If you have onsite technicians working with any type of a hard drive, DSP-based CT resource card or CPU card, they should already know that electronic components are extremely susceptible to electrostatic discharge and that they should handle electrostatic discharge sensitive devices (ESDS) only in an ESD protected area (EPA), which is an area in which ESDS can be handled with accepted risk of damage as a result of electrostatic discharge or fields.

The electronic industry has placed a growing emphasis on reliability and quality assurance. ESD damage to parts and assemblies can be minimized by using ESD control measures. For example, make sure that your technicians are using static straps – these will greatly increase the lifespan of your equipment. The most irritating component of ESD is that it doesn't generally cause immediate catastrophic damage. Instead, ESD damage, which is not immediately noticeable, leaves the component operable, but damaged, such that it will fail at a later time under normal stress. Two weeks after your technician leaves a facility that has never caused any problems, you may suddenly get reports of weird things happening. Very often that ESD bug is going to be the culprit.

An ESD kit is a minimal investment – only about $20. You can pick up one from vendors such as the Prostat Corporation (Bensenville, IL – 630-238-8883, www.prostatcorp.com) or you can buy a full blown field service kit and other related products from UltraStat 2000 Inc., (Rye CO – 719-676-3782, www.ultrastat2000.com)

Workers who "carry" a charge into the work environment can rid themselves of that charge when they attach an ESD wrist strap (which contains a current limiting resistor which will discharge static but stop fatal currents from coming through) or when they step on an ESD floor mat while wearing ESD control footwear. The charge is routed to ground rather than being discharged into a sensitive component.

The biggest spark of all is a bolt of lightning. This can destroy any rackmount computer, save for an ingenious safety ground mounting block invented by Rittal (Springfield, OH – 937-399-0500, www.rittal-corp.com). Rittal developed this component for a Lucent company to address a Bellcore 1089 NEBS specification. NEBS testing is necessary for vendors wanting to sell equipment to the Regional Bell Operating Companies (RBOCs) and Competitive Local Exchange Carriers (CLECs). NEBS GR-1089-CORE testing is for electromagnetic emissions and immunity, surge standards and electrical safety.

NEBS demands a safety ground connection for any panel that has an I/O cable connected to it. Also, the PICMG Telecom Interest Subcommittee (TISC) requested a means for allowing the CompactPCI rear I/O module faceplate panel to be at protective earth potential (electrically grounded) when fully inserted into the cPCI system.

In the past, the rear I/O panels of a computer subrack were of a single, monolithic rear panel design, so you'd typically have just one earth grounded cable assembly coming from the outside world to connect to a computer subrack system. The actual hard wiring took the form of a safety ground stud on the computer's rear panel where a wire is connected to the cabinet. This provided an electrical discharge path – what's called a "lightning strike discharge path" – from the subrack to the safety ground.

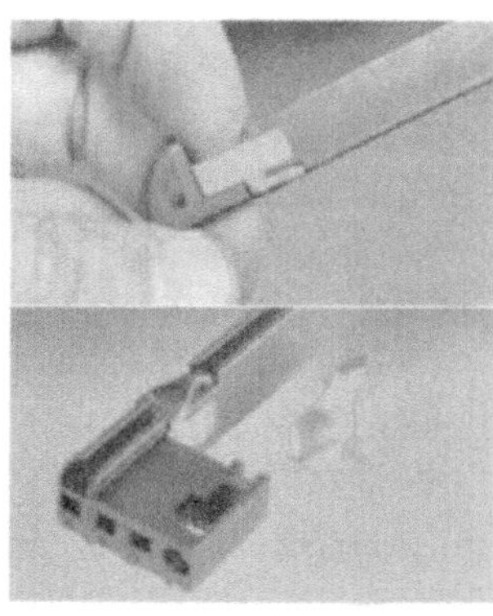

Elma's ESD clips

In the case of CompactPCI, companies designed their subracks with all of the I/O going out of the rear of the subrack. Because newer cPCI backplanes have lots of slots that hold lots of boards, you could have as many as 17 or 21 panels in the back of the box with I/O cables connected to each one. This also means that, in addition to the I/O cabling, you must provide a discharge path for the rear panel of every board. So, if each individual rear I/O module is hard wired with a traditional earth / ground connection, then, aside from the data I/O cabling for 21 boards, 21 pairs of ground wires would also have to be attached to 21 panels, resulting in a truly unmanageable mass of cabling.

One early solution: a standard card guide in a chassis can provide for an ESD path using what's called an ESD clip. Some very fine examples of these are made by Elma (Fremont, CA 94538 – 510-656-3400, www.elma.com) for connecting a board with a panel or outside case.

Generally speaking, non-heavy duty electrical contacts are suited only for eliminating any static charge build-up that may occur on the plug-in board as a result of its having been handled outside of the system. Standard ESD clips are fine if you're reasonably sure that your electrical path isn't going to be playing host to a whopper of an electrical surge, but when they're subjected to a NEBS lightning strike test (600 to 800 volts at high amperage), they often melt or blow open just like a fuse, are completely destroyed, and thus demand immediate servicing.

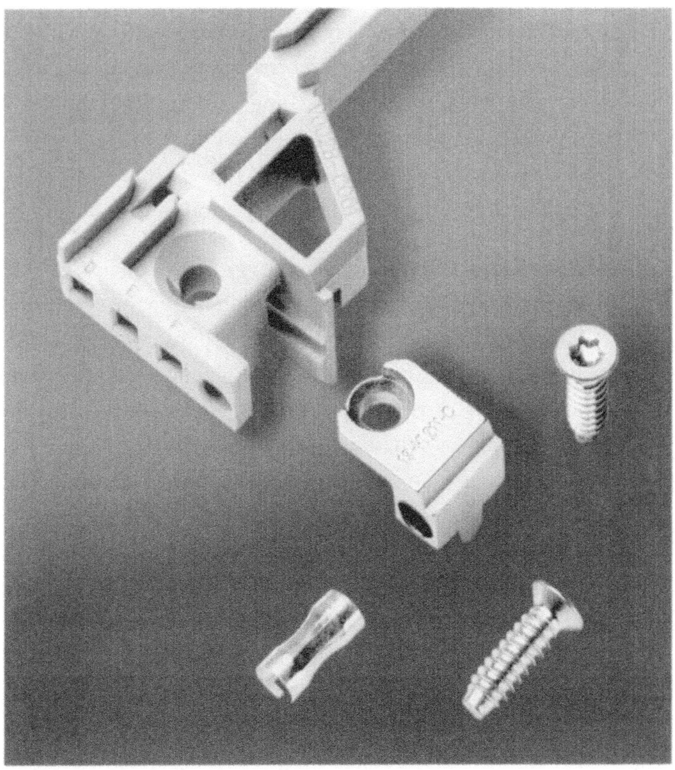

Rittal's safety ground mounting block protects rackmount computers from 600 to 800 volt discharges of up to 30 amps.

What Rittal did to improve this situation (and give rear I/O modules more of the same kind of modularity and fast interchangeabilty to the front mounted, hot swap cPCI modules) was to develop a safety ground mounting block casting that bolts down into the cPCI card guide and connects to the metal extrusion below that's bolted to the cabinet or the frame. The block also connects via a handle assembly to the rear faceplate panel where the I/O cables are attached to the board. This configuration provides an individual lightning strike discharge path for each panel, and the mass of the block is great enough to pass the NEBS lightning strike test without vaporizing. It will sustain 600 to 800 volt discharges of up to 30 amps.

THE POWER PROTECTION DEVICE FAMILY

There is a vast arsenal of products available to protect you and your business from potentially punishing spikes, surges and blackouts. If you work in a branch office, say, or a home office, where you're the remote call center agent ***and*** IT manager, it's up to you to make sure your hardware is safeguarded against power problems and the potential damage they can cause to your equipment and your business, so familiarize yourself with these devices:

Power Conditioners

Power conditioners are designed to sit between your electrical source and your equipment, delivering a steady current, with or without waveform noise reduction. It's like a filter on a fish tank, running all the time, cleaning up the electrical environment.

Line/power conditioners are used frequently in areas where there may be brownouts or irregular power but not blackouts. They can run all day long without using battery technology, providing a low-cost solution for correcting brownouts. Laser printers and copiers are especially good candidates for surge suppressors and something akin to a line conditioner that regulates power. Not only spikes but low voltage can be taxing for these machines. A UPS might be overkill for laser printers and copiers because (1) they pull so much power you'd have to get a huge UPS and (2) only a law firm racing against a deadline would be crazy enough to want to run a printer during a blackout.

AC Power and Telephone Line Surge Suppressors

A more common and less expensive defense against fried equipment is an AC surge suppressor.

Surge suppressors/protectors are perhaps the most basic form of protection. Essentially, they are designed for one purpose: to absorb high voltage spikes or surges before they reach connected equipment. If this is carried out effectively, a relatively small investment can save you a lot of money against damage to your electronic equipment.

Most of us are familiar with surge protectors as small devices into which you can plug multiple AC powered pieces of equipment, rather than going directly into the wall. Contained inside these devices are wires and fuses set up to redirect high voltages before they reach (and destroy) your valuable equipment. Most work by suppressing incoming electrical surges before they reach your equipment by shunting the surge to a ground, most commonly via a metal oxide varister (MOV). Other suppressors add a filter or choke which, rather than unloading the entire thrust of a surge onto the ground wire at one time, releases a "captured" surge to the ground wire in small and controlled increments.

A surge suppressor has a "clamping level," or the level at which the device limits the maximum voltage allowed to the system(s) it protects. The lower the clamping level, the better the protection. Once the particular level of surge is produced, the suppressor kicks in to prevent the disturbance from potentially damaging your equipment.

Connected equipment includes PCs, printers, phone systems, even some (but not all) UPSs themselves – virtually anything that runs on AC power and might be susceptible to a surge.

The classic (though by no means exclusive) example of a surge/spike is caused by a lightning strike. In such a case, the surge protector acts like a bodyguard: it takes the hit and often blows up (internally – most decent units will have a casing designed to protect against burning or melting), but will save your equipment from being fried. Since the device itself is in effect disposable, you want to look for warranties and replacement policies when purchasing. Above all, be sure that what you buy has a connected equipment warranty: no surge protector is bulletproof and you want to have a safety net.

Smaller companies find this alternative an easy expense to justify. It's the first line of defense you should consider if you have unprotected equipment already in place and operating. Surge suppressors don't regulate flow as such: they prevent overflows. Surges, spikes – events that can kill your equipment in a heartbeat can usually be prevented if you're protected with a surge suppressor. Suppressor design ranges from simple lower-level protection (i.e., not ***too*** many volts hitting it) like Kensington Microware Ltd.'s (San Mateo, CA – 650-572-2700, www.kensington.com) line of SmartSocket surge suppressor/power strips, to the MAX 8 Tel from Panamax (San Rafael, CA – 415-499-3900, www.panamax.com) which has a "power monitoring" system that protects AC equipment against prolonged over-voltages, as well as surges. In the event of a prolonged over-voltage, the

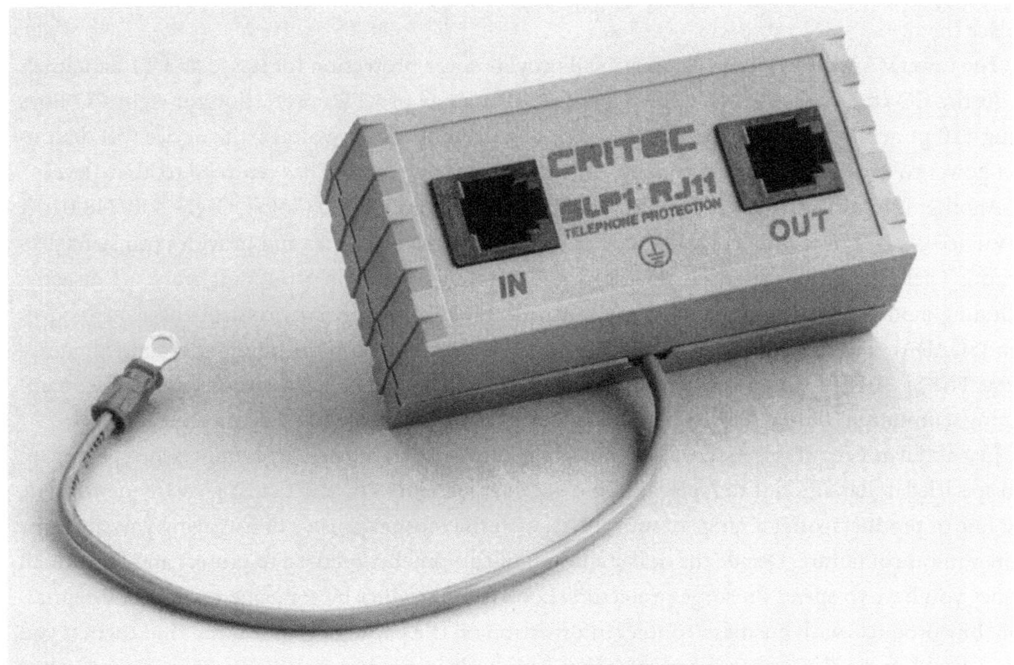

ERICO's CRITEC-RJ11 device connects to telecom lines and protects against electrical surges.

device's proprietary "SurgeGate" circuitry will disconnect power to any equipment plugged into its eight AC outlets, then automatically reconnect it once power has returned to a safe level. Panamax also offers the POWERMAX LAN UTP which provides rack-mountable data line protection for your network connections (most popular 10Base-T networks) and other rack equipment.

For whole-facility surge protection, Atlantic Scientific's (West Melbourne, FL – 800-544-4737, www.atlanticscientific.com) ZoneMaster PLUS installs at the main AC entrance and is a remarkably flexible device. Upgradeable and configurable on site to meet the exact needs of any facility, this interesting product is engineered to provide from 150,000A to 600,000A of surge capacity. Options include an extended range filter and surge counter.

Less obvious to some is the need for surge protection on phone/data lines, as well. The urban myth of Aunt Betty killed while talking on the phone, as lightning strikes the telephone pole in front of her house is an extreme example. More often, line surges have killed machines at the customer premise. But even low-level line disturbances can screw up transmissions or be just plain annoying, like the crackle you hear on the line as lightning strikes in your area.

In any case, you should install surge supressors on telco and data interfaces entering your communications server as well as filter the power going into the server. One cannot put enough emphasis on this. As simple as this may appear, many people overlook putting surge suppression on telco interfaces, be it digital or analog. They'll put UPS systems on the power side of a computer or networking device, but they won't protect the PSTN interface side, so very often lightning will hit the power pole outside their building and blast the computer telephony or other resource boards, which are usually under a warranty that precludes lighting strikes. It can be a very expensive proposition to replace them.

The towerMAX LL(T1) from Panamax will provide surge protection for leased and T1 data lines.

At the CO end, the towerMAX CO/8-110 from Panamax provides protection for eight CO lines using 110 punchdown connectors. In the event of a prolonged over-voltage, the device will disconnect power to equipment, then automatically reconnect it once power has returned to a safe level.

Another interesting device is the CRITEC SLP1-RJ11, from ERICO (Solon, OH – 440-248-0100, www.erico.com), a neat little component that connects to telecom lines and provides transient protection. There's a cornicopia of applications and equipment to help a business stave off disaster, including modems in programmable logic controllers (PLCs), supervisory control and data acquisition (SCADA) equipment, fire and security alarms and industrial monitoring and control equipment. The SLP1-RJ11 uses a hybrid, multi-stage clamping circuit to ensure protection while maintaining a minimum of line interference. Installation is simplified via RJ-11 connectors.

Just about any reputable manufacturer offers a lifetime warranty on surge suppressors, paying up to a specified dollar amount to replace equipment damaged if the product fails. They frequently will, in a line of products, offer a range of warranties, with the most expensive units paying you the most money in case of failure. Decide the dollar amount of equipment you need to protect and how much money you have to spend on surge protectors. Look for a product by someone with a good reputation. Buy products with business/contact information on the packaging so you can find them if you have a problem. Look for surge protectors that have built-in modem protection. This is often called "backdoor protection" because lightning takes the line of least resistance and whether that's your AC outlet or your phone line, be prepared. Look for the UL listing: ***UL 1449 A&B protected.*** In 1996, The Underwriters Laboratories, Inc. set a standard for the safety of transient voltage, ensuring various endurance and performance levels for surge suppressors. The rating also distringuishes the better suppressors from the $6.99 klunkers you can pick up in a drug store. The rating allows for what's called a 330V let-through, which is the lowest let-through voltage specified when a test device is hit with 6,000 volts in a UL testing lab. This voltage (or a lower one) should be your target when shopping for a suppressor.

Often it is not high-voltage spikes that cause problems for offices and call centers. Low-voltage sneak currents (below 50V) can wreak havoc too. These low-energy spikes cause locked screens and erroneous numbers that damage equipment over time, like a slow virus.

PowerSure products from Liebert (Columbus, OH – 614-888-0246, www.liebert.com) are among those that provide surge protection that's rugged enough for communications servers and small PBXs. They also protect against telephone line surges, and can stop current as low as 10V by using sinewave tracking to detect these transients and cut them off.

UPSs

UPSs come in three basic flavors: Stand-by, line-interactive, and online. ***Stand-bys*** are the lowest rung on the food chain. When utility power fails (or rises or falls suddenly in voltage), a stand-by UPS will switch to battery power to protect your connected equipment. Typically these units will also filter out some of the line noise or interference that can affect utility power output. A ***line-interactive*** UPS also switches from filtered utility power to battery power in the case of a blackout (total loss of power), a severe "surge" (short-term voltage increase) or "brownout" (short-term decrease). For most "spikes" (sudden, short increase in voltage) or "sags" (brief voltage decrease), however, a line-interactive UPS does not need to draw on battery power. Instead, the unit will provide a voltage regulation window , which boosts or lowers the output voltage in response to changes in input voltage from the utility, within a specified range. Finally, ***online*** UPSs provide "bulletproof" protection for truly mission critical systems. These units use what is called double-conversion technology – running input AC power through the battery and converting it to DC, and then converting back and running it out as AC. Unlike stand-by and line-interactive systems, there is no delay whatsoever in switching from utility to battery power. Online UPSs also produce only pure sinewave output, the cleanest form of AC power possible, as opposed to the pseudo-sinewave or squarewave outputs that some line interactives and standbys put out.

These categories can help to group UPSs and distinguish one from another. They should not, however, be used as simple or exclusive indicators of quality. Just as you shouldn't buy a cheap stand-by UPS to protect your company's back office servers, you probably don't need the most expensive and sophisticated systems available for every application in your business. Some pertinent questions to address when shopping for UPSs are:

- **What are my size requirements?** UPSs are sized in terms of Volt Ampere (VA) supply ratings – the same ratings are used to describe current drawn by equipment. In the simplest terms, you add up the VA figures drawn by the equipment you want to protect (the number is usually written on the back of your equipment), then buy a UPS rated somewhat higher than the sum (generally add 20%, so volt-amps = amps X 120), but what "somewhat higher" means depends on several variables, including the basic technology of the UPS in question (e.g., ferroresonant UPSs are inefficient unless you load them close to capacity). Still, when you size your UPS, you're tabulating the maximum load all of your systems will draw, which they rarely do. So when a UPS lists the run time at half load, this is not as lame as it sounds. Then again, new power supplies are "Power Factor Corrected," which means they draw low distortion current from the AC source, and are incredibly efficient. So it's always been a good idea to somewhat overprovision your UPS. In any case, many UPS vendors provide sizing tools on their Web sites, where you can quickly calculate the size of an appropriate unit.
- **How much backup battery time do I need?** Batteries are not an alternative to AC power. A UPS can't keep your systems going forever, but it can provide you with enough time to perform a "graceful" shutdown. Definitions of graceful, however, can also vary. In a revenue-generating call center, for instance, downtime is unacceptable and you may need to keep a system running on batteries for several hours. In this case you want a power supply to which you can add external battery packs – the more scalable, the better. You also want to check the guaranteed rating of your battery (both in terms of "cycles" and years), and find out what kind of battery monitoring tools the UPS offers.

MGE UPS Systems' (Costa Mesa, CA – 800-523-0142, www.mgeups.com) has an interesting product line called the Comet eXtreme UPS. Introduced partly as a result of the famous power crisis in California, the Comet eXtreme can, despite its small size, provide up to 100 minutes of online, double conversion power protection in a compact tower or rack configuration thanks to optional add-on battery extension modules.

Minuteman UPS's new digital on-line UPS tower models are available in six models (up to 7 kVA) and three footprints.

The Comet eXtreme series includes five power configurations (4.5 kVA, 6 kVA, 9 kVA, 12 kVA and 40 kVA). For maximum availability, you can service these devices' inverter module and/or replace batteries without interrupting the supply of power to the connected equipment. Also, you can combine two Comet eXtreme UPS modules to achieve a totally redundant power solution.

The units communicate with the outside world via RS232 or you can use a selection of optional cards, such as the Ethernet/SNMP card, to monitor, manage and conduct automatic orderly system shutdowns of remote UPS-protected devices via any SNMP monitor. The UPS units also include MGE's Solution-Pac power GUI-based management software for Windows 95/98/2000, NT, Novell NetWare, SCO Unix v.3.0, SCO Openserver 5.0, and Linux.

John Pooler, senior product marketing manager at MGE, says "Comet eXtreme's claim to fame is that it's the first UPS in its category to deliver over an hour of backup time and in a very compact design. A lot of customers are asking for it, especially in a rack configuration. They don't want the UPS to take up twice as much space as the equipment it's backing up."

Another maker of marathon UPSs is Minuteman UPS (Dallas, TX – 800-238-7272, www.minutemanups.com), which primarily focuses on UPSs that are 10 kVA or smaller. They've built their reputation on providing backup to computer and phone systems, and networks. Their top-of-the-line Minuteman MCP 10001 with four MCP BP1 Battery Packs is a 10 kVA online UPS that can supply 161 minutes of power at full load.

The Minuteman staff has also introduced the MCP-E line of tower UPSs which can produce continuous true sinewave AC power. Available in 700 VA and 1, 2, 3, 5 and 7 kVA models, each MCP-E unit incorporates constant output voltage regulation, input power factor correction, customizable input and output configurations and a zero transfer time to battery power. Both tower and rack-mount models are available. All models are network IP and SNMP capable and include Minuteman's Web-based Sentry II power management and diagnostic software.

Rod Pullen, Minuteman's president and general manager, says: "In the past people wanted enough power to save the files gracefully and shut down the equipment in case of an outage. Now, however, we're seeing a focus on productivity. They want extended runtime capabilities, and we've focused on this."

"Since the World Trade Center disaster we've gotten many requests for backup power for the security systems (intrusion alarm) market," says Pullen.

Not to be outdone in the my-battery-is-better-than-yours department, Tripp Lite (Chicago, IL – 773 869-1111, www.tripplite.com) sells a 3 kVA true online dual conversion UPS unit that's just 3U high. Called the SU3000RT3U, it's the latest member of the SmartOnline UPS Series, filling a critical niche series which stretches from 1 kVA to 10 kVA and includes 208-volt and 240-volt on-line solutions as well. The unit's runtimes can be increased indefinitely just by adding more external battery packs (BP72V12-2U). The system also supports front-access and rapid hot-swap battery replacement.

Tripp Lite's SmartPro SU2200RT is an amazingly small 2U high rackmount UPS that can deliver and output of 3 kVA (about 2,250 watts).

Two built-in communication ports (1 "smart" RS-232 and 1 contact closure) manage up to two servers simultaneously. There's also an SNMP/Web accessory slot. The bundled PowerAlert Software can control all networked UPSs regardless of their brand or what operating system is running on the network. Interestingly, Tripp Lite's PowerAlert management software is the only UPS software to receive Palm certification for wireless applications. PowerAlert can also be accessed and controlled via remote WAP phone communciations.

An even smaller 2U version, the SU2200RT, has nearly all the same features.

- **How easy is it to replace the batteries?** Given that a battery is, without question, going to give out at some point, you want to make sure you can replace it yourself, simply and with minimum system downtime.
- **What and where is the voltage regulation window?** Specifically applicable to line interactive UPSs, these numbers will tell you how much (or how little) input voltage the system can handle without going to battery, and how much fluctuation occurs in the output voltage (the closer it stays to nominal utility voltage – e.g. 120 volts – the better).
- **What kind of management capabilities do I require, and what management software does the unit include?** Software has become one of the key factors in differentiating among UPSs on the market. Before becoming enchanted with long lists of sexy features, however, it is important to determine what management tools you can and will actually use, and to make sure the UPS software is compatible with the systems you plan to connect to it.

Server-Room and Site Power Protection

Beyond a certain point, power protection becomes more serious than just plugging a UPS between the system and the wall socket. When you have to keep an entire server rack, or two, or a roomful of servers and support equipment running, it's time to take a more planned and coordinated approach to power protection. The needs are the same as with individual-system protection: Keep the signal noise, transients and spikes from getting through to disrupt the systems, and keep the supply voltage steady and flowing. It's the scale of the equipment and the elements of the protection scheme that change.

Special Considerations for the Low End Marketplace

The greatest variety of power conditioning and UPS equipment can be found at the low-and medium-end of the market. At the lowest end (300VA to 650VA), these UPSs protect the PCs and residential gateways of homeowners. The power needs of the PCs, key systems and routers of small offices, branch offices, or enterprise departments can be handled by the slightly larger systems (700 VA and above).

If you work in a large office, with lots of people and computers and phone lines, and you've got a skilled IT manager on hand who runs things so smoothly that you never see a problem, you may

➤ Uninterruptible Power Supplies — A Tester's Perspective

The primary purpose of a UPS is to keep your electronic systems working when the utility power drops or disappears. Most UPS also provide surge protection, to prevent damage to your equipment from lighting transients coupling onto the power lines. Depending on the type of equipment you wish to protect, you will be concerned about different UPS characteristics.

Small Office/Home Office (SOHO)

- How long do you need to run on back-up battery, for an orderly shutdown? Check your IT load, and look for a UPS with adequate capacity.
- Are you in a frequent lightning zone, such as Florida or Texas? Line voltage surges from nearby strikes can damage components and destroy memory devices. Be sure that the UPS limits power line transients ("let-through voltage") and not just total energy or Joules.
- Will the UPS cause interference to other electronics, such as portable radios or TVs? A UPS contains high-speed switching circuits which must be designed correctly to avoid interference. Look for an FCC label indicating compliance with Part 15 Rules.

Mid-sized Office/Commercial

- Does it meet the applicable electrical safety standard? (Also important for the home, of course). The UPS should have a label from a nationally-recognized testing laboratory indicating conformance to ANSI/UL 1778 for the US. For the EU, the CE-marking should be supported by compliance with EN 50091-1, and EN 50091-1-1 for operator-accessible areas.
- Does the UPS cause its own power surges when it switches from utility power to battery and back again? These switching spikes can cause memory damage or loss. UPS performance specifications such as IEC 62040-3 rate switching spikes by duration from none to greater than 10 milliseconds. Check UPS ratings for the smoothest switching.
- How "clean" is the AC power from the UPS running on battery? Highly-distorted AC power can drive delicate electronics nuts, garbling mass memory. A typical figure is 8% or less distortion for "clean" power.

Large Building or Plant

- These are 400 – 500 kVA capacity systems and above, for three-phase. How efficiently does the UPS convert battery to AC power? The answer determines operating cost and back-up time.
- For critical applications such as RAID, does the UPS support redundant module operation?

These are some of the key features you should consider in selecting a UPS, but of course there are many others. Become familiar with them by checking out several specifications, and consider what options such as computer ports and reporting protocols you might also want. The right choice of UPS could just prevent a real data disaster.

— Roland Gubisch; Chief Engineer, EMC and Telecom; Intertek Testing Services NA Inc. (Laguna Niguel, CA — 949-448-4110, www.itsglobal.com)

not ever think about what could happen to your computers, phones, and servers – and the work you do with them – in case of a power outage or surge. But if you work in a branch office, say, or a home office, where you're faced with the responsibilities normally borne by skilled IT technicians, ignorance is definitely not bliss. You must take responsibility for making sure your hardware is safeguarded against power surges and spikes and the potential damage they could cause to your equipment and your business.

A teleworker's home office, or a SOHO, is obviously not going to be as well supported by IT staff as would be a large organization, and equipment at this level must be sufficiently versatile to make up for this loss. Many informal users are under the impression that a surge suppressor or a UPS at a workstation will provide all the protection necessary to keep systems running. While protecting the obvious AC power side, few small office personnel or teleworkers suspect that even in the comfort of a peaceful suburban SOHO, surges have various ways of entering sensitive equipment that can damage them. Misapplication or makeshift approaches to power and data quality owing to an informal home environment, such as a home office, can actually result in more severe problems for small networks than no protection at all. Telephone lines, networks (LAN/WAN), signal, security, cable and/or building management systems are examples of surge pathways that equally affect any kind of electronic equipment, from a $700 PC to a $3000 server. Surge-related damage to I/O ports or communications boards can disrupt the tranquility of a small office just as much as it can infuriate some IT boss in a multinational enterprise.

In a teleworking small office or a corporate temporary office setting, many power-surge problems can be associated with improper wiring and grounding. Grounding must adhere to the requirements of manufacturers, local code, and the National Electrical Code (NEC) – for safety as well as for proper operation of any surge protection system. Commonly overlooked problems include improperly grounded main disconnects and circuit devices. Some companies recommend that multiple grounds (such as those at telephone and CATV service entrances, satellite poles and dishes, metal well casings, and antenna towers) be bonded together to reference a common ground. Also, if your location is in a lightning-prone area (such as parts of northern Florida) utility companies can supply you with "surge diverter" devices that defend your premise at the electric service entrance against power surges up to around 300,000 amps.

Today, customer configurable, upgradable surge protection systems are now available for telephony, data and signal systems of all sizes. These products are UL listed and made of high quality materials that assure long life and safety. They are also transparent to normal operations with components designed not to dampen transmission quality.

Panamax offers a huge line of surge and transient protection devices for the home and small business, such as their towerMAX line. With Panamax equipment you can customize the signal line protection with an unlimited number of add-on towerMAX modules. Modules are available for the following applications: Analog, digital, ISDN or T1 lines, networks, RS232 lines and miscellaneous accessory lines.

Tripp Lite (Chicago, IL – 773 869-1111, www.tripplite.com) also offers a premium surge suppressor – the Isobar – of which more than 14 million have been sold.

If you simply want a really inexpensive (under $100) UPS that will be dedicated to powering a non-telephony device (a large monitor, for example), Minuteman UPS offers a small 300 VA standby UPS with three outlets. Called the MBK 300, this is capable of powering a computer and perhaps a flat screen. Its clean output waveform is a simulated sinewave.

APC's powerful 700VA Back-UPS LS 700 is a UPS that will protect any multimedia desktop PC and monitor.

American Power Conversion (West Kingston, RI – 800-800-4272, www.apc.com), known as APC in the industry, offers the Back-UPS LS 700, part of its premium line of desktop solutions. This unit gives desktop PC users a hefty 700VA UPS for new high-performance computer systems. Desktop computer users who have a larger monitor or need longer run-time should be able to protect their system and get ample runtime with the high-powered Back-UPS LS 700. The unit includes Windows XP and Apple Macintosh compatibility, user friendly PowerChute Personal Edition software for Windows 98 / ME / XP (which ensures a safe, automatic shutdown in case users are away from their computer during an extended power outage), along with a Data Recovery warranty. Based on APC's Back-UPS Pro USB family, this higher power model is easy to use and manage, offers users a space-saving, sloped faced shape with seven power outlets on top and alarm indicators visible from any angle.

Complete hardware protection is ensured via two AC outlet banks, including two transformer block spaced outlets, along with fax/modem jacks located on the rear of the unit. The first bank of four outlets provide battery backup power, AVR (automatic voltage regulation) and surge protection, while the second bank of three outlets provide surge protection for non-data sensitive equipment like scanners and printers. The fax/modem jacks enable a user to protect a system from surges entering that system via a two-line phone/fax/modem line. This protection is also fully compatible for DSL high-speed Internet users.

Moving up the power scale a bit, Fortress.TeleCom from Best Power (Necedah, WI – 800-356-5794, www.bestpower.com) is a UPS having features tailored especially for protection of key switches, routers, hubs, CSU/DSUs and modems against the effects of spikes, sags, surges, noise, lightning and power outages. Fortress.TeleCom delivers smooth, sine-wave output, for communications equipment that requires pure, continuous power; and is designed to sit on a rack or on a wall using special brackets.

Fortress.TeleCom also has built-in Transient Voltage Surge Suppression (TVSS) to safeguard phone or network connections (RJ11 or RJ45) against surges and spikes that can enter through the "back door."

Models range from 750VA (450 watts) to a 2250VA (1600 watt) model.

Vendors taking care to position their UPS products in the marketspace may try to offer slightly more expensive yet dependable technology that they have perfected for several product cycles.

Controlled Power (Troy, MI – 800-521-4792, www.controlledpwr.com) offers everything from a small 700 VA UPS all the way up to 25 kVA single phase ferroresonant-based isolation transformers. They stick with proven technology. They give short shrift to exotic concepts like supercapacitors or flywheels.

As Ken Krause, director of new business development, says "With our equipment you buy one unit and it's got everything in it that you'd ever look for in a UPS. Unlike some other companies that use a TRIAC (a semiconductor used for controlling AC – you can find them in light dimmers) that uses the incoming AC voltage wave form as a reference, Controlled Power's equipment uses its own internal circuitry to generate a perfect sinusoidal AC waveform, free from harmonics."

"I'll take any of my 3 or 5 kVA single phase AC UPS units and stand them up against anybody's out there in America," says Krause. "I'll beat them in terms of input current distortion, probably by a factor of 2-to-1. We'll even build larger single phase systems for mainframes, since they don't use polyphase AC. We can do 18 kVA systems if you want."

Incidentally, the author's home office consists of a Panamax power filtering unit, two 500VA APC UPSs and two 640VA APC UPSs, as well as a 700VA IBM Office Professional UPS, all used to power three PCs, an HP desktop scanner, a Nikon film scanner, a broadband modem, a Linksys router and assorted ZIP drives, and CD burners.

Special Considerations for the High-End Marketplace

When providing power protection, line conditioning and backup power at high VA levels, a number

➤ PowerDsine's Power-Over-LAN

And now for something altogether different. PowerDsine (Petah Tikva, Israel -97-23-9347663, www.powerdsine.com) develops and manufactures power supplies for telecom and datacom OEMs. They offer a forward-thinking power solution that targets IP telephony systems as a primary application. Called Power-over-LAN, it's a multi-channel hub that hooks up to a 10/100BaseT switch on the network server, and uses existing data lines to supply power to any Ethernet connected device. Because most existing devices on your LAN are not built to accept power in this fashion, a DC-to-DC splitter is needed to enable the device to receive power over the same line as it receives data. PowerDsine has also built a good deal of intelligence into their product on the switch side, and can automatically detect whether a connected device has the ability to accept LAN power before sending it out over the line. When a device receiving power-over-LAN is disconnected, the power supply to that terminal will shut off after two minutes, preventing another device from being connected and potentially damaged.

This "distributed" method of supplying power has two major advantages. First, and most obvious, is that connected devices such as IP telephones can use the same line for data transmission and electrical power. Second, because the hub has redundant DC inputs on each channel, it can draw on a central UPS for its power, and thus obviates the need to employ an individual UPS at every terminal in the network.

PowerDsine offers Ethernet Switch vendors an embedded IEEE 802.3af compliant, Power over LAN Single Inline Package (SIP) solution, which integrates into Ethernet switches. The PD-IM-7124 Power-over-LAN Integrated Module is designed for 24 port Ethernet switches. It's brother, the PD-IM-7148 supports 48 ports.

of issues crop up that either don't apply, or apply only lightly at the lower end. Here's the major factors that need to be considered when implementing heavyweight power protection:

Single-Phase vs. Three-Phase Power

Most end-user electrical power is single-phase, meaning that there is a single sinewave power signal relative to neutral. When higher levels of power are needed, a three-phase AC power signal is used, with three power sinewaves that are 120 degrees out of phase with each other. Three-phase power is common for industrial motors (such as those for heavy shop equipment and elevators), and is sometimes used to deliver general site power when a lot of power at high efficiency is needed, such as a central office air conditioning system.

Unless you're installing the biggest server room in the Northern Hemisphere, you're likely to be working with single-phase power (or split-phase power, which is essentially two single-phase circuits with a common neutral). If you're wiring a completely new site with higher than average power requirements, you might want to talk with your power utility or a licensed electrician about the advantages of three-phase power.

Most UPSs and line conditioners are single-phase. However, many three-phase models are available, especially in the heavyweight range.

Line Frequency

Most power protection equipment is available for either the 60 Hz line frequency used in North America and Japan, or the 50 Hz frequency used in the rest of the world. Smaller equipment will often work with both frequencies, adjusting or compensating automatically. At least one line of midrange UPSs (Clary's GT series) can accept almost any input voltage or frequency and produce clean, stable 120 VAC/60 Hz power at its output.

At higher power levels, equipment must be optimized for one supply frequency or the other. Although it may seem like a trivial difference, running 60 Hz equipment on a 50 Hz supply or vice versa can cause major problems – including massive overheating, fire and explosions. In this chapter, we've mostly assumed that we're talking about 60 Hz power. If you're designing protection for a site in Europe, South America or another location that uses 50 Hz power, the information and application is all the same – you just have to use the right-spec models.

Differential Ground Problems

A tricky problem that can mess up network, serial and telecom communications is varying ground levels. If there is more than a slight difference in the ground potential at different points in a communication circuit, it can muddy or obstruct network or digital telephony signals. It can also cause operating problems for the servers and systems themselves. Any equipment in the power line, including UPSs, line conditioners and isolation transformers, can produce differential ground levels.

Most UPSs and line conditioners have provision for straight pass-through ground. If they don't, there are alternative ways to ensure a consistent ground level throughout a site. If the protection equipment manufacturers do not include specific instructions for ensuring consistent ground levels, you may want to bring in a licensed electrician with power protection experience to wire the room properly.

AC Signal Purity

UPSs regenerate the 60 Hz AC signal from the DC battery supply. Since AC signals from the power grid are near-perfect sinewaves, the ideal output from a UPS would in theory be an equally perfect waveform. Many of the heavyweight UPS manufacturers emphasize the "true sinewave" output of their UPSs. However, it's not completely cut and dried as to whether this power signal perfection is needed.

At high power levels, many AC devices function as tuned circuits, and frequency and waveform variations can cause overheating, power loss, and malfunctions. On the other hand, most electronic equipment such as computers, routers, monitors and the like are powered by switching power supplies. Switching supplies are not nearly as picky about their input as linear (transformer-based) supplies. Since they chop the incoming power to bits anyway, a rough approximation of a sinewave doesn't bother switching power supply circuitry at all. Many lower-rated UPSs have very rough power signal shapes without causing any problems for the systems they protect. A clean, perfect-sinewave output from a backup power supply is certainly a plus. However, don't put undue emphasis on it when choosing your server-room UPS if all of the systems being protected use switching power supplies.

High-End Line Conditioners

If your goal is to clean up the power coming in to a rackful or roomful of servers, the device you want is a true ***line conditioner*** with high capacity. Line conditioners do not provide the backup power of UPSs, but they do provide all of the filtering, boosting and limiting that protects the downstream equipment from damage and erratic function. Many use a combination of spike protection devices and ferroresonant transformers to eliminate noise and spikes, reduce surges and overvoltages, and boost line undervoltages. The result at the output is clean, stable power under almost any condition except a total power loss.

Line conditioners (and their simpler kin ***noise-suppressing isolation transformers*** which comprise many of the simple "power conditioners" mentioned earlier) can be used in two ways to protect sensitive equipment. One way, obviously, is to route the power for the sensitive items through the line conditioner. Line conditioners can also be used in "defensive" mode by screening transient- and noise-generating equipment from the rest of the site. For example, HVAC systems, blowers and elevators are all brute-force devices that are notorious for generating line garbage and voltage fluctuations in site power. Instead of trying to carefully screen every potentially sensitive system from their assaults, the problem can be reduced by placing an appropriately hefty line conditioner or noise-suppressing transformer between the blower or elevator motors and their incoming power. Presto, the noise and fluctuations are reduced or eliminated at the source.

Defensive line conditioning is not usually a solution by itself, but it can simplify and reduce the cost of protecting the more sensitive systems on the site. If the worst culprits are screened off, the level of protection the server room needs will be reduced.

Liebert makes one of the most complete lines of high-VA line conditioners. Their DataWave line has models that provide from 15kVA to 200kVA support, with protection against line variations from 40% undervoltage to 40% overvoltage. The units can be paralleled for additional capacity or redundancy, and their status can be monitored from a central location.

Oneac (Libertyville, IL – 847-816-6000, www.oneac.com) makes single-phase line conditioners with ratings from 200VA to 9.6kVA, and three-phase conditioners rated from 5.8 to 45kVA. Their isolated outputs and stabilized virtual grounds are significant features for users looking for exceptionally clean and stable power.

MGE has a line of power conditioners and noise-suppression transformers with ratings from 500VA to 225kVA. Noise-suppression transformers can be excellent choices for "defensive" line conditioning.

High-Power UPSs

The best level of power protection in any environment is a UPS. In the server room, level, UPSs are no longer a simple plug-in box, but a larger, rack-mounted or pad-mounted element. The rackmount versions are plug-in, with a bank of output sockets on the rear to power the systems in the rack. The pad-mounted units are usually wire-in models that must be installed by a licensed electrician.

➤ Energy Utility Power Conditioning and Support

One of the best sources for help and information when you're trying to build a clean power system is the power utility itself. Most people are aware that electrical utilities have programs under which they'll provide free or low-cost audits of energy use, as well as low-cost loans and rebates for upgraded, energy-efficient equipment. What fewer people know is that most utilities also have programs under which they'll come out to a business site and perform a power quality audit.

A power quality audit uses the utility's tools and expertise to evaluate the quality of line power at the customer site. In addition to plug-in monitoring, a power quality auditor will examine equipment at the site and recommend steps and solutions for potential problems. A thorough audit can uncover insufficient power capacity to a particular room or area, inadequate grounding, and more. Even more importantly, potential troublemakers such as blower motors, chillers, HVAC systems, elevators and other heavy, electrically noisy equipment can be identified, and specific recommendations made for fixing or blocking the problems they cause to more delicate equipment. If you're going to invest significant capital dollars in a conditioned-power setup, a power quality audit by your utility is an excellent first step.

As far as the quality of incoming power goes, utilities are required to meet certain standards for voltage, frequency and electrical noise at the power drop. Ask your utility to certify that the power at your incoming drop meets these standards – that means make them come out and put a monitor on the lines. This will stop some problems at the source, particularly if you have nearby neighbors who run heavy noise- and spike-generating equipment (it's up to the neighbors and the utility to fix those problems if they exist).

Finally, there's the issue of power supply – if the power goes out, all the line-conditioning equipment in the world doesn't do any good. Again, your utility has an option that can reduce your backup power costs. As an alternative to investing in massive, long-runtime UPSs, you can contract with your utility for what's called "guaranteed" or "minimum downtime" power. It's not available everywhere, even if the utility offers it to some customers, and it's not particularly cheap. Under a minimum downtime contract, the utility guarantees that power will never be interrupted to your site for more than a specified number of minutes. The shorter the maximum interruption, the higher the cost. It's worth checking into. If your utility company provides it, guaranteed power can sharply reduce your backup power needs and costs.

Whereas standalone units have ratings from 200VA to about 1.5kVA, rackmount UPSs tend to be heftier, with ratings from 1.5kVA up to about 5kVA. Pad-mounted units may have ratings as low as 2-3kVA, but are more often in the range from 10-25kVA. When very large sites or corporate-level server rooms need backup power and protection, models with single-phase or three-phase power ratings of 100kVA, 225kVA, 600kVA and more are available. The larger units can often be paralleled to produce whopping ratings in the mega-VA range.

Most larger UPSs can be paralleled for higher capacities, or to provide "N+1" redundancy for enhanced protection. As with other uses of N+1 redundancy, this means that any one of the UPSs can fail or be taken down for maintenance without reducing the overall level of protection. Coordination between the units can be through network or dedicated connections, or through active monitoring of the load conditions, as is the case with the Hot Sync-equipped UPS models of Powerware (Raleigh, NC – 919-713-5300, www.powerware.com). These UPSs are available in the 10-40kVA range with the powerful Powerware 9330, which allows up to four modules to be paralleled with no system-level single-point-of-failure.

Scalable UPS models are typically more expensive than non-scalable types. If you're not going to expand from one UPS to multiple units, don't waste money on a model with scalability or redundancy features. (On the other hand, if your server load is going to increase past the rating of one unit, be certain that redundancy and scalability are available in the model you select.)

Most large UPSs are online, double-conversion models that regenerate power for the downstream devices. At high power levels, standby UPSs (which switch from line to backup power on demand) may not cut in fast enough to prevent equipment glitches. The drawback to double-conversion is that it consumes more power when compared to standby or line-interactive approaches. Some manufacturers offer a choice of technologies at a variety of rating levels; Chloride Power (Libertyville, North Chicago – 847-816-6000, www.chloridepower.com) even offers dual-mode models that allow the user to switch modes for increased efficiency or increased reliability.

Liebert is one of the masters of the large-scale UPS market. They've sold more than a million installations. They offer single-phase and three-phase UPSs, environmental control systems (which are often integrated with power protection for server-room and other mission-critical sites) and comprehensive site security, environmental, power and emergency monitoring equipment. Their 60Hz UPStation S3 is a three-phase, true on-line UPS for networks, mid-range computers and telecommunications. The system offers prepacked battery packs, distribution modules and maintenance bypass switchgear that can be configured to match the need of any application. Models range from 12-24 kV. Liebert's Series 300 delivers on-line UPS protection for three-phase loads. 10-125 kVA models are available. Npower, the next generation of Liebert's medium size UPSs, has true double-conversion online technology that protects against the full spectrum of input and output power disturbances. It can make ultra-fast adjustments to changing loads. Npower has all-digital, DSP-based controls and a new patent-pending switch technology that delivers a nearly flawless output AC waveform. Even in worst-case situations such as unbalanced loads and non-linear loads Npower can maintain a total harmonic distortion (THD) of less than 2.5%. The Npower system is available in seven models ranging from 30kVA-130kVA.

Chloride Power, the parent company of Oneac, has international expertise with power protection. While Oneac handles most of the smaller scale UPS market and the line conditioning arena,

Comet eXtreme an MGE's UPS line. You can add battery modules to get up to 100 minutes of backup power.

Chloride handles single- and three-phase UPSs. Their Synthesis line incorporates both line-interactive and double-conversion technologies, allowing users to choose between power-efficient operation (up to 97% efficient) and the premium power protection offered by double-conversion systems.

APC has both midrange and high-end UPS lines. The SmartUPS RM rackmount models handle up to 5kVA, while the SmartUPS DP pad-mount line offers up to 10kVA. For full server-room and site support, the single-phase Symmetra models feature scalable and redundant operation in the 4-16kVA range, while the heavyweight Silcon DP300 series backs up three-phase power in the 10kVA to 4300kVA range.

Clary Corporation's (Monrovia, CA – 626-359-4486, www.clary.com) Continuous Power Systems are midrange units with up to 10kVA ratings. Their product lines are certified for a number of highly critical applications, including traffic light support, shipboard power systems, military and scientific power backup, and medical system support. All of their units have digitally controlled double-conversion power regeneration for exceptional power stability and reliability.

For three phase AC, MGE offers the 10 kVA to 30 kVA AC Galaxy 3000 for critical Internet Data Center operations. The Galaxy 3000 series supports 10 to 40 network servers. Through automated load testing of the batteries, the Galaxy 3000 can now detect battery performance problems before they can affect the UPS's backup time.

High-Power UPS Issues

Small UPSs and their bigger kin do the same fundamental job, just at different magnitudes. But when the technology is scaled up a host of noisome issues appear.

Hot-swappability can be extremely important. If a UPS must be shut down or taken offline to change out batteries or other major components, routine maintenance and repair will be more difficult. If batteries can be swapped while the unit remains in service (i.e., "hot swapped,") then everything can be kept running with fewer service interruptions. This is a much more significant issue with large UPS installations, since shutting down an entire rack or server room can be a major crisis for a company. (For very high reliability and continuous protection, consider an array of paralleled or N+1 redundant UPSs instead of a single unit.)

For example, Compaq Computer Corporation (Houston, TX – 800-282-6672, www.compaq.com/ups) has emphasized modularity and hot swappability even in their smaller "XR" (for "extended runtime") models that are the new generation Compaq tower UPSs: the UPS T1000 XR, T1500 XR and T2200 XR. They all support optional Extended Runtime Modules (ERMs), hot swappable batteries, an enhanced front panel display, and Enhanced Battery Management (EBM) software. The T1000 XR is rated up to 1000 VA / 700 watts, the T1500 XR is

Compaq's T1000 XR, T1500 XR and T2200 XR UPSs accept optional extended runtime modules that are hot swappable.

rated up to 1500 VA / 1050 watts, and the T2200 XR is rated up to 2200 VA / 1600 watts. They're designed for non-rack optimized servers (ProLiant ML servers), storage, workstations, and networking equipment like hubs and routers.

Cooling and ventilation can be an issue with larger UPSs, particularly online models. Power conversion at high VA levels can generate a lot of heat, more than most server-room equipment. Adequate cooling will have to be provided, otherwise the fans in your PC will be intaking hot, not cool, air. Although most UPSs use sealed gel-cell type batteries, some battery types and large concentrations of gel-cells under continuous charge and discharge can produce hazardous emissions. Venting these gases to a safe external discharge will be necessary. UPS manufacturers know the cooling and ventilation requirements for their systems, and it's often listed in the unit's specifications.

Enclosures take the idea of cooling and ventilation idea one step further – why not provide the equipment with its own enclosed, controlled environment. While computers and other sensitive electronics have changed over the past years, one thing remains constant: excessive heat, poor power quality and unauthorized access can damage or impair the operation of vital systems and peripherals. Companies such as Liebert have integrated enclosure systems include self-contained air conditioning, UPS and wiring management in a sturdy, lockable cabinet. For example, Liebert's Little Glass House is a totally integrated solution that combines precision environmental control, uninterruptible power supply (UPS), access security, SNMP communications, and cable management in a sealed 19" or 24" rack enclosure. In the Little Glass House, Liebert has combined all of the elements necessary to ensure the long-term viability of your network components such as servers, routers, hubs and switches.

Along the same lines is APC's NetShelter VX series of enclosures, all designed to address major trends within the data center and computer room environments, including rack-optimized thin servers with increased depth, higher power densities that cause thermal issues, and the proliferation

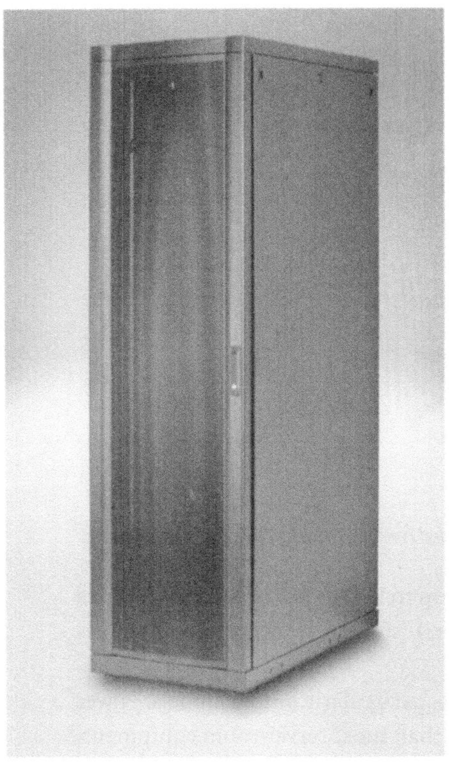

APC's NetShelterVX, an environmental enclosure for data centers and computer rooms.

of cables that demand management within the enclosure. Enhanced thermal management features of the NetShelter VX series fully perforated front and rear doors is designed to maximize airflow by promoting the natural cooling properties of installed servers and networking equipment. Also, ventilation requirements for major servers are maximized with 830 square inches of open ventilation area in front and 839 square inches in the rear of the rack to address increasing power densities. The NetShelter VX series is available in three heights: 25U, 42U and 47U.

Power consumption becomes a significant issue with large, double-conversion UPSs. Smaller UPSs draw a relatively insignificant amount of power to drive their charging and control circuitry. Even the losses involved in double power conversion are minor with UPSs in the under-1kVA range. At the 10kVA level and up, though, the power drawn by the UPS itself, over and above the load it is feeding, can be significant. Many UPSs draw 10% or more of the total downstream load in circuit power and conversion losses. For a 100kVA unit, then, the overall power load of the UPS and its protected systems can reach 110kVA, or about 80,000 watts for a net load of about 70,000 watts. Be certain your site designer or electrical contractor is aware of the additional load of the UPSs themselves.

Exterior UPS Placement must occur for certain larger applications. Alpha Technologies (Bellingham, WA – 360-647-2360, www.alpha.com) designs and manufactures systems that perform overall power conditioning and backup power for COs, ASPs, ISPs, and on the cable television or broadband networks at the "head end." Many of their devices are also in the "outside plant" infrastructure: On the telco side they power remote terminals (RTs) and on the cable broadband side they power "system actives" such as signal amplifiers, and other external network components having an attached power device. Their remote systems are housed in cabinets that can survive very extreme environmental conditions such as wide temperature fluctuations, and all weather-related elements associated with an exterior installation."

For example, Alpha's Novus XT CFR Series systems for exterior cabinets allows for the quick addition of power capacity as new services are integrated. Power ratings of 3000VA to 5000VA are available.

➤ Size up your UPS!

Don't know how big a UPS system you should buy? Don't worry, Minuteman UPS maintains the www.SizeMyUps.com Web site. Just enter eight pieces of information, click on the "Size It Now!" button, and you'll get a precise recommendation.

Many powerful Alpha Technologies UPSs, like this one, can be found in the telco outside plant. Alpha offers complete installation, turn-up, test, and preventative maintenance programs for its communication power systems.

Alpha's extensive catalog of products range up to very large capacity DC power protection for COs, as well as AC UPS applications, both in a single phase and three phase AC. These ferroresonant-based UPSs are generally not standard off-the-shelf generic AC UPS solutions, but are usually designed for particular applications running in a demanding environment.

Alpha's VP of Marketing, Eric Wentz, says "We've found recently that communications service providers are demanding more than just a couple hours of backup, many want six or eight or more hours of backup time. So, we offer solutions such as extended battery backup to reach that; to go beyond, we also offer an array of combustion engine, or propane powered generator systems that are fully integrated into the overall power system. Those are what we call 'indefinite runtime' solutions. They're small 3.5 kW systems that are part of a cabinet installation."

The Smart UPS

Many smaller as well as larger UPS systems by various manufacturers have in recent years begun to differentiate themselves by incorporating embedded processors and software that support various ways of communicating with PCs such as cards running the Simple Network Management Protocol (SNMP), serial and other kinds of ports to local PCs; or they may have their own Ethernet network cards and IP address, allowing them to be managed by legacy network management software such as HP OpenView. And if the data network is down, some UPSs have modems so you can dial in and check them. You can even command the remote UPS to perform self-diagnostics, and to send out e-mail or pager alerts when it needs servicing or when a power event has occurred. While this information can help prolong the life of your batteries and ensure that the UPS will be ready when it needs to be, it really proves crucial ***after*** a power failure by allowing you manage system shutdowns in an organized and effective manner. On some systems you can even toggle individual outlets on a single UPS, allowing you to shut down certain connected systems while keeping others up and running.

David Slotten, director of product management at Tripp Lite, says UPS management software simply started out as a way to monitor UPSs at remote sites with a single GUI interface and provide some capability for unattended shutdowns.

This makes an IT manager's life easier for the obvious reason that he doesn't have to be physically present in order to shut a system down when a power failure occurs. Most systems have now added GUIs that let you, for example, schedule automatic shutdowns and monitor the power supply's status (which otherwise required looking at LEDs on the exterior of the unit itself). In most cases you can also tell the UPS in what order to shut down and bring back up the connected systems, ensuring that the most critical are given priority. When you have multiple servers connected to one UPS (a common scenario), you should be certain that the software can shut down and bring up all the connected systems. This either requires hooking up a separate box to the UPS, into which you can run serial lines from all the systems, or setting up the servers in a master-slave configuration, where the "smart" master server sends out command signals over the network to its slaves.

"Our IP-based software management technology soon allowed us to do a lot of cross-platform, even cross-manufacturer monitoring of UPS systems globally," says Slotten. "Now, we've become more action-oriented, so you can remotely toggle outlets on and off. If power becomes scarce you can shed loads that don't need to be backed up and extend the runtime for those that do. Or, if you've got a router hung up out there, you can toggle a UPS outlet and get it going again."

Tripp Lite's PowerAlert v. 11.0 software provides network power control from any Web-enabled device, including the industry's first support for wireless phones and Palm Pilot PDAs. A new Wireless Application Protocol (WAP) Web server converts PowerAlert's normally GUI interface into a text form compatible with all wireless Internet browsers. Now, instead of logging into a workstation, you can control check UPS status, reboot outlets or power down devices from a WAP phone or wireless PDA. And PowerAlert can automatically contact administrators during power events via e-mail, pager, PDA or Web phone.

And if your Tripp Lite UPS is going to be left unattended but you'd still like it to have 100% availability, the company has a Watchdog System that ensures high availability by monitoring and restarting critical services and the operating system if necessary. The Watchdog System is comprised of a special Watchdog application used in conjunction with any Tripp Lite UPS System that features Watchdog compatibility.

The Watchdog System can provide fault tolerance for retail kiosk and e-commerce Web servers. It can preserve a high level of security when used with security systems which permit entrance/exit based on audio/visual or keycard recognition. The Watchdog System can also help maintain constant corporate, direct marketing and customer support communications when used with remote e-mail servers.

Using the properties menu in Tripp Lite's Watchdog application, systems administrators can set CPU usage thresholds and query performance for every service on their system. A benchmarking capability allows systems administrators to collect and view service performance in order to set usage thresholds and proactively address recurrent problems.

One thing to watch out for when comparing UPS software is operating system compatibility – most support at least Windows NT/95/98/2000 and UNIX, but you want to know that there won't be a conflict if you're running multiple servers or multiple OSs in one network (or, in some cases, on a single UPS).

The level of sophistication that UPS processors and software have achieved, impressive as it is, should not obscure the fact that this software's primary purpose is to allow an IT manager to worry less, rather than more, about power protection.

PowerChute *plus* is a good example of full-featured UPS management software. It was developed by one of the great names in power protection, APC, a company that offers a huge selection of accessories, which include environmental, Web-based, and out-of-band management capabilities.

In addition to scheduled self- diagnostics, graphical displays, and battery status updates, PowerChute gives you the ability to choose response actions for any number of possible power events. You can select from a menu of predefined shutdown actions, or even trigger the server to run an executable file in response to a particular event. Optional SNMP integration and a Web-accessible browser interface are also available. Out-of-band, dial-in communication is provided for when the network goes down.

There are actually three versions of PowerChute.

First, APC PowerChute Personal Edition v1.0 runs under Windows 98, 98/SE and ME, can be set up in four mouse clicks, provides unattended shutdown and can automatically save open Microsoft Office files. APC PowerChute Business Edition v6.0 is for small or medium-sized businesses which have up to 25 UPSs in a networked environment. It has a Windows NT/2000 based Management Console along with a simple mass configuration wizard that lets you configure settings for e-mail or broadcast outage notification, UPS parameters, and shutdown behavior for multiple UPSs simultaneously over the network.

APC PowerChute Inventory Manager v2.0 is enterprise-level UPS management software built upon an SQL database that can manage up to 10,000 devices connected on a large corporate network, including APC's Smart-UPS, Symmetra, Silcon UPS, and MasterSwitch-managed outlet strips.

APC will offer a new driver to ensure that their serial Back-UPS, Back-UPS Pro and Smart-UPS products work with Windows XP.

As for hardware on which to run PowerChute software, APC's Smart-UPS 1000 and Smart-UPS 1500 VA tower UPS units sport eight outlets. They can connect to PCs via USB or serial port connections. New features to enhance battery life include a temperature compensated battery charger that continually adjusts for optimum performance, and an extended range automatic voltage regulation (AVR) design that can deliver steady computer grade power during brownouts or over voltage conditions without resorting to batteries.

The Smart-UPS 1000 and 1500 tower models are available in both 120-volt and 230-volt versions.

Finally, APC also offers what they call the Network Management Card EX that lets you configure, control and monitor APC UPSs via standards based management such as the Web, SNMP, WAP, PPP/Slip, and Telnet. The auto-sensing card operates at either 10 or 100baseT connection speeds, and has a revamped, modular design that lets you quickly add options, such as environmental-monitoring, out of band control and network standards based security, without using a second card slot.

Various in-band and out-of-band communications paths can be used to monitor and control Liebert's Npower 3-phase UPS, including RS-232 direct serial connection and Liebert's SiteScan

APC's Smart-UPS 1000 and 1500 tower models come in both 120-volt and 230-volt versions.

software interface. Optional communications equipment also enables simple network management protocol (SNMP) capabilities, the SiteScan centralized monitoring system, and remote monitoring panel and alarm status contacts. The unit also has self-diagnostics and retains an event log with 512 time-and-day stamped alarm events to simplify maintenance and troubleshooting.

When a UPS Needs Help

The rash of lengthy power outages in California during 2001 underscored the fact that while most power outages last less than 20 seconds, a surprising number can last up to eight or ten hours.

"There is a cost tradeoff at about the eight to twelve hour range such that maintaining a diesel or gas generator makes more sense than having a bank of batteries," says Rod L. Pullen, President and General Manager of Minuteman UPS (Dallas, TX – 800-238-7272, www.minutemanups.com). "Supplying five kiowatts for over an hour during a blackout using batteries demands, well, a truckload of batteries."

Still, the good old battery-based UPS will be with us for some time to come. It's not too expensive, it's reliable, and the amount of legacy equipment out there is immense.

EXOTIC FUTURISTIC STUFF

That's not to say that new, more bizarre ways of storing energy aren't continually being developed.

Supercapacitors (SCaps), ultracapacitors or electric double layer capacitors (EDLCs) store energy by maintaining two charged plates (usually carbon) at opposite polarity, separated by an electrolyte such as an aqueous acid, a salt solution, or an organic electrolyte. Carbon can be made into a cloth with an incredibly high surface area (2000 square meters per gram) and can hold a charge powerful enough to run a PC for up to a minute, enough time to ride through most small power outages.

Companies in the SCap game include Power Cache Ultracapacitors (San Diego, CA – 877-511-4324, www.powercache.com) and PolyStor (Livermore, CA – 925-245-7000, www.polystor.com).

Another interesting energy storage resource is the flywheel. Electricity drives a composite rotor to spin up a flywheel, storing energy kinetically. To draw power, the process is reversed, and the spinning wheel drives the motor/generator to convert the kinetic energy back into electricity. Some of the more powerful flywheel systems are quite compact and can produce tremendous (500 KVA) output from 5 seconds to a minute, giving a microturbine or engine/generator power source time to come on line.

Companies in the flywheel marketspace include Active Power (Austin, TX – 512/836-6464, www.activepower.com), Beacon Power (Wilmington, MA – 978-694-9121, www.beaconpower.com), Precise Power Corporation (Bradenton, FL – 888-522-1600, www.precisepwr.com) and United Power (Richmond, VA – 804-359-6500, www.unitedpowercorp.com).

In particular, Active Power has introduced a new "mechanical battery." It's the High Inertia Turbine 6 (HIT6), a six-kilowatt, quiet, low-emissions alternative to both batteries and engine gener-

Active Power's HIT6 is a 50,000 rpm turbine, a generator/motor, and a counterbalancing rotor mass. It can provide uninterrupted DC power to business or telcos.

ators. HIT6 is designed to replace or supplement the lead-acid battery bank on the +24V or -48V DC bus of the telecom power plant.

The brainchild of Active Power's CEO and founder Joseph Pinkerton, HIT6 can provide uninterrupted DC power during utility outages of any duration as long as fuel (or a heat source) is available, and the unit can supply optional AC voltage for auxiliary loads (such as climate control) after a few seconds of utility disruption. HIT6 systems can be paralleled for capacities up to 24 kW.

The HIT6 system is normally powered by the utility to maintain a charged state and operates in parallel with the DC bus that provides power to the telecom equipment. The system maintains the DC bus at set voltage levels during a power interruption.

HIT6 is a highly integrated combination of things: a 50,000 rpm turbine, generator / motor and flywheel. The unit can accept 185 to 264V AC current to spin the internal components of the system so that the telco can maintain a charged state and operate the unit in parallel with the DC bus that supplies power to the telecom equipment. In case of an electrical outage, a special gas circulating in a closed cycle can be instantly heated by a propane burner before being sent through the turbine, although in theory any heat source (such as solar or geothermal) can be used.

Because of its low maintenance design and the fact that it fits in an enclosure the size of a central air conditioning system, HIT6 should find a home in critical remote and outdoor power sites. Marconi Communications, Inc. a global provider of advanced communications solutions, is the exclusive distributor of the HIT6.

An even more promising storage technology are fuel cells, which combines oxygen and hydrogen (or natural gas) to generate electricity with heat and water as byproducts. As early as 1999, the First National Bank of Omaha, Nebraska, decided to protect its credit card transaction system by installing an 800 kW system made by Sure Power Corporation (Danbury, CT – 203-790-8996, www.surepowersystem.com). Fuel cells operate best at full capacity, so the bank runs them all the time and sells power back to the local utility. Also, the system generates enough heat to keep the sidewalks around the bank snow-free in the winter.

CHAPTER 6

Wireless — Staying Connected

NEEDLESS TO SAY, THE ROLE THAT cellular phones and their providers played during these trying times have created a renewed interest in all things wireless, in general, and cellular, in particular. While the utility of cell phones and other wireless devices in shaping the response to a crisis had never before been examined, it was made crystal clear in September 2001 that modern, robust communications systems can play a major role in successful disaster management. Devices such as, laptops, handheld computers, PDAs, interactive pagers and cell phones can be the lifeline needed for governmental agencies, disaster relief organizations and the business community to stay afloat during a crisis.

Workers across the nation were forced to morph into "mobile workers" – making cellular connections do the kinds of jobs normally performed by the landline PSTN, scrambling to get their work done on laptops, handhelds and PDAs, working their e-mail and short messaging services as never before. People found themselves not only at disaster's doorstep, but also in a state of paralysis, stranded at airports and in trains, hotels and other places on the road.

Businesses can learn much from this experience. It can help them to prepare for the future – a future where their employees can keep in touch and stay productive under even the most trying conditions.

WIRELESS TECHNOLOGY TO THE RESCUE

The American Red Cross Disaster Service (ARCDS) understands the value of wireless technology during a crisis. Like any other organization, it still tries to meet its objectives during the aftermath of a disaster, but unlike any other organization, its objectives ***depend*** on a continuing confrontation with disaster in all its forms, wherever it may strike. Working forever in disaster recovery mode, the ARCDS' very mission puts it in harms way. Its operations take place under extreme conditions: heavy storms, power and telephone outages, earthquakes, fires and floods; yet it must always be ready to take on the logistical difficulties any destructive force might pose.

When an organization is suddenly forced to temporarily relocate because of a disaster, it usually doesn't know ahead of time what the new office space will look like – it might be a school gym, a church, an empty warehouse or even a barge floating on a river. But the organization probably won't be able to alter the temporary space, nor in all likelihood even want to incur the expense of doing so. During such crisis there generally isn't time to pull a lot of telephone and LAN cables to enable even the vague appearance of a traditional corporate IT communications infrastructure. Thus, an organization's communications network needs to be easy to set up and capable of working in a wide variety of building spaces.

The key requirements for a disaster recovery communications network are mobility, reliability, security and, finally, ease-of-use for both the IT staff and the end users. Such a network also must have the ability to provide six to eight hours of continuous battery operation (via battery swaps, if necessary) in the event of rolling power outages, which are common in the aftermath of a disaster. In addition, the system needs to be capable of locally maintaining data and also transmitting that data to a repository of some sort located out of harm's way.

From a communications standpoint, wireless is the optimal choice for keeping a handle on things during a disaster-related crisis. It can communicate across large areas and can transmit (if thoughtfully planned) through walls and other obstacles. Remember that corporate staff will be, in all probability, mobile – scurrying about, dealing with matters outside their normal job description in order to get things back to some semblance of normalcy.

Since simple set-up and ease of use are the hallmarks of any good disaster-based network system, a handheld portable system may be the answer for outfitting many of the corporate staff so they can quickly resume their daily chores. It could be a system much like Dauphin Technology Inc.'s (Palatine. IL – 847-358-4406, www.dauphintech.com) mobile, handheld, pen-based computers, as well as other equipment capable of incorporating a PCMCIA ("PC Card") wireless LAN card (more on these devices later in this chapter).

For example, you will need a few portable printers, such a Canon's (Lake Success, NY – 516-488-6700, www.canon.com) portable bubble jet printers. These little printers are fully equipped with desktop-class capabilities, including Canon's Drop Modulation Technology (which provides both fine image quality and high-speed output), Image Data Correction capability (for automatic correction of low-resolution Web data at retrieval for printing), and IrDA infrared interface (indispensable for a mobile printer), full-scale photo-processing and scanner capabilities.

There's also the Hewlett-Packard (Palo Alto, CA - 650-857-1501, www.hp.com) DeskJet 350 with an optional IrDA-compliant infrared adapter. The DeskJet 350 has a built-in NiMH battery that will print 130 pages of black text in the normal mode and somewhere around 20 to 30 pages in color before you need to recharge the battery. The battery takes about one hour and 40 minutes to recharge and charges automatically when the power supply is connected to an electrical outlet. Add the optional sheet feeder that holds 30 sheets at a time and you're in business. The DeskJet 350 accepts plain paper, transparencies, labels and more in letter, legal and executive sheet sizes.

One or more laptop computers equipped with wireless LAN cards will also be needed to act as the network servers and a few PCMCIA hard disk drives for data backup and doing the initial system load. Then all you need is to tie together the restored backup data sitting safely in a remote location and the staff can begin to once again respond to customers' and clients' needs.

The beauty of this emergency set-up is that the handhelds can run the same application software and can operate, along with laptops, as standard workstations when connected to the wireless LAN. In addition, any notebook or handheld PC can be quickly configured as a server, should a critical hardware failure occur.

A PCMCIA-based wireless solution allows users to easily interchange cards, including battery backup; and making future upgrades by swapping out PCMCIA cards is inexpensive and easy, compared to replacing major components. The ability to quickly add new components, such as notebooks, printers, and handhelds is critical. With a wireless LAN, adding a workstation to the network is only a matter of copying some configuration files onto the new device from a floppy disk or PCMCIA hard drive, plugging in a PCMCIA wireless LAN card and bringing the new station within radio range of the rest of the network.

There are so many laptops that I'm not going to give any specific recommendations. Many times in a disaster situation, you will simply go with what's available. However, portable hard disks don't quite fall into that category – yet. So, I will relent and give the reader a couple of recommendations:

Calluna Technology Ltd.'s (Methil, Fife, Scotland – +44(0)-1592-630810, www.calluna.com) 1040 MB Callunacard offers a host of features that can supply a temporary corporate network with the most advanced performance available in supplemental laptop storage. At only 3 ounces and 10.5mm thick, the card provides a remarkable 1 GB of storage capacity, and also allows high-speed back-up and data transfer to any device equipped with a Type III PC card socket. Requiring no complicated installation procedures or configuration, the card is both easy-to-use and free of the need for power converters or messy cable connections.

Or perhaps Maxtor Corporation's (Milpitas, CA – 408-894-5000, www.maxtor.com) portable hard drive – offered in 40 GB and 120 GB sizes, it's ideal for users who demand maximum capacity and great performance. With data transfer rates at up to 40 times faster than USB 1.1, the USB 2.0 interface is backward compatible with virtually all existing USB ports. Another Maxtor product you might find desirable is the 3000XT. This 160 GB hard drive is specifically designed for tremendous capacity and the high performance demanded by a corporate network. But, to use this device, you need a 1394 (i.LINK or FireWire) interface. Still, the 3000XT can help you run applications, download files from a central corporate data location, and backup files on the temporary network.

For more information on portable hard drives and PCMCIA drives, just plug these terms into your favorite search engine, you'll end up with enough product data to keep you busy for days.

WIRELESS LANS – PRACTICAL AT LAST

We've alluded briefly to how laptop-equipped wireless LANS can help you cope with disaster and keep you and your business up and running during a crisis. Let's now take a closer look at the Wireless LAN (WLAN), it's technology, benefits and disadvantages.

The Wireless LAN, once considered an exotic, before-its-time technology, is finally going mainstream. In both the home and the enterprise, setting one up has never been easier. One of the "ultimate temptations" for any experienced IT person (or even an inexperienced SOHO owner) is to set up a WLAN. Just imagine the freedom of wireless data connectivity (at least when you're not using such a network to counteract the effects of a disaster): the ultimate mobility for laptops, printers and now PDAs, which has the side benefit of both reducing and simplifying the administrative overhead of equipment "moves, adds, and changes." There's also no need to pull cable through walls and ceilings while hoping and praying that nothing becomes damaged in the process.

From Fantasy to Reality

In the past, WLANs were a hard sell simply because they were based on proprietary technology and didn't provide much practical bandwidth. Since 1997, however, a family of wireless specifications, grouped under what's referred to as 802.11, has undergone refinement by the Institute of Electrical and Electronics Engineers (IEEE) along with various manufacturers.

There are three main 802.11 transmission specifications: 802.11a, b and c. All of them use the Ethernet protocol and, best of all: any LAN application, network operating system or protocol, including TCP/IP, will run on 802.11 compliant WLANs as easily as they run over the Ethernet.

In the US, the most popular, "universal" WLAN standard at the moment is 802.11b, now called "Wi-Fi." It operates in the 2.5 GHz frequency range and can transmit up to 11 Mbps. Wi-Fi certification allows for interoperability, so if necessary you should be able to integrate different manufacturers' products into one system.

The Business Case For WLANs

The price of wireless equipment has dropped considerably. When your crisis has passed and you are faced with re-establishing a cabled environment and a conventional LAN set-up, take a moment and consider that a WLAN is now also a good idea for "peacetime" business use as well. Most WLANs designed for larger businesses can easily function as an extension to pre-existing wired LANs, transforming the conventional LAN into a sort of "wired backbone" (as we'll see later).

A recent study of 300 companies having 100 or more employees using WLANs entitled "Wireless LAN Benefits Study" by NOP World-Technology (London, UK – +44(0)-20-7890-9000 www.nop.co.uk), reveals that using wireless LANs allowed end-users to stay connected 1.75 more hours more each day, amounting to a time savings of 70 minutes for the average user, increasing their productivity by as much as 22 percent.

The study shows that WLANs also have an impact on return on investment (ROI), with organizations saving an average of $164,000 annually on cabling costs and labor, more than 3.5 times the amount IT staff had anticipated. There are also, undefined benefits derived from increased productivity. Through a combination of cost savings and productivity gains, per employee savings are estimated at around $7,550 annually.

The Components

The three main components of a WLAN are: WLAN client adapters, Access Points (APs), and Outdoor LAN bridges or routers.

WLAN Client Adapters

Adapters – complete with adorable little antennas – will get your laptop, handheld PC, printer, PocketPC, PDA or other device onto the WLAN. They are Network Interface Cards made essentially to the same specs as their wired brethren. Since mobile laptops are a natural for use on a WLAN, 802.11b client adapters tend to be PCMCIA or "PC Cards," as they're called these days.

All wireless LAN adapter cards have an "ad hoc" mode enabling them to communicate directly to

Peer-to-Peer WLAN

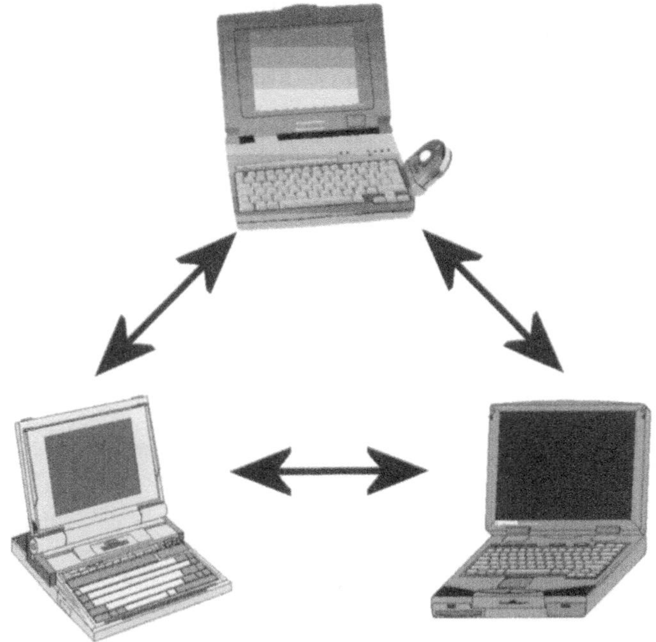

each other. So, if you're building a small, impromptu WLAN (such as in a crisis situation), you can quickly set up a peer-to-peer wireless network.

There are also wireless-enabled PCI cards for the desktop computer. If you are using desktops either in a crisis situation or that aren't within reach of a walljack (when used in the typical corporate environment) these devices are a godsend. Some allow for a PCMCIA/PC Card to plug in to a supporting baseboard that sits in a PCI slot, such as the Wireless PCI Adapter (Model WDT11) from Linksys (Irvine, CA – 949-261-1288, www.linksys.com); while others have been specifically designed as wireless PCI cards such as the Linksys Instant Wireless PCI Card (Model WMP11).

And, if you are using desktop PCs that don't have a spare PCI slot, you can use a Wireless USB adapter, such as the ORiNOCO USB Client Adapter-Gold from Agere Systems (Santa Clara, CA – 800-372-2447, www.orinocowireless.com), the Linksys Wireless USB Adapter Model WUSB11, the Wireless USB Client Model WLI-USB-L11G from Buffalo Technology Inc. (Austin, TX – 800-456-9799, www.buffalotech.com), the USB Client from Avaya (Basking Ridge, NJ – 908-953-6000, www.avaya.com) and the USB Wireless Adapter Model DWL120 from D-Link (Irving, CA – 949-788-0805, www.dlink.com). These all cost between $95 and $150.

The new Wireless Networker from Symbol Technologies (Holtsville, NY – 631-738-2400, www.symbol.com) is an 802.11b CompactFlash I/II card for PocketPCs . It can also be used in the Casio E-125, Compaq iPAQ, and HP Jornada 520/540.

Another interesting device in the same vein is Linksys' Instant Wireless Network CF Card (Model

Linksys Wireless LAN Adapters

CompactFlash **PCMCIA / "PC Card"**

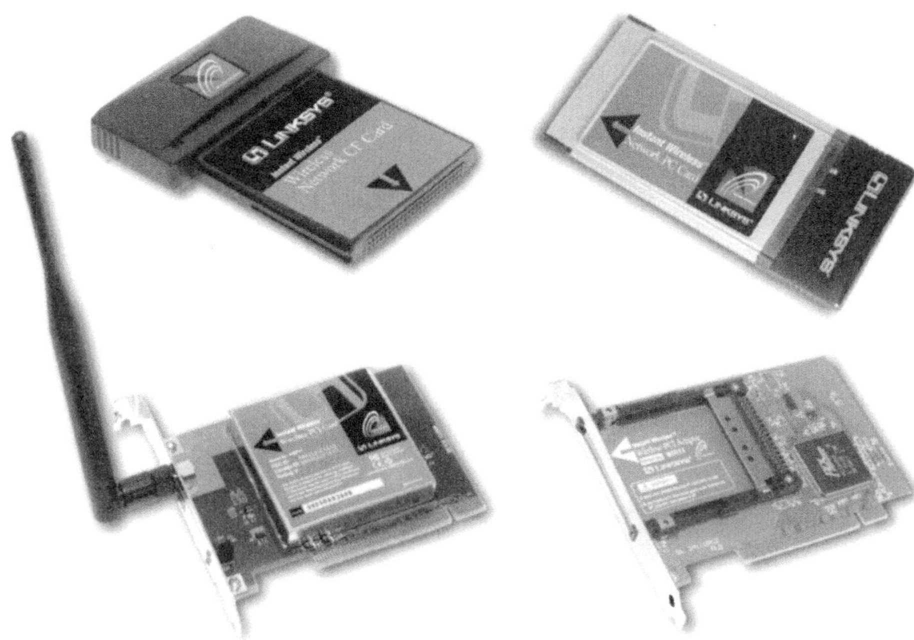

Wireless PCI Card **PCI Adapter (PC Card plugs into this)**

WCF11). It's a Type II CompactFlash card that connects directly to your PDA. With it, your little PDA can now send and receive data at speeds up to 11 Mbps at distances of up to 1,500 feet. Compatible with Windows CE 2.1 and 3.0, it can also be quickly configured from your PC. It should be available by the time you read this.

Access Points (APs)

Access Points (APs) are radio transceiver devices attached to the wired LAN that take the place of the usual wall outlet data ports. APs act as Ethernet bridges, passing data between the wireless and Ethernet parts of your network transparently, both parts being in the same subnet.

Generally, APs are more than just simple bridges, many tend to be "managed bridges" which means that you can do things such as allow/disallow clients via Media Access Control (MAC) lists, set the AP channel (frequency), control AP access methods such as (Ethernet, Telnet, PPP via serial port, and the Simple Network Management Protocol, or SNMP), and set encryption keys and configure client access controls.

Some APs don't need to hang off of a wired LAN and can communicate with each other directly, via AP-to-AP "wireless network bridging," a feature that comes in handy if you need a signal repeating device to extend your wireless network to places where wired LAN cabling can't go (large open areas like warehouses and retail stores) or if you're building a totally wireless LAN.

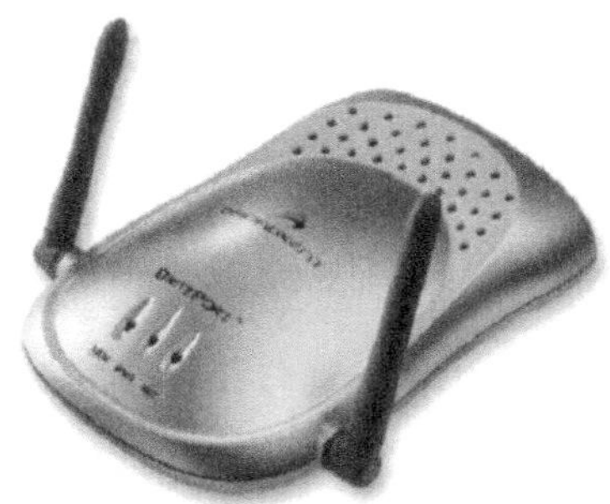

An example of a 11 Mbps 802.11b access point that can be used indoors or out, the BritePort Wireless LAN Access Point from Broadxent (Milpitas, CA — 408-719-5100, www.broadxent.com) lets users roam within the wireless range and has built-in 128-bit encryption.

APs convert a location into something resembling a miniature cellular data network, with each AP acting as a "base station" that covers a "cell" – or in this case, a "microcell" between 100 and 1,000 or more feet in diameter. With WLANs, wireless-enabled laptops and PDAs can roam seamlessly from one microcell to another, as long as the APs are within the same network.

If the location consists of one large space on one floor, you can use a single AP that may not support wireless AP-to-AP bridging but offers a lot of coverage, such as the Wireless LAN Access Point 6000 from 3COM (Santa Clara, CA – 408-326-5000, www.3com.com) which supports 65 users within 100 meters (328 feet). Of course, by adding more access points you tend to decrease network congestion and enlarge the coverage area.

WLANs can support many devices and large physical areas simply by adding access points to boost or extend coverage. By strategically distributing APs around your offices or facility, WLANs can be made simple (one AP) or amazingly complex, boldly going where no wired LAN has gone before. Of course, attention must be given to the distance and obstruction limitations imposed by the physical infrastructure (open floor plan, maze of corridors, reflecting metal walls, big empty warehouse, church with imposing pillars, etc.).

In complex environments one would think that wireless-enabled LAN devices would become confused over which AP to use, but 802.11b's modulation method, known as complementary code keying (CCK), is less susceptible to such "multipath-propagation interference" than the phase-shift keying (PSK) used in the first, low-bandwidth (2 Mbps) version of the 802.11 specification.

Once you and your business have relocated to your new location you'll find that many APs can support "Power over Ethernet" or "Active Ethernet" which sends DC power along with LAN signals over Ethernet cabling (or at least many of them can – check before purchasing). Doing this eliminates the need to add AC outlets and power supplies near APs, so you can now provide networking in hard-to-wire facilities, such as Historic buildings and structures sealed because of asbestos.

Although 802.11b "Wi-Fi" certified equipment is interoperable, many IT people tend to buy their equipment from whatever company they find comfortable to deal with. To make things easy for them, Cisco's (San Jose, CA – 408-526-4000, www.cisco.com) Aironet 340 Series is a comprehensive, integrated family of access points, easy-to-install PC Card, PCI and ISA client adapters, Ethernet clients, as well as line-of-sight outdoor bridges. Client driver support includes Windows 95, 98, Windows NT 4.0, Windows 2000, and Novell Netware.

The Aironet 340 Series Access Point can function as the anchor of a stand-alone wireless network or as a bridge between wireless and wired 10/100 Ethernet networks. Cisco says their Aironet

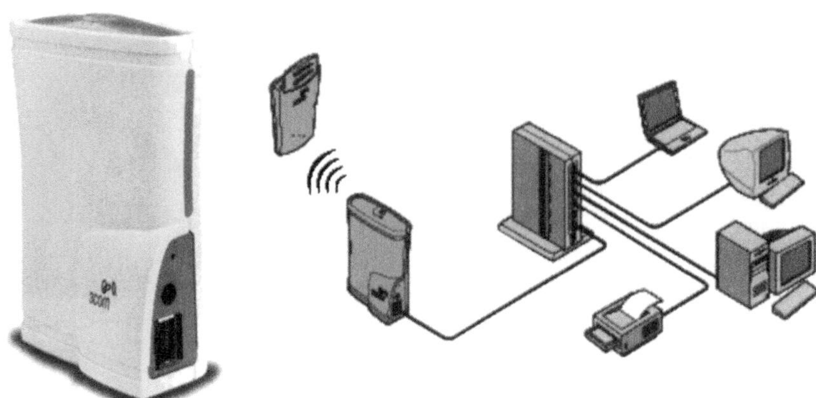

With 3Com's 11 Mbps Wireless LAN Workgroup Bridge, you can first connect as many as four Ethernet-enabled LAN devices to a hub, then connect the hub to the new 3Com 11 Mbps Wireless LAN Workgroup bridge. The bridge works with any Wi-Fi (802.11b) certified access point.

wireless bridge products can do point-to-point or point-to-multipoint connections between buildings as much as 25 miles away, depending on the antenna. Naturally, Cisco offers a wide range of antennas, cable and accessories, enabling customized deployments.

Not to be bested by the competition, 3Com's latest take on the bridging aspects of APs is their new 3Com 11 Mbps Wireless LAN Workgroup Bridge, which was said to be the industry's first wireless workgroup bridge with Wi-Fi certification.

Aside from being a wireless bridge, the 3Com Wireless LAN Workgroup Bridge acts as a sort of aggregator for up to four wired Ethernet devices. Rather than buy a separate Wi-Fi certified adapter, PC Card or PCI card for each LAN device, you can simply connect up to four Ethernet-enabled PCs, printers, 3Com NBX VoIP phones or LAN/Internet appliances to a hub (such as 3Com's own OfficeConnect Dual Speed Hub 5) and then connect the hub to the Workgroup Bridge. The Workgroup Bridge then transmits the signals of all four LAN devices wirelessly to a standard Wi-Fi AP, such as 3Com's own Access Point 6000 mentioned previously, or the 3Com AirConnect 11 Mbps Wireless LAN Access Point (Model 3CRWE74796B) that can serve 63 users within 100 meters (328 feet).

The devices aggregated in this way can even include non-Windows-based computers running Macintosh and Linux operating systems.

Some readers may recognize that the 3Com Wireless LAN Workgroup Bridge is reminiscent of their four-port NJ100 Network Jack, which allows up to four networking devices to be connected to the network via one Ethernet port.

Interestingly, Cisco offers a similar device, the Cisco Aironet 350 Series Workgroup Bridge which can connect eight Ethernet-enabled laptops or other portable computers to a WLAN, linking them (actually, the hub they're connected to) to any Cisco Aironet Access Point (AP) or Wireless Bridge.

Bringing WLANS to the Great Outdoors

When the average IT person thinks of 802.11b they think of wireless LAN-type products for building interiors. Yes, interior APs can be used to bridge separate buildings in close proximity to each other, but there's a whole other set of "b" products – "outdoor LAN bridges" or routers – that work in the Wireless WAN (WWAN), from building to building.

These outdoor-type products are still 802.11b, they use the same frequencies and modulation, but they employ a different "handshaking" protocol between them. If you took a WAN bridge and placed a laptop next to it, they would ignore each other, since the laptop is looking for an access point, and the bridge is looking for other bridges. Of course, you could change the driver and the configuration parameters in one of the devices, which then would allow them to communicate.

With tall (50 to 120 foot) masts and focussed Yagi or parabolic-like antennas, even 802.11b WLAN bridges can achieve a line-of-sight transmission distance of about 18 miles over perfectly level Iowa corn fields (Cisco says they can do 25). The practical limit is about five miles.

The 3Com 11 Mbps Wireless LAN Building-to-Building Bridge (Model 3CRWE90096A), for example, will give you point-to-point and point-to-multipoint connectivity options. You can also choose the antenna option that best fits your needs: You can achieve transmissions up to 1,300 meters (3,609 feet) with omni-directional antennas, and 4,100 meters (14,435 feet) using sector panel antennas.

The bridge supports both 40-bit and 128-bit WEP security. The wireless bridge is also transparent to VPN protocols, using a special encapsulation protocol to avoid conflicts with other tunneling protocols.

Security or Lack Thereof

There's been some hysteria over WLAN security issues, some justified – some not. First of all, these security anecdotes relate to "interior" LAN products as opposed to WAN versions of those products. Second, since various forms of encryption may decrease network performance by 20%, many manufacturers ship their products with this option defaulted to "off". Your average IT guy then installs the equipment out of the box and discovers he (or anyone else lurking around) can telework from the company parking lot.

In any case, "pretty good" security use to involved simple link level security, such as provided by WEP (Wired Equivalent Privacy), a protocol that was part of the original 802.11b formulation and adds encryption to prevent others from seeing your data as it's transmitted. WEP uses RC4 encryption and 40 bit keys that must match between the mobile device and the AP.

Continued attacks and deficiencies found in WEP, however, have encouraged companies to move up to a "better" security system. Wireless LANs *can* be made relatively secure with the use of encryption technology and stringent management to ensure that the system isn't broadcasting data to the world at large. But to up the security level, the wireless devices used to access the network need to be used under an umbrella of access control as well as security procedures that ensure the company's data is secure as it traverses public networks and RF spectrum outside the corporate firewalls.

The IEEE 802.11 Task Group I (IEEE 802.11i) is currently working on extensions to WEP for incorporation within a future version of the standard. The committee's 802.1X (or 802.1.x) standard provides a scalable, centralized authentication framework for 802-based LANs. It provides user-based authentication and automatically creates and distributes new 128-bit encryption keys automatically at set intervals. End users with 802.1X friendly Windows XP clients, for example, can be authenticated by a RADIUS server and supplied with a WEP security key.

The standard is flexible enough to allow multiple authentication algorithms, and because it is an open standard, multiple vendors can innovate and offer enhancements.

Still the IEEE 802.11 Task Group is rapidly making progress on three new security improvements for "legacy equipment." Known collectively as the Temporal Key Integrity Protocol (TKIP), these measures are intended to quickly fill security holes left by WEP.

Cisco beefed up its Aironet WLAN product security with LEAP (Lightweight Extensible Authentication Protocol). When you log-in a LEAP system, all clients have unique keys and clients dynamically generate a new WEP key instead of using a static key.

Other companies, such as Avaya, support all of this but believe that the best way to secure traffic entering the network from a wireless access point is to use the security and policy enforcement found in an IPSec-based VPN, which they also support in their products and in their VPNremote client desktop software.

One can use a VPN with other manufacturers' products, though this demands the addition of a designated VPN server to the system.

Encryption

The US government's Federal Emergency Management Agency (FEMA) deploys its fleet of 802.11b wireless laptops armed with MAC (Media Access Control). The FEMA guys, after experimenting with encryption, strongly feel that MAC, instead of standard encryption, can be used to prevent unauthorized access to their wireless data. On a local area network (LAN) or other network, the MAC address is your computer's unique hardware number (see the Security chapter for a bit more on MAC). Whether MAC can serve as the basis of a "perfect" wireless security system remains to be seen.

Fixed Wireless or "Wireless Broadband"

In the aftermath of the September attack, Pentagon officials opted to rely on a wireless LAN running over a "fixed wireless" network due to its increased security capabilities.

"Fixed-wireless" systems have been around for a long time. Point-to-point microwave connections have long been used for voice and data communications, such as in "backhaul" networks operated by phone companies, cable TV companies, and paging companies. Fixed wireless is essentially wireless broadband, and the technology uses wireless spectrum frequencies at frequencies that differ from mobile wireless systems, ranging 1 GHz to as high as 40 GHz. The higher the frequency, the smaller and cheaper the antennas that can be used. Also, fixed, wireless broadband carrier-based services involve multiplexing many data streams onto a single radio-carrier signal.

Conventional fixed-wireless systems are mounted at some elevation, such as on a rooftop, using a large dish antenna, and provides local phone service and high-speed Internet access primarily for business customers. The terms wireless broadband and broadband wireless tend to be used interchangeably, but the term is also sometimes applied to privately deployed wireless networks.

Pentagon IT officials with the help of several companies, including Avaya and AT&T, set up a fixed wireless network on the grounds of the Pentagon following September 11, which connects to a service in Arlington, Virginia. This network allows the Pentagon's staff to send and receive e-mail and maintain access to critical government records and databases.

Within three days from the time the germs of the idea began, line-of-sight antennas were installed in Arlington (where access was available to the necessary networks) and at the Pentagon (where access was needed). Then, basically, all that was necessary was to set up the network – a little hardware here, some software there, and the staff in the world famous octagon building was back in business.

➤ A Man, A Van, A Plan

Paul Zucker works for integrator RMS Business Systems (Buffalo Grove, IL – 847-215-1661, www.rmsbus.com) but he's also a volunteer with Buffalo Grove, Illinois, Emergency Services, a volunteer organization that provides communications support to the village's local police and the fire departments.

Zucker and his company had noticed that when a real emergency takes place in the village and the emergency van pulls up, nobody can make cellular phone calls from the scene, since this kind of network voice traffic is the first to become overloaded by panicky citizens.

Thus was born the idea of putting the van on the village's municipal wireless WAN, then making Voice-over-IP calls (as well as video and data transmissions) over the WWAN back to headquarters and then over the PSTN.

"Buffalo Grove has antennas all over the place," says Zucker, "but in the center of town they have a pretty large tower at their village hall, and we put a 53-foot pneumatic mast with a dish on the command van, and we can transmit up to about three miles."

"We use Cisco Aironet 802.11b equipment at both ends." says Zucker. "By virtue of connecting this van back to our wired network using this link, we now can provide some very nice services to the police and the fire department. They can access their files on servers that are resident in their stations and give them access to the Internet if they need it, and give them access to e-mail if they need it, right out at the scene of the incident."

"We then realized that since we have this nice LAN connection, we can also provide VoIP over RF. For this we're using a pair of Cisco routers and a PBX in the van. Cisco equipment, besides routing calls over IP, can connect to the PSTN with a VIC (voice interface card), to a PBX with a WIC (WAN interface card) or to an ISP via PPP over ISDN. These options depend on where you route your calls."

"The Cisco wireless bridge is attached to a hub in the van that provides LAN services to all the van's devices, such as a PC, a videoserver and a Cisco router. This router doesn't do any routing – a PBX in the van connects to the router's WIC card, so the router is simply acting as an analog gateway for the voice network."

"So when I pick up the phone in the truck the dialtone I'm getting is coming from Village Hall over the RF link through the phone system. The packetized voice goes over the RF link back to village Hall, where the opposite happens, there's another router that takes those TCP/IP packets, converts it back into analog using a VIC card which is in turn connected to a POTS line."

"We can send voice, video and data. We can send a live video signal from our truck. We digitize it, turn it into TCP/IP packets, send it out over the wireless WAN, and it becomes available on any PC in the village network. So if the Police Chief is sitting at his desk at the Police Station and he wants to see what's going on at the emergency scene, he just has to go to his Web browser, punch in the URL address of the truck's intranet server and boom, he's looking at live video from our van," say Zucker.

Future Migration

802.11a is a form of wireless ATM that runs in the relatively interference-free 5 to 6 gHz frequency band, has lower power consumption and can transfer data at an impressive theoretical maximum of 54 Mbps. It also supports eight non-overlapping channels, yielding 13 times the capacity of its more popular brother, 802.11b. Because of the frequency difference, however, it's not made easily compatible with other wireless Ethernet technologies.

802.11a is, however, championed by companies such as Avaya. Their AP-3 access point, designed and built in cooperation with Agere Systems, has a dual CardBus architecture that allows for the cohabitation of 2.5 GHz and 5 GHz radio cards in the same box, which means you can slowly migrate your WLAN from 802.11b to higher bandwidth 802.11a by changing client cards when convenient.

The AP-3 has Spectralink VoIP support and can be remotely managed via a Web browser or standard SNMP management tools. The AP-3 also includes a new Wireless Distribution System (WDS) that enables a single radio in the AP-III to act as a repeater station or wireless bridge to expand a network across a facility and between buildings. The AP-3 is also supported by Windows XP.

On the horizon is 802.11g which operates at 2.4 GHz, supports speeds ranging from 11 to 54 Mbps and is backwardly compatible with 802.11b. But many in the industry now feel that 802.11a has too great a head start and will end up as the "winner" for high bandwidth WLANs.

The business community, as a whole, will eventually build a wireless infrastructure to extend its reach beyond the four walls of the enterprise. But businesses shouldn't limit themselves to a single

➤ Specialized Stuff

Symbol Technologies (Holtsville, NY — 631-738-2400, www.symbol.com) has staked out a major claim in the WLAN world, thanks to their unique vision of a single handheld device that literally does/has everything: The SPT 1800 Series of pen-notepad computers, which is a Palm OS platform that has a one-dimensional bar code scanner, and WLAN or WWAN 802.11b connectivity. A WWAN GSM version even has cellular voice capabilities.

With a SPT 1800, as shipments are unloaded, warehouse management can easily record new inventory levels to networked host systems. Hospital physicians and caregivers can have the latest patient data at hand, and law enforcement officers can scan and verify vehicle registration or drivers' licenses.

In retail, when used with a magnetic stripe reader and portable printer, the SPT 1800 becomes a portable point-of-sale (POS) device. For scan-intensive tasks, an attachable pistol grip handle converts the SPT 1800 into a more ergonomic design. This "pistol" configuration is being used by four ski lift operators at the Park City Mountain Resort near Park City, Utah, a ski resort that hosted the 2002 Winter Olympics. Skiers wear ski lift tickets having bar codes. The lift operator scans the bar code and the device transmits via the WWAN to a Windows 2000 SQL database to verify the ticket.

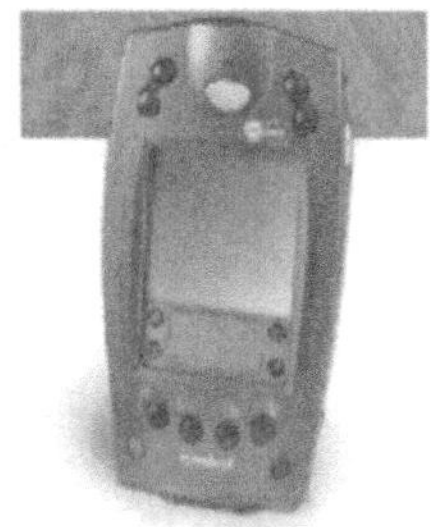

Recently Symbol announced that it will team with IBM to integrate mobile devices with IBM e-business applications and services for a variety of industrial applications, including field service automation, logistics supply chain, consumer goods delivery and postal and package delivery.

Symbol Technologies' SPT 1800 is a rugged Palm OS platform that combines computing power and bar code scanning with WLAN / WWAN connectivity and even GSM telephony.

wireless project; rather, they should go for a broader access initiative that takes into account such things as security, reliability, portability, ease of use and integration. In other words, a multi-channel delivery system that can handle *all* the ways people access a company.

WIRELESS VOICE SERVICES

Wireless voice networks can meet the challenge of just about any disaster. For example, cellular telephone providers installed temporary towers on wheels in New York and Washington to replace permanent towers that were damaged in the September terrorist attacks. The wireless voice providers can also relatively easily reroute traffic and redirect antennas in a stricken area to improve coverage.

Just as the service providers had towers on wheels that could take over when their more permanent brethren came tumbling down, the business community needs to provision their critical people with redundant communication methods. Even when thinking wireless, stay the redundancy course. Provision your employees with a two-way e-mail device, such as Research in Motion's (Waterloo, Ontario, Canada – 519-888-7465, www.researchinmotion.com) BlackBerry handheld wireless devices – most work on wireless networks that are separate from the typical cellular network – and a cellular phone. In this way your communications conduit now enjoys multiple redundancy, with landlines *and* cellular devices using different access networks.

IP Phones – Wireless?

Although it was not at the epicenter of the September 11 disaster, the telephone network for the New York State courthouse complex (located in downtown New York City) lost many of its T1 and frame relay circuits, as a result of the damage to Verizon's CO adjacent the World Trade Center buildings. These data lines are what connect the state court complex to courthouses around the New York region.

The innovative courthouse IT staff installed a fixed wireless connection from one of the court's buildings to a data center unaffected by the attacks. That center has an OC-3 link connecting it, in turn, to New York's upstate court system. IP phones were then installed. All of this, the fixed wireless connection, the IP phones and the data access was made possible by using Nortel equipment. The phones, for example, connect to an IP-PBX box situated in the previously mentioned data center (which is powered by what Nortel euphemistically calls the "Succession Communication Server for Enterprise"). Then, Nortel Voice-over-IP gateways located in that same data center and in another building in upstate New York provide a means for callers to access the public telephone network.

Evolution of the Wireless Voice

Before including "cellular" in a disaster recovery or business continuity plan, you should under-

➤ Not Strictly Wireless but...

If your telephone circuits get themselves fried, you can, in an emergency, take a walk down the street and see if your neighbors are feeling "neighborly." Many times, the events that brought your circuits down didn't affect your next door neighbor's telephone service, with the help of point-to-point laser transmission devices, a friendly link can be established between your damaged building and the one belonging to your fortunate, unscathed neighbor. This allows you to get your voice service up and running as fast as possible.

stand the technology, where it is now and what direction it's headed. Let's first take a look at the wireless network ecosystem, as it exists today. The evolution of wireless networks looks deceptively simple. You see numbers such as 2G, 2.5G and 3G, and you think it's all quite straightforward, as simple as 1, 2, 3. Narrowband begat Broadband. Circuits begat Packets.

In fact, developments in the wireless industry resemble a textbook description of how to write a screenplay: Give all of your characters – that is, wireless technologies such as CDMA, GSM, iDEN and TDMA – wildly different subplots and have them all converge at the shoot-em-up climax near the end of the film. What comes to mind is the madcap dash for the buried loot in ***It's a Mad, Mad, Mad, Mad, World.***

So break out the popcorn and Milk Duds, dear readers, as we unravel the Strange Case of Wireless, a true tale of convergence.

By The Numbers

Before we start, keep in mind that there are two main paths to the Promised Land of 3G – one involving GSM and the other CDMA. Both (hopefully) converge at the end of the story.

1G

The first generation consisted of "analog" mobile networks such as the US-based Advanced Mobile Phone System (AMPS), running in the 800 mHz frequency band. AMPS provides only one voice conversation per channel, uses analog Frequency Modulation (FM) to transmit unencrypted voice over the air and doesn't offer multiple services as do more advanced 2G services.

2G

A 2G phone uses digital encoding, allows for roaming over semi-global regions, and gives you data services as well as circuit-switched voice. The data services are also circuit-switched, so using them is pretty much like making a call with a low-speed analog modem.

The Japanese 2G Pacific Digital Communications (PDC) system failed to take the Far East by storm, but the Global System for Mobile Communications (GSM) conquered Europe, Asia, Australia and New Zealand, and is now spreading through North America too, thanks to its aggressive marketing by VoiceStream. GSM operates on the 900 mHz and 1.8 gHz bands worldwide except for the Americas where it occupies the 1.9 gHz band.

GSM almost singlehandedly brought the words "smart card" (or "smartcard") back into our vocabulary with its Subscriber Identification Module (SIM) – a credit card sized piece of plastic with an embedded microprocessor that contains user service information for accessing the GSM network.

GSM is easily the most popular mobile standard in the world. As of July 2001 there were 565 million GSM subscribers in 400 networks in 171 countries worldwide (as reported by EMC World's Cellular Database). Outside of Japan, GSM accounted for 78% of all new mobile subscriptions during the first half of 2001.

Other 2G systems include cdmaOne which uses the same frequency bands as AMPS, and the Integrated Digital Enhanced Network (iDEN) launched by Motorola in 1994 that runs in the 800 mHz, 900 mHz and 1.5 gHz bands.

Qualcomm owns most of the CDMA-related patents. Two major cdmaOne carriers are Verizon and Sprint.

Unfortunately, typical data transmission rates for 2G networks range between 9.6 and 14.4 Kbps. This isn't that great for Web browsing and multimedia applications, but okay for using Short Messaging Services (SMS) where subscribers can send and receive short text messages via their cellular phone and two-way paging devices. Things improved in November 1999 when cdmaOne IS-95B appeared, which allows for 56 Kbps data bandwidth instead of 14.4 Kbps.

2.5G

The major improvement with 2.5G wireless is the introduction of packet-switched data services, which means that when you use a data service you don't occupy the whole channel for the whole call (but the voice channel is still circuit-switched).

On the GSM path, the General Packet Radio Service (GPRS) is a GSM-based packet data protocol that can be configured to gobble up all eight timeslots that exist in a GSM channel. It can support a 115 Kbps data rate, though 50-60 Kbps is more likely in practice, especially since the packets must contend for the same bandwidth as GSM circuit-switched voice, and providers will tweak the bandwidth based on the number of subscribers that can be attached as they try to find a profitable mix of number-of-users versus bandwidth-per-user.

Hughes Software Systems (Germantown, MD – 240-453-2498, www.hssworld.com), employs 1,700 programmers in India who work on ready-to-go GPRS solutions for carriers who realize that 3G is still far away and want to grab some of the GPRS action. Ajay Kumar Gupta, VP and general manager for US software operations says, "Instead of deploying GPRS to everybody (100,000+ subscriber base), some providers are zeroing in on business campuses and more lucrative areas of likely GPRS users (10,000 subscriber base)."

For even more bandwidth, GPRS can be upgraded to use a modulation technique called EDGE, which stands for Enhanced Data-rates for GSM (or Global) Evolution. EDGE allows GSM operators to use existing GSM radio bands to increase the data rates within GPRS' 200 kHz carrier bandwidth to a theoretical maximum of 384 Kbps, with a bit-rate of 48 Kbps per timeslot and up to 69.2 Kbps per timeslot in good radio conditions. Existing cell plans can remain intact and there is little investment or risk involved in the upgrade.

AT&T has announced it will move its entire network to GSM/GPRS and thence to EDGE. VoiceStream is also converting to GPRS, and Cingular Wireless has announced it will take the GPRS/EDGE route too. Cingular originally planned to launch GPRS in Seattle, where AT&T recently run test trials of its GPRS service using Nokia phones. At the moment, however, Cingular is said to be doing trials in the San Francisco Bay Area, to be followed by rollouts in Nevada and the Carolinas.

The GPRS equivalent in the CDMA world is CDMA2000 1XRTT. Instead of one code for access, this uses five codes within the 1.25 mHz radio channel as well as more sophisticated modulation to boost bandwidth for individual users up to about 144 Kbps. It also involves a new phone and demands a change to some of the base station equipment, doubling voice network capacity and allowing data to be packetized and sent without the need for a circuit to be established.

By mid-2002, the entire Sprint PCS all-digital nationwide network will offer CDMA2000 1XRTT. It's the first phase in what they refer to as a "3G" service. Even now Sprint will sell you a silver SCP-5000 Sanyo phone with a full color display for $399. Like Japanese carriers, Sprint will charge for screen savers and ring tones.

And although they currently offer Motorola iDEN phones and haven't officially made an announcement, many in the industry feel that Nextel will also install a CDMA2000 1XRTT overlay on its nationwide integrated digital enhanced network.

CDMA2000 1XRTT has two new descendants, CDMA2000 1xEV-DO and CDMA2000 1xEV-DV.

CMDA2000 1xEV-DO (Evolution-Data Optimized) is about to be deployed in bandwidth-crazed South Korea. Faster than CDMA2000 1xRTT, 1xEV-DO is not officially 3G but has 3G-like bandwidth, supporting fixed and mobile applications at 1.2 to 800 Kbps on average and 2.5 Mbps peak.

In the US, Verizon may do 1xEV-DO deployments at the end of 2002, as will an operator in Japan. Airvana (Chelmsford, MA – 866-344-7437, www.airvananet.com) specializes in building 1xEV-DO end-to-end IP infrastructures. Its president and CEO, Randy Battat tells the author that: "In North America certain regional operators are looking to use 1XEV-DO to come up with a fixed wireless technology that delivers Internet access at about 200 to 250 Kbps without line-of-sight transmission and without a truck roll. Semi-urban areas with little DSL deployment are candidates."

3G

3G is the long-sought juncture at which the CDMA and GSM evolutionary paths converge into a single official, globally-roamable system. Many European carriers wanted 3G to be based on GSM, while Americans pushed for descendants of cdmaOne. The CDMA camp, however, suffered from a schism. The European Telecommunications Standards Institute (ETSI) and the Japanese operator NTT DoCoMo wanted Ericsson's Wideband (W-CDMA) to be the basis for 3G, which demands a large swath of new spectrum. Qualcomm and the Korean carriers, however, wanted backward compatibility and found they could achieve the same objective as Ericsson's W-CDMA by simply aggregating the existing codes and channels, and so Qualcomm promoted the series of incremental upgrades leading to the various flavors of cdma2000, any of which they hoped would be adopted as an official 3G system.

Finally, in 1998, European and Japanese standards bodies agreed to create yet another standards group, the frightfully official and definitive Third Generation Partnership Program (3GPP). 3GPP will bring forth a common, global wireless standard called the Universal Mobile Telephone Standard (UMTS) which should appear by 2005. To keep the Americans in the loop, a separate Third Generation Partnership Project Number 2 (3GPP-2) was founded

UMTS will have circuit-switched voice and packet-switched data. It's a combination of some GSM technology and the W-CDMA "air" (radio) interface. 3G networks must be able to transmit wireless data at 144 Kbps at mobile user speeds, 384 Kbps at pedestrian user speeds and an impressive 2 Mbps in fixed locations (home and office).

This flexibility derives from UMTS' two complementary radio access modes: Frequency-Division Duplex (FDD) which offers full mobility and symmetric traffic, and Time-Division Duplex (TDD) which offers limited (indoor) mobility and handles asymmetric traffic such as Web browsing.

Ultimately, UMTS itself will evolve into an "all IP" or "end-to-end IP" network, or at least a network where IP is used as much as possible.

UMTS is Europe's answer to an earlier (and ongoing) project, the ITU-T's International Mobile Telecommunications 2000 (IMT-2000) that stakes out frequencies for future use. Amusingly, the independently-minded allies of US and Japan refer to 3G as IMT-2000 (not UMTS), even though the

Europeans, to keep the Americans in the loop, established a separate Third Generation Partnership Project Number 2 (3GPP-2) body.

Going out on a limb for 3G

Once the 3G frequency band was identified, European operators went on a crazy bidding war to grab a piece of the spectrum – UK operators alone squandered over $35 billion, while Sonera, Finland's largest carrier, spent $7.7 billion for a license in Germany – and no money was left for R&D or actual construction of the 3G infrastructure.

In the US, just allocating the frequencies is a problem. The United States Army and Navy, which occupy some of the UMTS band, refuse to move to other frequencies, citing security risks. The United States Air Force, however, will be more than happy to *sell* its portion of the UMTS spectrum for $3.2 billion.

Despite these roadblocks, UMTS is being developed by a consortium that includes Alcatel, Sony, Motorola, Italtel, Bosch, Nortel, Siemens and others. When their work is finished you'll be able to roam around the world using an UMTS compliant 3G phone or a W-CDMA, EDGE or cdma2000 phone.

Why Use 3G?

After all is said and done, however, it may turn out that the standard approach to presenting wireless data solutions on mobile devices is wrong. Simply porting the PC's browser-based wired Internet paradigm to a mobile device may not work, since the PC with its large screen and keyboard is far different than a PDA or smart phone. Also, providing high-speed mobile wireless Internet access alone doesn't appear to excite users – as can be seen from the collapse of Metricom and it's Riccochet service.

Multimodality Saves the Day?

Part of the problem may not be so much the bandwidth as the interface. To make 2.5G and 3G interfaces truly "friendly" an exciting new technology has appeared, called multimodal capability. It allows for simultaneous voice and data over wireless and dovetails neatly with locality and "presence" services.

For example, standing in the middle of an unfamiliar town, you could find a hospital by using speech-rec to verbally ask a wireless device for directions to one from your current location. A map could then appear on your color screen and directions to the hospital could be read to you via text-to-speech technology. Or you could elect to receive an SMS message with text directions along with a graphic, or use a touchscreen to trigger a display of each step of the directions as you walk along.

Whereas today's technology demands that you finish a voice session to enter a data mode, multimodal, as its name implies, enables you to input data using speech and/or a keyboard, keypad, mouse or stylus, and retrieve data as audio, synthesized speech, plain text, streaming MPEG4 (or equivalent) video and/or graphics. Modes can be used independently or concurrently.

Multimodality, therefore, allows you to relate to your phone more as an intelligent "personality" than just a phone, and it may just make wireless data delivery a mainstream phenomenon once and for all. This is a "good thing" when you're operating in disaster mode. After all, the Internet was originally conceived by the RAND Corporation as a distributed communication networks for military command and control. Today the 'Net can "take a beating and keep on ticking" and, thus, serves as a mechanism for information dissemination during a crisis.

Cisco, Comverse, Intel, Microsoft, Philips and SpeechWorks recently formed the Speech Application Language Tags (SALT) Forum to create a formal specification to accelerate standards for multimodal applications (www.saltforum.com). SALT is developing a device and network independent *de facto* standard that will rely on extensions to HTML, XHTML and XML.

Action Engine Takes Action

Still, the mobile phone is a small device that's not easy to use when you're walking along. It would be helpful if the device itself had more intelligence or ran more flexible, intelligent, "multidisciplinary" applications.

A fascinating platform-independent technology that's right along these lines is Action Engine from Action Engine Corp. (Redmond, WA – 425-498-1500, www.actionengine.com).

Action Engine's Mobile Services Framework is composed of a "Smart Client / Smart Server" software system that enables a company or a service provider to build intelligent agents to simplify and speed up wireless informational, transactional and enterprise services. MyCasio.com, which enables the Casio BE-300 to access rich media content, as well as business and personal services, was built using Action Engine's framework.

Action Engine's Smart Servers take your request (an XML packet sent from the Smart Client) and can call upon enterprise data like a CRM sales database, informational data like weather, or transactional services like travel booking via a Service Provider Plug-in Layer. This layer can integrate any data source, including existing Web sites (via "scraping"), XML interfaces, SQL and other databases, and custom APIs.

As for the agents, their one guiding principle is to minimize the number of data exchanges. If you book a plane ticket to Lisbon, for example, before any data is sent you're asked to provide "up front" all of your desired flight and seating information. While waiting for your choices to display, you can see screens filled with news headlines or content tailored for you. When the flight options appear, you choose one, enter the payment method, and send the data off once more. The agent now knows you're going to Lisbon, so while you wait your screen is filled with information about Lisbon. You get your confirmation.

But that's not all! The genius of this system is that the intelligent agents can talk to each other, working like little goal-oriented elves to make things easy on you.

So, after you book your flight, the details have already been added to your personal calendar. Another agent then kicks in and asks you if you want to book a rental car and a hotel room. Yet another agent asks you if you want to e-mail the information to your secretary, boss or "better half."

The Action Engine Mobile Services Framework can now sit on mobile devices running the Palm OS, PocketPC/WinCE, and RIM operating systems. The framework is also available for Windows PCs (via a native Windows client) and on any HTML 3.2 Web browser. In early 2002 the framework will be available for J2ME devices and perhaps Linux or Symbian/Epoch.

Hopefully, every wireless (and wireline) carrier and service provider will buy into the Action Engine concept.

To 3G or Not To 3G?

Certainly it would be cool to have a phone with a color screen that can display streaming video, do video conferencing with tiny cameras, tell you where the nearest hospital or restaurant is, or what the traffic like is up ahead. But keep in mind that a globally roaming system won't appear for years, though some useful services not requiring copious amounts of bandwidth will appear piecemeal on slower, 2.5G services.

There are many carriers and service providers out there that are moving cautiously, since there is a lot of money at stake. So don't hold your breath for a sci-fi holographic, telepathic, handheld device anytime soon.

IN SUMMARY

It's a dynamic world and for businesses to survive, they need the ability to instantly communicate with (and receive communication from) anyone from anywhere. Wireless technology is one way to meet these rapidly growing demands.

Wireless initiatives, like those described in this chapter, can help the business community to maintain critical operations during and after a crisis, even when the scale of the disaster is overwhelming. But only if you ensure that every wireless component that enables the stated disaster recovery or business continuation strategy can deliver extremely reliable and cost-effective performance under challenging conditions.

CHAPTER 7

Security

IN THE ERA BEFORE THE 20TH CENTURY, "security" consisted mostly of "physical security" of putting physical objects of some worth in a secure room or box and posting live people around to keep an eye on them. As the computer era dawned, "security" became the protection of data and data systems – initially of not letting the data escape via a disk in somebody's pocket, later preventing "hackers" from gaining access to a system and trashing it or transferring valuable data (e.g. credit card data) over data networks.

Anyone who attempts to either access specific information that resides on a corporate network or use network resources – when unauthorized to do so –is by definition an intruder – regardless of their intent. Network intruders can range from a bumbling employee to hackers, "crackers" (the malevolent form of hackers), and disgruntled employees. Actions by any nefarious individual, group or even bumbling employee can spell disaster for many businesses. If the intruder has accessed the network with malicious intent, they can steal information; disrupt businesses by modifying data files, software and network operations; and pirate resources like long distance data services for their own use – often without being detected.

Businesses that are attacked through network intrusion can suffer large losses in financial and intellectual assets, along with deterioration of business capabilities that can have wide-reach adverse affects, including dealing with the aftermath-public embarrassment and damage to their reputation that can lead to a loss of customer confidence.

VOICE NETWORK SECURITY

As the reader can tell, terrorists, acts of war and Mother Nature aren't the only disasters a business has to contend with. One good example is the voice system. In recent years the voice system itself has become a security problem, as new features and technologies are tacked onto it.

Someone hacks into a major airport's phone switch, ringing up thousands of dollars of unauthorized usage over a two-day period. Someone listens to an employee's voice mail messages without authorization, then uses the information for blackmail. Someone dials into your PBX's maintenance port and reprograms its call-routing at random. An employee repeatedly calls relatives in Peru from his office phone, charging the calls to your company.

These are real examples of the many ways phone systems are hacked and abused – costing businesses millions (and perhaps billions) of dollars a year in phone charges, litigation, lost trade secrets, and repair.

Nor are phone systems and voice mail the only vulnerable items in your infrastructure. You also have to watch remote access to your LAN (Voice-over-IP systems) and voice systems, calling cards, IVR platforms, and cell phones. All these systems are prime targets for industrial espionage, cyber-thieves, "crackers", disgruntled employees, and stupid pranksters.

Toll Fraud

Your company has never done business in the tiny middle eastern country of Eritrea, and your offices are locked up every night by 9. Strangely, however, your phone bill shows hour-long calls to Eritrea's capital city of Asmara after midnight every night for a week. Or you may have a toll-free number for the convenience of your customers, but when your phone bill comes, you notice 50 calls in a row from the same number, lasting less than a minute each.

Congratulations, your a victim of toll fraud, the theft of long distance service. More precisely, toll fraud is the unlawful, unauthorized use of another's telecommunications system to make long-distance phone calls. It typically involves compromising or tapping into a telephone customer's equipment, but it can also involve tricking an employee into giving away access to outside lines.

Every year, companies pay billions of dollars for phone calls that are illegally charged to their accounts by people who raid their phone systems. Estimates by the telecommunications industry place annual losses due telephone toll fraud at over $2 billion. It's quite a lucrative business, and it's becoming more sophisticated each day.

Basically, there are two main types of these criminals. The first are organized groups who sell stolen long-distance for profit. These are known as "call-sell" operations, where calls are sold on the streets, with no questions asked. The second type of criminal is the cracker who hacks into your system for fun or profit. Crackers have found a ready and enthusiastic market for their toll-fraud know-how, and sometimes team up with criminal organizations, drug dealers and call-sell operators who pay huge sums of money for the special technical services. Once crackers find themselves getting steady pay for breaking into systems and doing harm, they tend to perfect their abilities. These people combine technical skills, tools, and money, and can be quite successful, hard to trace, and difficult to defend against. Often, the cracker is either trying to crack your authorization codes (to sell them) or attempting to gain personal access to further his or her own illegal activities.

Toll fraud is attractive to these criminals because the stolen calls cannot be traced.

After you or your business has become a toll fraud victim, there's little you can do about it. Hopefuly, the damaged doesn't put your business in jeopardy.

One discovered, you can contest toll fraud charges, but the law says that the owner of a PBX is responsible for all calls made from his system, legitimate or otherwise, so don't get your hopes up. Whether it's an experienced hacker who intends to post your code on the Internet to sell cheap to hundreds of other bad guys, or a disgruntled former (or even current) employee who wants to make a few long-distance calls on your dime, you need to protect your PBX from this high-tech assault.

You don't want to deprive your business of telephone services like remote access, voice mail, and automated attendants; you need to keep up with competitors and increasing customer demands. These state-of-the-art villains stay informed so as to gather the latest information on how to attack these systems and gain access to your telecom equipment. Once they're in, they have the run of your system: they can set up voice mailboxes, make long distance calls, even reprogram the system so you and your employees can't access it.

According to telecom security expert and author, John Haugh, there are three distinct varieties of toll fraud:

"First Party" Toll Fraud, which is helped along by a member of the management or staff of a victim-company. An example would be the telecommunications manager at the Human Resources department of a company in New York City (an "insider") who sold his agency's internal code to the thieves, who in turn ran up unauthorized long distance charges exceeding $500,000.

"Second Party" Toll Fraud, which is facilitated by a staff member or subcontractor of a long distance carrier IXCs, vendor or local exchange telephone company selling the information to the actual thieves, or their "middlemen." An example would be a "back office clerk" working for one of these concerns who sells the codes to others.

"Third Party" Toll Fraud is facilitated by unrelated "strangers" who, though various artifices, either "hack" into a user's equipment and learn the codes and procedures, or obtain the needed information through some other source, to enable them to commit Toll Fraud.

As for actual techniques, today's most common forms of toll fraud are DISA, voice mail infiltration and shoulder surfing, but there are others, all of which are examined below:

Simple Unauthorized Dialing

The least sophisticated kind of toll fraud are the placing of unauthorized calls by staff, visitors, maintenance workers and others. These calls can be made at any time of day, and are often made on site. The most typical situation is one where the thief is visiting the company, often under seemingly legitimate circumstances (maintenance, delivery or courier, etc.) and asks to use the phone. These calls are often made in the presence of company staff!

Solutions/Prevention

- Assign authorization codes from each extension to enable long distance dialing outside. Codes should be at least four digits long, and should not match any user extensions.
- Phones with public access, such as those in lobbies or conference rooms, should be restricted to only internal and local calls. If this is not possible, never let a non-employee make a call, offer to dial for the person. Train your employees not to allow outside personnel access to phone lines.
- Block the area codes of countries that your staff has no reason to call. Consider blocking even certain area codes that you need to call, but only periodically. The FBI identifies the Medellin area of Colombia and the Virgin Islands in the Caribbean (809 area code) as prime targets for fraudulent calling. Crooks are using these numbers as pay-per-calls to get around US regulations and 900 block. Every time you call that number, they get rebates from their foreign phone companies. In this way they get paid, but they don't have to obey US 900# regulations. The US regulations require them to warn you of the charge and rate involved, and also to provide a time period dur-

ing which you may terminate the call without charge. Consumers often get paged or get a message on their answering machines asking them to call, say, an 809 number. If they call the number, they hear a long recorded message and later find huge charges on their phone bill for the call. To minimize fraud, block those area codes, then require staff to use calling cards to reach those areas.

- Pay-per-call numbers should be screened and disallowed. Examples of these are 1-900 numbers or 976 prefix numbers. Chances are, there are no legitimate business uses for these numbers.

Direct Inward System Access (DISA)

Remote access through the office PBX (Private Branch Exchange) or "switch" is perhaps the most common cause of toll fraud today, generally involving Direct Inward System Access (DISA), a telephone system feature that allows an off-premises employee to dial directly into a company telephone system and to access all the system's features and facilities as if one were dialing from within the office. The remote access features enable an employee phone-in dictation for the typing pool, or, if the employee is traveling, to dial in, and then make outbound long distance calls using a company's less expensive long distance lines, like WATS or tie lines. Because of DISA's capabilities, many PBX owners use DISA in place of calling cards, a convenience for staff but an opportunity for call-sell operators.

To use DISA, you must input from your touchtone phone a short string of numbers as a password code. The problem with this idea is that "unauthorized people" often acquire that number or figure it out and run up expensive long distance phone calls.

How? "Social engineering" is the way many systems become compromised. Given the prevalence of automated-dialup/login software, people frequently forget passwords and access codes (since they don't have to enter them every time they log on). As a result, network administrators, ISPs, and other service providers have become accustomed to providing this information over the phone to desperate

➤ You versus Toll Fraud

There's plenty you can do to guard your systems, but first you need to know where you're vulnerable and how to spot potential problems before they get out of hand. Here are some of the indicators you're being hacked:

- Legitimate employees call in at night trying to access voice mail but can't because the ports are busy.
- There are too many busy lines at all times, night and day for the amount of traffic your business should generate.
- Call Detail Record (CDR) reports show lots of calls during times you normally don't have that type of activity, such as over holidays, weekends, etc.
- Switch reports show multiple, failed log-in attempts to a system.
- Unexplained surges in system use.
- Long holding times.
- Reports showing calls to unrecognizable area codes, or your long distance carrier flags a sudden start of call traffic to, say, Madagascar.
- Your voice mail codes have suddenly, and unexplainably, changed.
- Call tracing shows incoming calls being transferred to a particular extension that's been set up for call forwarding, which shows unauthorized outside calls dialing out.

callers who present even marginal identification. It's not hard to subvert this process.

Such "tele-thieves" have posed as systems administrators and conned employees into telling them PBX authorization codes. Consumers have told the FCC that they have received calls from persons claiming to be FCC agents or telephone company employees investigating calls placed to other states or countries from the consumer's telephone line – or are checking on possible technical problems with the consumer's telephone line.

The caller typically requests the consumer's cooperation in the investigation by just saying "yes" and hanging up the telephone receiver when a second call is placed to the consumer by a "supervisor." Other tactics include requesting the consumer to provide their credit or calling card number or to dial a specific series of numbers before hanging up the telephone receiver. For example, they will conclude by saying: "In order to finish these tests I need you to transfer this call to extension 9-0-# (nine, zero, pound), or 9-0 (nine, zero)". By transferring such as call, the caller can then make long distance telephone calls which will be billed to your phone extension.

All of these types of calls are made to trick consumers into providing information or taking actions that will enable the caller to place fraudulent calls using the consumer's credit or calling card number or telephone line.

If social engineering doesn't work, tele-thieves will often search a company's trash ("dumpster diving") for directories or call detail reports that contain 800 numbers and codes.

More sophisticated crackers use personal computers and modems to break into databases containing customer records showing phone numbers and voice mail access codes. Others locate the DISA feature with the use of a "war dialer" or "demon dialer." The device dials telephone numbers randomly, generally 800 numbers, until a modem or a dial tone is obtained. (If an incoming 800 line has been used, you may be billed for the 800 call as well.) After the cracker finds a number, he or she tests random authorization codes until the correct one is found.

You might scoff at this, thinking that your pass codes are safe. But according to the laws of probability, the raw odds of correctly guessing a pass code of two digits are 1 in 100; the odds of correctly guessing a pass code of four digits are 1 in 10,000; a six-digit code, 1 in a million. This sounds relatively secure. Except it doesn't work that way. Default passwords (e.g., "SUPERVISOR") are frequently left in place, making the hacker's job easy. Even when passwords are changed, legitimate users (employees, maintenance personnel, etc.) frequently employ patterns to make passwords easier to remember. Repeating numbers. Geometric patterns on the phone keypad. Clever words. Likely combinations are relatively few in number, and can be generated by automatic hacking software. Indeed, research indicates that crackers using random generation programs can go through the 432 most common pass codes in less than one hour.

Once found, the authorization codes may then be distributed via bulletin boards and pirated voice mailboxes. Once these thieves have the numbers and codes, they can call into the PBX. Once in the system, they're able to place or sell long-distance calls, often international calls, which of course show up on your bill. When making these illicit calls, call-sell operators can even hide their activities from law enforcement officials by using "PBX-looping" – using one PBX to place calls out through another switch in another state. Moreover, in many cases, the PBX is only the first point of entry for such criminals. They can also use the PBX to access the company's data system.

Solutions / Prevention

- Once again, prevention is the key when protecting your DISA feature. To decrease your vulnerability to fraud, it's best to restrict DISA to trusted people and use complex passwords or authorization codes to control access. Authorization codes should be as long as possible, and not easily defined (i.e., sequential, 12345, or all one number, 77777, etc.). Ideally, use randomly generated pass codes (not birthday, home phone numbers, social security numbers, etc.). You should change the passwords frequently at regular time intervals. Also, add an additional layer of security with authorization codes required after pass codes. And don't write pass codes down (even on order forms for new voice mail boxes and extensions) then throw them out in the trash.
- Consider limiting use of the DISA after regular business hours, or using an attendant to intervene.
- Limit 800 number access from non-essential areas of your markets.
- Restrict the use of outbound trunks as much as is feasible without restricting authorized company business.
- Train your operators and employees to be on alert for suspicious calling patterns, and to deny requests for outside lines to callers. Receptionists have unwittingly connected to an outside line, rude and forceful tele-thieves who have claimed to be the president of the company. Beware of callers who repeatedly request invalid extensions or names of non-existent employees. Lookout for excessive hang-ups, obscene calls or wrong numbers – these are often a ploy to get the employee to hang up, which could then leave an outgoing line vulnerable. Telephone company employees checking for technical and other types of telephone service or billing problems and FCC personnel will not call and ask you to "just say yes" when receiving a telephone call, provide a credit card or calling card number, or dial a specific series of numbers before hanging up the telephone receiver. If you think a call is a toll fraud attempt, be sure to ask the caller for their name and their telephone number, tell the caller you are going to call the telephone company to determine whether or not there is a problem with your telephone service, immediately hang up, find the telephone number for your telephone company business and security offices on your telephone bill or from directory assistance or your telephone book and call that number (rather than the number provided by the caller) and provide details of the call you received to the telephone company representative. Finally, contact law enforcement.
- One of the signs of a cracker at work is if calls are coming into the switch and going out a lot, especially on one extension. Check "call path tracing." A cracker with admin privileges can set up one extension to forward calls to an outside line (or maybe they already know a particular extension is on call forward). They then call into the switch, DTMF (or ask a receptionist for) that particular extension and get dialtone at your expense.
- Set up your system to disconnect after a preset number of invalid attempts. For example, the Model 125 Site Monitor from Gordon Kapes (Skokie, IL – 847-676-1750, www.gordonkapes.com) has a feature that lets you decide how many times a user can enter name-password combinations to log onto the system in a given time period, and restricts access to the system for up to 99 minutes when that number is exceeded. After repeated unsuccessful attempts to access the system, an alarm is generated and the unit can send a report to your service center within minutes. It can even page you.
- Check the numbers called and bills generated, or use a telemanagement / call accounting system

that does so. More and more call accounting systems are incorporating real-time toll fraud detection and prevention features. These subsystems can pick up anomalous calling activity while doing data-collection. You define what "unusual" is for your company: international calls after business hours; calls to a specific area code; etc. The system can then compare real-time call activity using such call activity data as "calls per hour," "call-minutes per hour" and abnormal conditions against the preset thresholds you define. When the system detects possible fraud, an alarm manager will be notified for execution of appropriate reporting and/or corrective action procedures. Alarms can be sent to multiple PCs and pagers, and you can configure some systems so that if someone unplugs it an alarm goes off at your service provider's or your own office!

The Model 125 Site Monitor from Gordon Kapes looks for repeated attempts to access a phone system — a sign that hackers are at work.

- To minimize the vulnerability of Nortel's immensely popular Meridian 1 system to unauthorized access through DISA, the company suggests the following safeguards:

1. Assign restricted Class of Service, TGAR and NCOS to the DISA DN.
2. Require users to enter a security code upon reaching the DISA DN.
3. In addition to a security code, require users to enter an authorization code. The calling privileges provided will be associated with the specific authorization code.
4. Use Call Detail Recording (CDR) to identify calling activity associated with individual authorization codes. As a further precaution, you may choose to limit printed copies of these records.
5. Change security codes frequently.
6. Limit access to administration of authorization codes to a few, carefully selected employees.

If You Become a Victim of Toll Fraud

You can be responsible for as much as $50 in unapproved calling card charges under the Truth in Lending Act and Federal Reserve Board regulations. Call the telephone company that issued your calling card as soon as you become aware that you are a toll fraud victim and request the company to cancel your calling card number and issue a new one to you.

Toll Fraud Complaints

Toll fraud complaints involving calls placed to another state or country are within the jurisdiction of the Department of Justice. You should contact the Federal Bureau of Investigation at the following address if you are a victim of toll fraud involving calls placed to another state or country:

Federal Bureau of Investigation
7799 Leesburg Pike
South Tower, Suite 200
Falls Church, VA 22043

You should contact your state attorney general's office if you are a victim of toll fraud involving calls placed within your state. You can obtain the telephone number and address for this office from your local or state consumer offices or the government section of your telephone directory.

Talk with your local telephone company about having your telephone line blocked from third-number and collect call billings.

Stay on top of telephone toll fraud. The Alliance to Outfox Phone Fraud, a group of telecommunications industry and related companies, provides fraud information to consumers. Visit the Alliance's Web site at http://www.gnat.net/outfox or contact the Alliance at: 2890 Fairview Park Drive, 10th Floor, Falls Church, Virginia 22042 – or call 1-800-9-OUTFOX (1-800-968-7369).

Note that toll fraud is within the jurisdiction of the Department of Justice or your state attorney general's office.

Billing Complaints Resulting From Fraudulently Placed Calls

You should first try to resolve a billing complaint resulting from fraudulently placed calls with the companies that billed you for the calls. If you are unable to resolve your billing complaint with those companies, you may file a complaint with the proper regulatory agency.

➤ Toll Fraud Protection Buying Tips

When shopping for a toll fraud system, keep this check list in mind:

1. Look for a system that continuously monitors calling activity, 24 hours a day, seven days a week.
2. Make sure you can tailor the system to the calling patterns of your company. It should be able to look at several parameters, including the time of day and day of week. Don't get something with limited, preprogrammed monitoring.
3. Make sure the system doesn't issue false alarms. And make sure it can get alarms to you several ways – pager, e-mail, printer, etc.
4. Some systems can shut off facilities such as a specific trunk group if an unauthorized call has been detected. But be careful when setting this up, if only one phone is making the undesired call, cutting off all other phones using that trunk group may cause fallout.
5. Make sure the system can handle a variety of switches if you have different switches in several locations. Make sure it properly handles different time zones.
6. Many buffer boxes and remote access devices that look for toll fraud rely on ugly, clumsy character-based programming, i.e., press Ctrl-B *X5 to dump data, etc. Luckily, more of these are using Microsoft Windows-based programs, with easy-to-use drop down menus. Go GUI.
7. Watch calling reports, tracking calling and traffic patterns. Some of your toll fraud requirements and what you consider abuse might change due to new calling and traffic patterns. Have self audits. Review your setup regularly to make sure everything is correct and the system is doing what it's supposed to do.

If your complaint involves calls placed from one location to another location within the same state, you should address your complaint to your state regulatory agency. You can obtain the telephone number and address for your state regulatory agency from your local or state consumer offices or the government section of your telephone directory. The telephone numbers for the state regulatory agencies can also be found on the FCC's Web Site at http://www.fcc.gov/consumer_news/state_puc.html. You may file a complaint with the FCC if your billing complaint concerns interstate or international calls. There is no special form to fill out to file a complaint with the FCC. You can simply send a typed or legibly printed letter to:

Federal Communications Commission
Common Carrier Bureau
Consumer Complaints
Mail Stop 1600A2
Washington, D.C. 20554

Your complaint letter should include the following information:

- Your name, address and a telephone number where you can be reached during the business day.
- A brief description of the complaint.
- The telephone number involved with the complaint.
- The name, address and telephone number of the person or persons who you believe placed the calls.
- The names and addresses of the telephone companies involved with the complaint.
- The names and telephone numbers of the telephone company employees you spoke with to try to resolve your complaint, and the dates you spoke with them.
- Copies of the telephone bills listing the disputed charges. The disputed charges should be circled on the copies of the bills.
- Copies of correspondence you received from the companies involved with the complaint and from state or federal agencies you contacted in an effort to resolve the complaint.
- Copies of other documents involved with the complaint.

Voice Mail Fraud

The two main types of voice mail fraud are "squatting" in a pirated voice mailbox, and using a voice mailbox to acquire a dial tone, which is then used to make long distance calls. Voice mail normally enables callers to leave messages for the owner of the voice mailbox. Using the "demon dialer," the cracker accesses a company's voice mail system and commandeers a voice mailbox. Both the password and the outgoing message can then be changed at will, and the pirated voice mailbox can now be used to receive messages from and leave messages for others in a crime or terrorist ring.

In some systems, there is a feature that provides a link to a PBX remote access feature or gives a caller a dial tone after the main voice mail function has finished. Sometimes, this feature allows one to find an outgoing line. Using this, the cracker has the ability to call into your system (possibly using an 800 number) via a Remote Access port, access the voice mail system by finding a busy or unused mailbox, and access the transfer feature to dial long-distance, or even internationally.

Solutions / Prevention

- This is another type of fraud that can be prevented with some advance planning. Passwords should be extremely well-guarded. Do not allow administrative access via the telephone. If an employee can add, delete and change boxes via the telephone, so can a hacker. Passwords should be at least six digits long, and should expire frequently (once a month or so).

- One sign that a cracker has hit your system is that voice mail codes are suddenly changed. A cracker could have obtained the code (through several ways, including "trashing" your garbage, or via programs that randomly generate codes until they get one right) and taken over the voice mail system for nefarious means.
- Remove voice mail boxes when an employee leaves or is on long-term disability. Thieves can take over an unused mailbox very easily, and you wouldn't even be aware of it. Never install more mailboxes than you need. Some companies install several mailboxes, in anticipation of hiring additional staff in the future. These unused mailboxes are a treasure trove for thieves.
- Review lists of authorized mailbox users periodically (every two or three months). Verify that miscellaneous users no longer employed by the firm; e.g., summer interns, former part-timers and retirees; no longer have mailboxes assigned to them.
- Review reports periodically that highlight inactive mailboxes, and disable those that remain inactive for longer than one month, unless the user can substantiate the need for a mailbox.
- Review "bad password attempt" reports on a daily basis. No mailboxes should collect more than three during a 24-hour period. If any do, you should be looking for signs of hacker activity.
- Never publish a list of phone extensions with voice mailbox passwords.
- Disable the transferring out feature within the voice mail system. Limit use of the system to receiving and retrieving messages (or limit the transfer feature to only 4 digit extensions, so that any transfers occur only within the company).
- If a voice mailbox remains unused for an extended period of time; e.g., vacation, out-of-office for two or more weeks, disability; consider forwarding calls for that employee to a secretary rather than the voice mail system. If you were a hacker and listened to a voice message that said the user would be not using the mailbox for an extended period of time, wouldn't you be tempted to "borrow" the mailbox?

Infiltrating Automated Attendants

Some companies replace the switchboard operator with an automated attendant or "auto-attendant". An automated attendant has an interactive voice response (IVR) front end that accepts touch-tones from the caller's phone and might give a caller a menu of several options, such as picking various departments or choosing staff by name or extension. A typical example would be, "Hello, you've reached Newton's New Magazine. Please press 1 for Subscriptions. Please press 2 for the Accounting Department. Please press 3 for the Book Division. Press 4 for Harry. Press 5 for Robert Johnson in the Complaint Department. If you know your party's extension, please enter it now."

These auto-attendants can also be used to forward after-hours calls to a 24-hour work center, an answering service, or even an employee's home. Some of these leave behind a dial tone. Thieves take advantage by dialing the forwarded number, waiting for someone to answer, then using the dial tone after the party hangs up. Some attendants allow the use of dialing "9" to access outside lines, which gives hackers the opportunity to dial out without waiting for an answer. Thieves who know how this works can abuse your system on a grand scale. This can also be done with a receptionist or switchboard operator.

In a typical scam, the tele-thief calls a company extension at random. When the call is answered, the thief apologizes for misdialing and asks to be returned to the operator. When the operator answers, the thief pretends to be the employee and claims to be unable to access an outside line. The operator, believing the call to originate from within the company, switches the hacker to an outside line. The thief is then free to make (or sell) long distance service at the company's expense.

➤ Protecting Auto Attendants and Voice Mail

If you manage an auto attendant/voice mail system, keep this checklist in mind:

1. Block transfers to numbers not defined as extension numbers or auto attendant choices. Don't let callers get an outside line by pressing "9" or some other trunk access number. Also, set up your voice mail lines to only answer calls, not dial out.
2. Prevent unauthorized use of server storage space. Don't let callers leave messages in uninitialized mailboxes (boxes that have been created but aren't being used). This prevents unauthorized callers from using mailboxes and hearing sensitive company information sent out over voice mail by logging onto an unused mailbox. It also prevents hackers from doing cute things like leaving messages for other hackers, messages that could contain access codes to your switch, personal information about employees heard on their mailboxes, etc.
3. Check reports frequently, especially port traffic statistics and log-in attempts. If one of your ports is constantly busy, this could indicate someone trying to hack in. Check for failed server and mailbox log-in attempts, and set up your system so callers are disconnected if they fail to reach a mailbox after a specified number of N attempts. Too many failed log-in attempts could mean you're under attack. Also look for other unusual activity in your reports, like calls made when the business is closed, over holidays and weekends, etc.
4. Change pass codes periodically, to keep new ones in use as much as possible. Make sure pass codes aren't guessable by a four year-old, which means don't match mailbox numbers, or use the mailbox numbers in reverse, "1-2-3-4," employee names, birthdays, or words like "admin", "supervisor", etc. Also make sure employees don't leave access codes lying around on scraps of paper tacked to their wall, taped on phones, etc.
5. As an addendum to #4 above: as holidays approach, or long vacations, change the passwords and authorization codes used for remote access, voice mail or administrative functions. Unauthorized people may have already discovered your current passwords and codes, and are waiting until the holiday season, when staffing is at a low level, to use the pass codes. If possible, route the incoming calls to a live operator or receptionist instead of an auto attendant. Otherwise, disable or restrict the remote access, voice mail, and other features of your system on holidays, weekends and other times when employees don't need them.
6. Server security. Use multiple levels of security access for your voice mail server. Give low level managers access to what they need to do to the system, and higher ups more access. Many systems can be programmed to require additional pass codes before letting major, high-level changes occur.
7. Watch out for callers who claim they're technicians, or use other ruses to talk their way into learning pass codes or other information about system. Keep a record of your technicians/VARs/interconnects on file, and if you're leery of someone, ask for name and phone number and say you'll call back.
8. Keep your voice mail/phone system itself in a secure area with a locked door. Make sure only authorized people have access to your telecom closet/war room.

Solutions / Prevention

- If possible, disable certain features, such as having the option to press 9 for an outgoing line. If available, install a verify extension capability in your PBX.
- Configure the call attendant so only valid extensions are transferred back to the PBX.

Maintenance Ports

One interesting means of entry into phone systems are the PBX maintenance ports or voice mail "terminals," which have modem access for administrators or vendors to access call system hardware and software in order to troubleshoot the system and to perform normal administrative moves, adds and changes remotely. This poses a great threat as it sometimes leaves your system wide open to demon dialers. Once a hacker has access to your system through the administration terminal they could disable all security measures, but enable outdialing features that a business had shut off, then create voice mail boxes and sell them to the underground hacker community, and even coordinate illegal trafficking operations such as stolen credit cards.

Crackers have been known to alter a PBX's configuration at 8:00 p.m. to allow fraudulent use, then reconfigure the system back to normal operation before business hours to hide their illicit activities!

Solutions / Prevention

- Some kind of security access unit (SAU) should be placed in front of the maintenance port. SAU's provide another level of user identification and password protection. This should be controlled by the PBX owner, and not the vendor. SAU's can be set up to further validate a user via callback or numerous token devices. For example, the Spectif Security Access Module (SAM) from Peregrine Systems (San Diego, CA – 858.481.5000, www.peregrine.com) is a security access management system that secures the maintenance port of your PBX and provides controlled access to the service providers and contractors who need access to your network equipment by directing them through your security system. SAM records, in detail, every transaction on the maintenance port. SAM is a versatile, friendly security management tool that operates with Peregrine's Spectif Events Manager.
- If you can't afford an SAU described above, at least ensure that you don't have any maintenance ports on Plain Old Telephone Service (POTS) lines, since random dialing programs crackers use can discover them. If you do put them on regular lines, hide them behind internal extensions to make them harder to find and break into. If possible, disable the maintenance port unless your equipment requires service (some vendors or services may not allow this).
- When the maintenance ports are not in use, deny all access.
- Password selection is important here. Make passwords difficult to figure out, and use all available digits. Change them often.
- A red flag should go up when technicians are locked out of certain systems or administrative functions, or when calling data can't be retrieved.

"Shoulder Surfing" Calling Card Numbers

Calling cards have become nearly indispensable to everybody: businesses with traveling employees, retirees, college student, foreign nationals wanting to call home, terrorists clandestinely planning their next feat of mass destruction, etc.

Using a calling card is easy – you just walk up to a pay phone and input an 800 number and your personal identification number. Unfortunately, it's also easy for tele-thieves to filch these numbers. The most popular technique is "shoulder surfing" or watching over peoples' shoulders as they use a phone or other communications or information system. Examples include videotaping people as they enter their passwords, observing air travelers as they use their computers and review documents while in flight, and observing users in normal operations who are trying to learn standard operating procedures. Call-sell operators routinely loiter around areas where there are usually large banks of pay phones, such as airports, train stations, or hotels. However, not all shoulder surfers need be close by. They can be located in an adjacent building using binoculars, a telescope, or a parabolic listening device to pick up the sound of touchtones. Tele-thieves may even masquerade as security officers and call you and ask for calling card numbers, claiming the need to deactivate its use to protect it from abuse. Organized crime rings run call-sell operations involving the theft of calling card authorization numbers.

Solutions / Prevention

- The first rule is to commit your calling card number to memory. This reduces the chances of the card and/or code falling into the wrong hands.
- Company employees must be careful when at airports. New York's Kennedy Airport is a notorious den of shoulder surfers.
- Employees should be alert to those around them while placing a phone call in a public location. One should always assume that someone nearby is listening for an authorization code. When keying in the authorization code, stand directly in front of the phone to keep others from seeing your number. If you've memorized it, input can be done quickly and securely. If you need operator assistance, speak directly into the mouthpiece in conversational tone and look directly at the phone when relaying the authorization code to the operator.
- Never give your authorization code to anyone. Guard your calling cards as you would a regular credit card. Never disclose the number to anyone, or else they will be able to use it.
- Never use the calling card to verify identification. As with a credit card number, this is not a legitimate way of identifying someone.
- If the card is lost or stolen, as with a regular credit card, report it immediately. The sooner the missing card is reported, the sooner the number can be deactivated, and the sooner a replacement can be issued.

50 General Tips for Preventing Voice Processing System Misuse, Abuse and Fraud

On February 3, 1994 my former boss dialed long distance through a voice mail system belonging to a company that should know better. Had he been malicious, he could have inflicted hundreds of thousands of dollars of unauthorized calls on this company. Fortunally, he isn't malicious. But hundreds of others are. Some people even set up "weekend phone companies," re-selling calls to foreign countries dialed through voice mail / auto attendant / IVR systems. It's a serious problem. No one wants to walk in on Monday to a $250,000 weekend phone bill. It's happened. Here are some ways to prevent it.

These tips will help you create a program for combating misuse, abuse and fraud on your voice processing system. Like most preventative maintenance, the steps are only effective if you carry them out quickly and diligently. If you wait for disaster to strike, you've waited too long.

1. Don't publish 800 numbers or voice mail access numbers in company brochures, newsletters, or other "public" documents that may fall into the hands of hackers.
2. Many systems boast of an on-line voice mail directory for callers who don't know the extension number of the person they want to leave a message for. While this is a worthwhile service, it can also be a boon to hackers if mailbox numbers are the same as the extension numbers, giving hackers a list of every mailbox.
3. Passwords should not be a social security number or someone's birth date. These may seem like safe numbers, but such information can be gotten from numerous sources.
4. Many systems can enforce the minimum number of digits users select for passwords. This minimum is determined by the system administrator. One vendor suggests a ratio of 10,000 possible passwords for every actual mailbox. This means that if a system has only one mailbox, the password should be four digits long (9,999 possible passwords). If the system has ten mailboxes, passwords should be five digits in length (99,999 possible passwords). One-hundred mailboxes would require six- digit passwords, and so on.
5. Limit access to PBX ports used for automated attendant service by using toll restriction. Callers can also be denied access to trunks using a class-of-service restriction.
6. Each application on the voice processing system – e.g., out-dialing, paging or networking – should have its own dedicated group of ports. This limits hackers from using one of these services as a gateway to the world.
7. System administrator passwords should be as long as the system allows. A passwork of 12 or 15 digits is not unreasonably long. Some systems provide several levels of password protection, depending on the activity or function performed. This is a good feature.
8. Change system manager passwords whenever system managers change.
9. System administrator passwords, if written down, should be kept in a secure place like a locked drawer or safe (not written on a Post-it note stuck to the screen of the system manager's terminal).
10. Trying to keep administrators from leaving their PCs or terminals running while unattended. A running PC makes a tempting target to hackers passing by.
11. Failing to achieve #11, at least use, a screen-saver program (very inexpensive) that causes the screen of the system administrator's terminal to go blank after 10 or 15 minutes of non-use. Hitting any key will bring back whatever was on the screen, but a hacker doesn't necessaril know that.
12. If your system provides for several "levels" of prompts, opt for a level that reduces feedback to callers logging on to the system. Omitting prompts that can guide hackers through the system may discourage less experienced hackers. There's no reason to "tell" hackers that extensions are four digits long, or passwords are five digits in length.
13. Using a "self-destruct" feature that erases all messages from a mailbox whenever the password is changed is an effective means for providing security. However, this may encounter resistance from some users (particularly those with a habit of forgetting passwords.) Let users know the consequences of forgetting passwords. This should improve their memory.
14. It may seem obvious, but instruct operators, receptionists and secretaries to use discretion when

giving out voice mailbox numbers. If a caller requests an individual's mailbox number, offer to transfer the caller to the mailbox rather than simply giving the number out.

15. Also obvious, pilot numbers (the special number users call to get directly into the voice mail system), should not be given out over the phone, regardless of who's calling. The same goes for access codes. If a caller claims to have forgotten how to access their voice mailbox, the secretary, receptionist or operator should offer to have the voice processing system administrator call them back and explain it. This will discourage all but the most tenacious hackers.
16. Passwords should be changed as often as possible. A bare minimum is four times each year, but monthly would be preferred. This may not go over well with users, but it is for their own good.
17. Some systems can automatically force users to change passwords after a certain period of time defined by the administrator. If your system has this capability, use it. If users complain, explain to them that it's the system and there's nothing you can do about it.
18. Restrict the use of outbound trunks as much as is feasible without restricting authorized company business. If there is no need to be calling Europe, don't allow the PBX to complete calls to those international area codes. If there is no need to allow long distance calling at night or on weekends, don't allow them.
19. During regular business hours, route any outbound calls to non-essential destinations (South America, Far East), to the Operator. If there is a valid reason for the call, the Operator can place the call.
20. Don't include executive mailbox numbers in any on-line directory. This won't make industrial spies happy, but too bad.
21. Check all authorization codes and passwords when the installation of a new system is complete. Factory default codes and passwords are often used. Experienced hackers know these codes. Don't make it too easy for them.
22. If you have a remote maintenance port, use a secure, two-line dial-back modem.
23. Warn users about shoulder surfers. Users should attempt to shield the key pad as much as possible to prevent a clear view of the numbers dialed.
24. Beware of dumpster divers (criminals who scout trash containers; e.g., dumpsters; looking for lists of passwords, authorization codes, mailbox numbers, whatever). If these kinds of records must be printed, be sure to shred them before discarding.
25. If the number of "failed attempts before disconnect" is a flexible parameter on the voice mail system, set it to three or less. If an authorized user can't get it right after two attempts, perhaps it's time for remedial training.
26. Program the voice processing system to wait at least five rings before answering the dial-back modem connected to the remote access port. This will discourage hackers and auto-dial programs.
27. Insist that the call-back line on your two-line modem is set up for out-dialing only, and cannot receive incoming calls. This must be requested from the local telephone company. Note: May not be available in all areas.
28. Instruct operators and receptionists to give outside lines to people requesting them only if they recognize their voice, or call the party back at their extension.
29. Some users have short memories and need to write down their passwords. Tell them to keep these little scraps of paper in secure places, such as their wallets, purses or the locked laptop

drawer of their desk. Explain to them that these little pieces of paper are as valuable as a credit card if a cracker gets a hold of them.

30. The out-dial feature for voice mail should not be used, unless absolutely necessary. Limit use to a minimum of extensions and monitor its use carefully.
31. Ask your voice mail, automated attendant system and PBX supplier about any loopholes they may know of that would allow callers to grab an outgoing PBX trunk from the voice processing system. Various features may be ***activated*** or ***inactivated*** when systems are shipped from the factory. Make certain that features you don't want activated are deactivated when the system is installed. After this, test the system yourself.
32. Check the list of authorized mailboxes against the system-generated list of active mailboxes (if one is available). They should agree. If they don't, investigate immediately. Someone other than the system administrator may be opening up boxes.
33. If you don't already have a call accounting system tied into your PBX, get one. Call accounting packages can provide not only records of what extensions made what calls, but they can also generate lists of calls that fit parameters that you select. These can included all calls over a certain length (60 minutes, for example), all calls to certain area codes (like the Caribbean), and excessive numbers of calls within a definable period of time. These exception reports can be generated on a daily basis and can detect most fraud long before the phone bill arrives.
34. Never allow "default" passwords to be used. Many systems can generate lists of mailboxes using default passwords. Contact each user and threaten them with the loss of their mailbox if they don't change their password immediately.
35. Insist that new users select a password as soon as they log on for the first time. (This should be within hours of when the mailbox is setup by the administrator.)
36. Many systems force new users to select a password, and some even select the password for them from a random number generator to prevent users from selecting "1111" or "1234" or some other obvious number. This is a good feature. Use it if your system has it.
37. Never leave a voice processing system unsecured. Even if it's a PC, lock it up in an equipment room. Restrict access to as few employees as possible (i.e., technicians, system administrators).
38. Secure manuals and lists of mailboxes and other specific system information safely.
39. Encourage users to check their mailboxes on a regular basis, even if they aren't receiving messages. If a hacker has taken over their mailbox, you want to know as soon as possible.
40. Change the number of the administrative mailbox. Often, this is a default number assigned at the factory, and used by every new system installed by that manufacturer or distributor. Hackers are well-informed on all default settings. This is one mailbox in particular that you don't want hackers cracking.
41. Monitor usage levels of the system. If available storage for messages suddenly drops, investigate. There may be crackers leaving extended messages somewhere in the system.
42. Limit the length of messages that callers can leave to a reasonable minimum (three or four minutes is more than adequate). Crackers will find it more difficult to use up system capacity this way.
43. Explain to users the danger of programming passwords into their telephone speed dial (particularly if they have a display phone). Anyone can walk by, press a key, and read the password on the LED display.

44. Limit the number of people given administrative privileges. Four or five is too many. Two is reasonable.
45. Consider additional security measures. There are keyboard locks on the market that disable the keyboard of the system administrator's terminal, making it impossible for anyone to do anything with the system. This may seem like overkill, but it is an inexpensive alternative to changing every password in the system after a hacker has commandeered the password file.
46. Many manufacturers are beginning to accept responsibility for alerting users of potential security problems. (Nortel offers a video dramatizing security and toll fraud issues.) Don't be afraid to ask what your supplier is doing to reduce the risk for abuse and fraud. If they respond with, "there is no problem," consider a different supplier.
47. Provide training in securing voice processing systems to all employees when a system is installed, and furnish remedial security training on a regular basis (annually).
48. Distribute fliers to all employees every two or three months covering tips for preventing security problems. Check with the voice processing system manufacturer or distributor. They may be able to furnish fliers.
49. Keep an emergency list of who to call near the system manager's terminal so that, in the event that a hacker makes his presence known, you will have the names and numbers close at hand. Make sure you have off-hours numbers too in case you must contact someone over the weekend or holiday.
50. If your present system doesn't offer an adequate number of safeguards, get a new system.

Get a Telemanagement System!

Call accounting and the more sophisticated telemanagement systems (that also include hardware management) are a good investment and are much more than something that can monitor and control phone system abuse by employees and hackers, and let you bill back calls to departments or clients (though they do both these things rather well). Besides keeping track of your company's telecom expenses, a good call accounting system can help you analyze those expenses and let you know how much you're spending based on industry trends, so you can establish budgets and allocate costs more efficiently and accurately. It can also monitor trunk traffic so you can determine if your setup is meeting your business' telecom needs efficiently – whether you have too much phone system or not enough. It can also tell you about employees' phone habits, and whether a call for change is in order. Why, it can even correct the mistakes your telco makes – are you absolutely sure your telco is billing you correctly? If you say "yes," you're probably wrong. A call accounting system (or adjunct bill auditing system) will help you run your call-detail data through financial models to help you check your phone bill and determine the best calling plans for which you qualify.

Call Detail Recording (CDR) is a telephone system feature enabling the system to collect and record information about each outgoing and incoming phone call: who made or received them, where they went or where they came from, what time of day they occurred, their duration, whether it was a misdial, etc. Some phone systems will collect and store the data internally; others will pump the information out of the box as the calls are made. The latter is usually accomplished via a Station Message Detail Recording (SMDR) plug which sometimes takes the form of an RS-232C port, into which one plugs a printer or a call accounting system. Whatever the exact teqhnique, the calling information should be recorded somewhether – either dumped directly into a printer or "captured" by a PC running call accounting or comprehensive telemanagement software – and subsequently processed into meaningful management reports. If you have a large phone system and need to store lots of records, check out the PollCat NetLink Call Accounting Terminal from Western Telematic (Irvine, CA – 949-586-9950, www.wti.com). It can store as millions of call records, depending on how much memory you get, and can be used with any call accounting software you may already have. It's an extremely reliable, PBX data recorder, designed for SMDR/CDR data collection and alarm monitoring. Collected call records can be retrieved via TCP/IP network, via modem, or by a local PC connected directly to the Console Port. The NetLink can also monitor call records for suspicious phone activity or critical alarm conditions. When an alarm is detected, the NetLink can immediately notify the proper personnel by pager, modem, or SNMP trap.

Call accounting software reformats and sorts the ASCII call data on the PC or downloaded from a buffer, and saves it as a database format. (Usually it can be imported into a spreadsheet program, like Excel, for easy reading.) The software then analyzes the data and generates reports based on templates included with the software; or you can create custom reports that include the specific information you want grouped the way you want it, depending on the software, and end up with a report of, say, the sales manager of the northern territories.

Ideally, your call accounting system should be able to monitor every piece of equipment attached to a phone line, such as faxes and modems. And for optimum efficiency, it should track more than minutes used. Look for a system that will collect and present all the information you need: cost allocation at your fingertips, state and country identification, trunk utilization. It should give you a complete picture of your network usage.

Even more sophisticated telemanagement systems also do facilities management, which tracks your telecom "physical plant" – helping you maintain records of equipment and cable locations, categories, terminations, and inventory. The best packages generate trouble-tickets, track adds, moves, and changes; and coordinate with purchasing to allow inventory control.

Most telemanagement systems are modular. If you need only call accounting and toll fraud detection, you won't have to pay for cable management and inventory. Factors affecting the price will be the number of modules, number of phone extensions, number of locations and other aspects of your workplace that will change your specific system configuration.

If your business is small, or you're just starting out with a handful of employees, you may not think you really need a call accounting or full-blown telemanagement system – but in fact you do, and there are many reasons. For small companies, telephone expenses can rank as a major cost item. Monitoring phone usage for employee misuse and abuse saves a company money not only directly on telephone bills, but also indirectly in improved employee productivity.

Any company can reap the benefits of call accounting. You don't have to be the size of IBM or Microsoft to need to monitor your telecom usage; in fact, you could make the argument that smaller companies have to keep an even closer eye on their telecom expenses and security because their resources are all that much more limited. The excessive unauthorized phone activity of an errant employee or crazed cracker could actually send a small company into bankruptcy. And just knowing the company is tracking calls can make the most brazen employee think twice about abusing his phone privileges. If you don't want to physically maintain a call accounting system on your PBX, or you don't want to make that initial cash outlay to buy your own system, you can have a service bureau do the work for you, usually for a per-call charge.

And always remember: Many times crackers try to get back into sites they've already hit, maybe six months to a year later, since they know managers sometimes relax and will again be off guard. Managers may even have disabled security features they've previously set up. In other words, stay on guard *constantly*!

Wireless Phones and their Vulnerability

As we've seen, telephone security should always be a consideration. But until now we've been discussing voice networks based on landlines and corded phones in the corporate environment. The variety of choices for telephones that has developed in recent years spans a huge spectrum of privacy capabilities. At the low end are ordinary cordless phones used in the residential, SOHO and teleworker environment. Calls over these phones may be picked up by walkie-talkies, baby monitors, and radio scanners. The most secure home-use telephones are traditional corded models, but there are a number of other options, summarized below.

Any of the telephones discussed in this section can be attacked by a knowledgeable adversary. One can, however, distinguish between those systems that require special knowledge and those that are vulnerable to casual scanning.

Cordless, 46 – 47 MHz. These commonplace cordless models are the most widely used among portable home phones. They are easily intercepted, and should be regarded as providing no security at all. Some models provide a "privacy" or "scrambler" feature based on audio frequency inversion. While this feature requires special equipment to defeat, the necessary equipment is easy to build, and plans are available on the Internet, so these phones should not be considered secure. If you do any kind of work from home, don't use this type of phone.

Cordless, 900 MHz. These high-end phones are not as susceptible to eavesdropping as ordinary cordless models, but they can be picked up using some radio equipment. Some models employ "frequency hopping" or spread spectrum technology, which uses a rapidly changing set of frequencies to scramble transmissions. Models that provide spread spectrum are reasonably secure for most uses.

Cordless, 2.7 GHz. These newer models are, like 900 MHz models, less likely to be intercepted than ordinary cordless phones, but should be considered reasonably secure only if they provide spread spectrum transmitting capabilities.

Cell phones. The most widely used older analog cellular telephones operate on frequencies around 800 MHz, which fall within the UHF television band. Radio scanners and television sets with UHF dial tuners can intercept such cell phone conversations, so these telephones are no more secure than ordinary cordless models. Be wary if conducting critical business over these devices. It's always been said that to maintain privacy you should trade your analog phone in on a digital (GSM, CDMA, TDMA) cell phone. Digital signals are scrambled (the actual technique depends on the particular digital transmission format being used) so that the calls are unintelligible to anyone listening in. Identifying codes that allow the cellular network to recognize your phone are encrypted as well, meaning that someone trying to intercept your phone's ID code would end up with nothing but gibberish. Because of these security enhancements, the incidences of wireless fraud drop dramatically in cell phone networks that adopt digital technology. However, an Associated Press dispatch of April 14, 1998 reported how, after six hours of work, two graduate students at Berkeley and a computer cryptologist were able to clone a "tamper proof" mobile phone using the world's most popular digital cellular technology – GSM. Defeating its security also revealed that GSM's code may have been deliberately weakened during its initial development so government agencies would have the ability to eavesdrop on telephone conversations. Thus, while digital cellular phones are a lot more secure than analog ones, with enough effort, the security of at least some can be compromised. Third generation or "3G phones" to be deployed by 2005 will be digital and will occupy additional bandwidths. These bandwidths have been allocated in Europe but not yet in the U.S.

As you might expect, cellular phones do cause grief for everyone – the individual, the teleworker and the business community. Let's look at some mind-boggling problems that can be encountered in the cellular communications environment.

Cellular Phone Fraud

Back in 1997 the author made his first visit to Portugal to give a presentation at SAT '97 (Servico de Atendimento Telefonico '97), a series of seminars held in Lisbon where experts give presentations on the latest developments regarding call centers, help desks, telemarketing, computer telephony, telebanking, and IP telephony. During my visit, I couldn't help but notice that, in spite of the impoverished populace – Portugal is Europe's poorest nation – ***everybody*** had cellular phones. There were more of them per capita than in the US. (The cell phone's popularity finally skyrocketed in the US too. They've become indispensable to business, and their value is also highly cherished by telethieves.) The ubiquity of cellular phones means that cellular-based fraud is on the rise.

Cellular fraud is defined as the unauthorized use, tampering or manipulation of a cellular phone or service. It was treated as a misdemeanor until 1993. Since then, it's a federal crime.

Tumbling Fraud

Aside from subscription fraud (where a tele-thief signs up for service with fraudulently obtained customer information or false identification) the first kind of cellular fraud to appear was a simple deceit known as tumbling fraud, where criminals use fake subscribers numbers from outside a particular calling area to access cellular service. Carriers validate cellular calls by checking the unit number of a phone, its electronic serial number (ESN), against the subscriber number, or mobile identification number (MIN) particular to the caller. These numbers are sent in the header of each call and must match the numbers issued when the phone was sold.

The problem was that carriers couldn't match the numbers instantaneously for a call if the caller is "roaming" outside of the subscription area . Thus, a phone programmed with a fake out-of-area MIN would be approved by the system. By the time a carrier is able to get verification (or lack thereof), the criminals could have changed, or tumbled, a single digit in the made-up MIN and be placing more calls.

Cellular Cloning Fraud

A more interesting (and lucrative) kind of fraud is cellular cloning fraud, where a tele-thief uses 800 MHz FM-capable signal interception equipment (typically a police scanner) that's tuned to one of the wireless reverse control channels. Such a device can grab or "snarf" the MIN and ESN numbers from the airwaves as they are transmitted by a user's cell phone to the wireless carrier's base station. A cell phone emits these identifying numbers when it is turned on, when it crosses from one cell to another and at regular intervals to maintain contact with the network. Tele-thieves are known to frequent airports, bridge overpasses, mall parking lots, or anywhere a large number of cellular phones may be in use to increase their chances to "snarf" a large number of MIN and ESN numbers.

Other ways to fraudulently obtain this information include unethical cellular phone dealer employees (in at least one case, information from a database was sold by an employee), and many security analysts caution that the greatest risks come "from the inside." Less sophisticated crooks engage in the time-honored activity of "dumpster diving" or collecting discarded carbon copies and receipts from trash dumpsters in back of stores that sell cellular phones.

A 1986 law, the Electronic Communications Privacy Act, made it illegal to intercept these calls – but didn't make it illegal to sell the scanners that can do so. A subsequent law, which went into effect in April 1994, made it illegal to sell scanners that are able to pick up cellular frequencies, though anyone with a modicum of electronics knowledge can modify these scanners for illegal use.

Once obtained and decoded, the numbers can then be programmed into another phone using a personal computer and a process that clones a new EPROM chip. This chip will make a phone appear to be a duplicate of the original.

Some cell phones allow the MIN to be changed from the keypad, often after entering a special access code. These codes are released only to authorized service facilities, but can be obtained from the manufacturer's technical manuals. Reprogramming the ESN is a bit more difficult, but obviously not impossible (since the criminals actually do it). Early phones had the ESN programmed into industry-standard ROM chips that were easily replaced. Some later phones stored the ESN in non-volatile memory that could be changed using special cables or connectors. In some models the firmware that controls the phone was also placed on ROMs, which could be later replaced by ingenious evil masterminds. Since a cellular telephone is essentially a radio controlled by a microprocessor, the tele-thieves would modify the portions of the phone's software that accessed the ESN, patching to use the pilfered number instead of the one installed at the factory.

Meanwhile, the unfortunate subscriber normally doesn't have a clue as to what's happening until he or she gets the bill. According to the ***Phoenix Gazette***, a bewildered Albuquerque, New Mexico subscriber once got a 20-pound phone bill delivered in a cardboard box after his phone's ESN was copied.

There are sometimes tell-tale signs that a cell phone has been cloned. A string of wrong numbers or hang-ups may occur. If you've having trouble placing or receiving calls, somebody else could be tying up your line. If you notice anything unusual, particularly on your phone bill, contact your carrier immediately.

Service providers have adopted certain measures to prevent cellular fraud. These include encryption, blocking, blacklisting, user verification and traffic analysis.

Encryption may turn out to be the most effective way to prevent cellular fraud as it prevents eavesdropping on cellular calls and makes it nearly impossible for thieves to steal ESN and PIN number pairs.

Blocking is used by service providers to protect themselves from high risk callers. For example, international calls can be made only with prior approval. In some countries only users with major credit cards and good credit ratings are allowed to make long distance calls.

Blacklisting stolen phones is meant to prevent unauthorized use. Some sort of equipment identity register enables network operators to disable stolen cellular phones on networks around the world.

User verification using Personal Identification Number (PIN) codes is another method for protecting against cellular phone fraud. Similar to an Automatic Teller Machine (ATM) PIN, the four-digit code is entered after dialing the destination number and pressing SEND or TALK. Since this number is sent via DTMF tones (the same touch-tones a landline phone uses) over a reverse voice channel, it will not be intercepted by a cloner listening to the reverse control channels. When PINs where first tested in the US in the 1990s, Bell Atlantic Nynex Mobile claimed an 80% reduction in fraud and Ameritech Cellular Services reported a 96% drop.

This method is vulnerable, however, to a cloner using two scanners and a DTMF decoder. But since the fall of 1995 cell phones have been manufactured with more advanced authentication capabilities based on authentication keys. In this process the cell phone and the base station exchange a "secret handshake" derived from a mathematical algorithm and a 20 digit number. A legal phone identifies itself by transmitting the answer to the algorithm. The keys to this process are stored in the telephone and in the cellular system database, and are never transmitted, so are not vulnerable to interception. This process is transparent to the user, and requires no additional dialing steps.

Traffic analysis detects cellular fraud by using artificial intelligence software to perform analyses similar to "data mining" applications, but in this case to detect suspicious calling patterns that deviate in some way from a customer's "profile", such as a sudden increase in the length of calls or a sudden increase in the number of international calls. The software can also determine whether it is physically possible for the subscriber to be making a call from a current location, based on the location and time of the previous call. In South Africa, the country's two service providers, MTN and Vodacom, use traffic analysis with the International Mobile Equipment Identity (IMEI) – a somewhat ungainly 15 digit number which acts as a unique identifier and is usually printed on the back of the phone underneath the battery - to trace stolen phones.

Radio Frequency (RF) fingerprinting is a more complex, and somewhat successful form of authentication. A "signature" is created and stored for each authorized cellular telephone, consisting of characteristic parameters that uniquely identify the transmitter. The theory is that even between identical cell phones, individual components and tuning variations create enough differences in the transmitted signal that a base station receiver can distinguish one from another. Thus, when a cloned cell phone sends an ESN/MIN pair, the cell system will notice that the transmitted signal doesn't match the signature stored in the subscriber database, and deny the call. This system has been deployed in many markets but some say it has not yet been fully perfected.

PhonePrint from Corsair Communications (Burlington, MA – 781-359-4000, www.corsair.com) is a powerful network-based fraud prevention system that uses an RF fingerprint, unique to each handset, to detect and disconnect calls from cloned phones in real-time. Already installed in many cell sites worldwide, PhonePrint is said to bring secure, highly reliable cloning fraud protection to any of the older analog AMPS or ETACS wireless network, and will soon be available for TDMA digital wireless networks.

Using signal-processing techniques originally developed for government signal intelligence applications, Corsair's engineers have achieved higher resolution data with better signal analysis algorithms than experts had once thought possible to deliver a cellular cloning fraud solution. This analysis generates the unique signal pattern associated with each handset, the "RF fingerprint".

When subscribers roam outside their home markets, PhonePrint protection is extended abroad. Corsair's PhonePrint Roaming Network allows carriers to partner with other PhonePrint markets for seamless, real time coverage between markets. When a system within the Roaming Network processes a call from a user outside its own network, it asks the home market for a copy of the RF fingerprint on file for that handset. In real-time, the print for the roaming handset is automatically made available to compare to the print in question. If the two prints don't match, the call is torn down.

PhonePrint is totally transparent to subscribers. It captures a series of signal intercepts from each active handset in a market to build an RF fingerprint. Once RF fingerprints are in the database for the majority of users in a network, the system matures and fraudulent users have nowhere to hide.

Corsair says that, compared to other wireless fraud solutions such as authentication key systems, PhonePrint is significantly more cost-effective, more user-friendly and easier to install.

Exchanging call detail records (CDRs) is another way cellular networks can protect themselves and their customers against fraud. TSI Telecommunication Services Inc. (TSI) (Tampa, FL – 813-273-3000, www.tsiconnections.com) has developed TSI DataNet, a fraud solution developed specifically for GSM operators. TSI DataNet is an automated service that enables roaming partners to exchange CDRs in near real-time. TSI DataNet's expedient approach is an alternative to the non-automated exchange methods currently used by GSM operators.

The TSI DataNet solution simplifies the management of roaming data collection, translation and routing to and from other wireless operators using TSI's GRX inpack network, based on TSI's general packet radio service roaming exchange (GRX) offering for GPRS that serves as a virtual private network (VPN) for other emerging services. Using DataNet over TSI's inpack GRX network, operators have the same quality and availability to transport CDRs as they have for their wireless packet data transport requirements.

The TSI DataNet system is an advanced central computer complex that receives roamer CDRs from visited operators, then sorts and routes them back to the home operator in near real-time using TAP3 or other operator-specified format. If operators already have a GRX provider, gateway agreements allow operators to use their GRX provider of choice to carry CDRs to TSI DataNet.

TSI's inpack packet data network solution is part of a suite of services, isolutions@tsi, offered to wireless operators, Internet content and wireless application service providers (WASPs). This suite of integrated services bridges the technical and business gaps between wireless operators and Internet content providers so subscribers worldwide can enjoy the benefits of convenient wireless IP services.

Solutions/Prevention

- The simplest, most obvious answer is to turn the phone off when not in use, which is the only time the phone can't be cloned. However, this is impractical for many busy businesspeople, and defeats the purpose of having a cell phone that can be easily accessed.
- Take advantage of the anti-fraud services provided by your cell phone company. Ask your cellular service for a PIN, which must be entered before placing a call. Although PINs can be captured through the airwaves as well, they are an additional safety measure.
- Another device offered by cell phone providers is a cellular-fingerprint, where a legal call is identified through a series of codes emitted by the phone. If the codes and number don't match, the call gets disconnected. A cloned phone, for example, would be unable to emit a code that

matches the phone number. This ties legitimate calls to the actual phone being used.

- Some companies are using anti-fraud software, which detects unusual calling patterns. Check with your provider to see what services are available, and what will work best for you.
- Finally, use common sense when it comes to cellular phone theft: lock the phone and remove it from plain sight in cars; protect your serial and phone numbers as you would credit card numbers; only allow reputable dealers to service your phone; and immediately report stolen cell phones to your carrier.

New in Town: M-Business Fraud

Vast new areas of fraud that the wireless industry hadn't even heard of prior to 2001 became an area of study and deliberation in 2002. Just as some say that "there's nothing new under sun," so these "new" forms of wireless fraud are simply new manifestations or extensions to old kinds of fraud. And so, e-commerce fraud has begat m-business or m-commerce fraud.

Since data can be sent over wireless transmissions, tele-thieves can now become more conventional thieves, using cell phones and new wireless data services to steal goods and services using falsified or stolen identities. It has already been demonstrated that supposedly secure "smart cards" (SIM cards) can be cloned, which means that it should be possible to extract personal data contained in a modern phone and duplicate it too for nefarious purposes. Since the cellular phone is becoming an extension of one's identity, technically brilliant tele-thieves of the near future will not just abscond with credit card numbers – they'll be making withdrawals with your bank account (or perhaps temporarily making a deposit of some illicit funds before transferring them elsewhere), rerouting short messaging service data, opening your garage door and doing who-knows-what-else.

Vendors are not standing still, although "open standards" for m-business security have yet to be conceived, let alone deployed. Still there are proprietary advanced technologies being developed to combat what's not yet arrived. For example, Diversinet (Toronto, Ontario, Canada – 416–756-2324, www.diversinet.com) is working on wireless Internet security technology. The company's Passport software package offers authentication for resource-constrained products, such as wireless phones and pagers.

Data Network Security

Today's businesses run on data. Expensive e-business, customer relationship management (CRM), enterprise resource planning (ERP), and date warehouse systems abound to help the business community to increase their profit margin. To earn their keep, all of these systems (and more) are

➤ Calling Card Tips

- Watch your cellular phone bill for unusual activity. If you find a call you didn't make, notify your service provider so appropriate measures can be taken.
- Don't give out PINs or other sensitive information over your cell phone.
- Don't flash or loudly give out your card number.
- Select a four-digit PIN you can remember, and make sure it isn't printed on your calling card. This obviously reduces the likelihood of fraudulent use if you lose the card or it's stolen.
- Challenge anyone from the phone company asking you for your PIN.
- Don't comply with requests to enter your PIN on touchtone keypads, because the numbers can be identified by their distinguishing tones.

dependent on accurate, clean data. Lose your data and you just might lose your business.

Maintaining the security of business networks and the data that run over them entails keeping networking devices, computer systems and information itself safe from the following:

- Unauthorized disclosure.
- Unauthorized destruction.
- Unauthorized modification.
- Unavailability of service.
- Unauthorized access to services.

The media frenzy over hackers has made data network security a pseudo-glamorous occupation. The public thinks of business LAN / intranet security (if it thinks of it at all) as a sort of castle surrounded by firewalls, beyond which lay the barbarian horde of hackers, crackers and an assortment of other bizarre and deranged miscreants. Through tiny peepholes in the walls, nasty viruses occasionally creep into the castle, defeated only after heroic battles resulting in many file casualties.

Indeed, in a perfect world, things would be that secure. But high and thick firewalls do not a secure business make, unless it is a SOHO, since small companies with small intranets and brochure-like Web sites have few critical segmented systems to protect. As it turns out, as many problems arise from the many bizarre and deranged miscreants dwelling ***within*** the corporate castle as from without. A whole fifth column of disgruntled employees (okay, at least one) with some technical expertise is lurking among the staff, ready to be transformed into hackers at a moment's notice.

Companies must take care of both the company perimeter and internal networking risks. Teleworkers, SOHOs and team-like smaller businesses running their own small network or even unnetworked PCs must focus on the perimeter, keeping that little firewall or proxy server up and running, while larger companies must add to this a paranoid stance regarding their employees.

Security in Distributed Environments

Companies are at greater risk than ever before precisely because of advanced networking technology. Distributed applications and computer networks make possible enterprise networks, since they operate between systems to allow users to run programs and access resources on multiple systems at the same time. They also make possible many points of entry for evildoers to attack or infiltrate such systems, since it is always the case that some users must access systems at remote sites or that database replication is necessary, and that this can occur over public data networks. The essential, decentralized nature of distributed networked environments also often results in slipshod or in some cases no consideration for the LAN security. The larger and more complicated the network, the greater the security problems and the more sophisticated and encompassing must be the ultimate response to put things back in order.

The essence of man-made data communications security "disasters" falls into two categories:

The first is the hacker or cracker. He or she wants to poke around systems, for fun, curiosity or profit (industrial espionage pays very well). There is no real "hacker archetype" Hackers range from the foolish to the unscrupulous, and their sometimes destructive appearance can never be predicted with any certainty.

The second are what we shall kindly refer to as non-technical employees, those unfortunate souls who regularly transport disks inadvertently laden with computer viruses ***and*** somehow manage to coincidentally deactivate their workstation's virus checker! These are bumblers who damage hardware (but somehow never themselves). One should more respectfully refer to these characters as "quasi-technical" employees, since they do know enough to get both them and the entire system in trouble, if just given half a chance (your job, to paraphrase W.C. Fields, is not to give the sucker an even break).

In any event, the following kinds of security measures should always be in place:

Identification, Authentication, Authorization

Indentification, authentication and authorization are all different steps to gaining access to the resources on a network.

Identification = claim to be a given individual
Authentication = "proof" that you are who you say you are
Authorization = privileges of a given individual

Authentication, for users and systems, can be achieved in three ways: Something you ***know*** (a password such as MyDogFido), something you ***have*** (a smart card), or something you ***are*** (voice, fingerprint, retinal or face recognition).

Essentially a hacker must try to masquerade as somebody with legitimate high-level access rights, which means they must come up with a logon and password for these prized user accounts. The company's internal "House Hacker" will go dumpster diving in the office trashcans, while the exterior-based "Foreign Hacker" will visit the actual company dumpster in the alleyway.

Often however, the hacker is helped by loose security, which accounts for most of the security issues on a network. Examples include easy access to key network hardware (including, but not limited to servers) and/or software, weak passwords, user work-arounds to existing security measures, non-monitored access to key user or corporate data (such as payroll files, corporate strategy plans, and product design documents). Loose security ups the risks of security breaches by both internal and external hackers.

Third party access control and security software is also available for most LAN environments. If the LAN is being used to process and store production data or confidential/sensitive data, you should consider additional security to that provided by the LAN operating system.

Thus, the first major consideration must be given to authentication, which commonly provides verification of users during an initial logon and a secure way for one server to be assured that a fellow server has accurately identified a user so that users need to log on but once.

The ways data networks are hacked by hackers are in many cases similar to the ways voice networks are cracked by crackers. They try to masquerade as somebody with legitimate high-level access rights, which means they must come up with a logon and password for these prized user accounts. The company's internal "House Hacker" will go dumpster diving in the office trashcans, while the exterior-based "Foreign Hacker" will visit the actual company dumpster in the alleyway. If a modem is connected to your LAN, a hacker will use a demon dialer to generate and test all possible entry authorization codes.

Authentication is a network operating system speciality, but passwords, like little locks, can only keep honest people honest.

If your LAN is being used to process and store production data or confidential/sensitive data, you should consider additional security to that provided by the LAN operating system. An industrial spy can even wear new disguises, including that of an outside cracker or that of an employee that's an inside cracker.

First, you can you can password individual workstations from its BIOS (basic input output system) utility screen. You can usually enter the BIOS by pressing either the F1 or the delete key as you boot up your computer. If the machine starts to launch Windows, or whichever OS you have installed, you waited too long to press the appropriate keys. Don't forget the passwords, or you will have to hack your way into your own workstation.

But a determined hacker can have assembled his own "toolkit." There exist several utilities hackers may use to gather data including screen grabbers which either save or send copies of workstation screens, TSRs which log crucial data like password keystrokes, and BIOS password bypassers which bypass ROM BIOS passwords and allow hackers to alter BIOS settings.

More enhanced authentication measures may be provided by any one of the following:

Dial-back systems. Authorized remote users who dial into a system are disconnected and then called back by the system to verify that they are calling from a recognized, trusted phone number, such as their home. This helps for authentication purposes only so long as no other dial-up software is installed on workstations attached to the LAN and that any password controls that are part of the dial-up software are in effect. Dial-up/dial-back password standards should follow those established for the LAN.

Kerberos. Kerberos is a network authentication protocol that's designed to provide strong authentication by using secret-key cryptography. A free implementation of this protocol is available from the Massachusetts Institute of Technology and it is available in many commercial products. Kerberos allows network applications to verify the identity of clients and other peer servers by using a third-party security server. The user is authenticated upon first logon, and other servers are provided with verification that users are legitimate.

Token authentication devices. Microprocessor-controlled "smart cards" can perform what's called two-factor authentication. During the 1990s "smart" credit cards incorporating tiny chips have seen considerable use in France and other parts of Europe. A set of standardized contacts on the front of each card supplements the card's previous coded magnetic stripe technology. The smart card can be made to generate one-time passwords good for just a single logon session.

GPS (Global Positioning System). This interesting approach uses a system that can verify the physical location of a user anywhere on Earth, thus at least ensuring that a call is not being transmitted from an unauthorized location. GPS is funded by and controlled by the U. S. Department of Defense (DOD). While there are many thousands of civil users of GPS world-wide, the system was designed for and is operated by the US military. GPS provides specially coded satellite signals that is processed in a GPS receiver, enabling the receiver to compute not just position, but velocity and movement through time as well. Four GPS satellite signals are used to compute positions in three dimensions and the time offset in the receiver clock.

Biometric devices. These scan eyes, whole faces or fingerprints to verify a user's identity so he or she may access the network and thus the computer systems attached to it as well as other facilities. For example, Advanced Biometric Systems, a division of Inforonics (Littleton, MA – 978-698-7300, www.inforonics.com) has an extremely powerful face-recognition system.

Authentication on the Internet is difficult to do since there is no previous (or perhaps any) real contact between parties. This requires authentication techniques that use such things as digital certificates. A digital certificate is an electronic "credit card" that establishes your credentials when doing business or other transactions on the Web. It is issued by a recognized certification authority (CA). It contains your name, a serial number, expiration dates, a copy of the certificate holder's public key (used for encrypting messages and digital signatures), and the digital signature of the certificate-issuing authority so that a recipient can verify that the certificate is real. A vendor trusts that the third-party CA has previously verified a user's credentials, in much the same way that retail stores trust that a financial institution will stand behind its credit card users. Some digital certificates conform to a standard, X.509. Digital certificates can be kept in registries so that authenticating users can look up other users' public keys.

There are other reasons why authentication is important, including a need to ensure that a message has not been altered in transit and that a message is indeed from the person who claims to have sent it. In addition, messages should be nonrevocable, i.e., the digital signing technique should prevent the sender from claiming that he or she did not send the message. These techniques are possible with digital signatures and public-key cryptography.

Authorization

Once a user is authenticated, he or she will be able to access resources based on authorizations. For example, a user may only be authorized to log in at night, or be restricted from accessing particular files or running particular applications.

Users should only be given rights to directories they need to accomplish their job. If a user needs temporary access to a directory, the access rights should be removed when they are no longer needed. Similarly, any administrative rights temporarily granted to users must be rescinded once the job requiring them is complete.

Files access attributes should be granted based on need. Obviously, files containing confidential or sensitive information should be restricted to a minimum number of users.

You should review system accounts on a regular bases and delete any accounts that are no longer required. Users should not be allowed concurrent signon privileges. The only exception to this rule should be the LAN administrator and his/her backup.

Users who are authorized to enter sensitive transactions or who perform sensitive and/or confidential work should be restricted to a specific workstation, preferably located in a restricted area. Access to the server itself should be restricted to the LAN administrator and his/her backup.

Consideration should be given to restrict user access to business hours only, especially for those users who are authorized to access and use sensitive and/or confidential data.

Network commands should be, in most instances, restricted to administrators and security personnel.

Data entered at a PC attached to a LAN can be transmitted over the network in ASCII text. Anyone able to use "sniffer" program will be able to to view and capture data transmitted over the LAN. It is therefore crucial to scrutinize the type of data being stored and processed on the LAN.

Encyrption Techniques

Encyrption protects data transmitted over a network from eavesdroppers. The Internet has presented new security challenges. Are you getting ready to send or open an e-mail with sensitive data? How can you be certain no one else has opened the mail intended for you? Worse yet, can you be absolutely sure that important file you just downloaded from your coworker was not intercepted by a third party and corrupted with a virus? Don't believe the hype that sending e-mail is a safe and secure

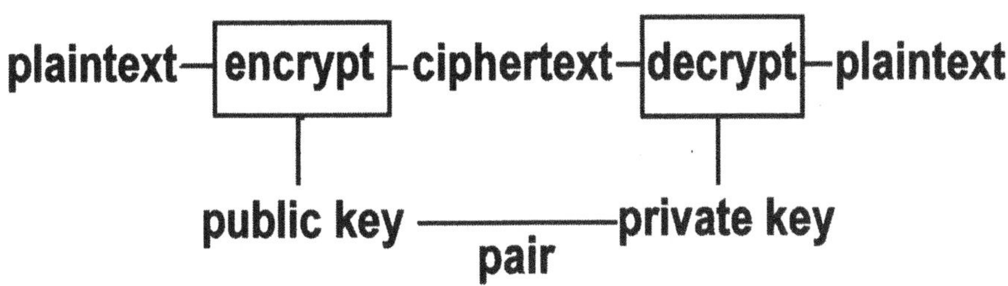

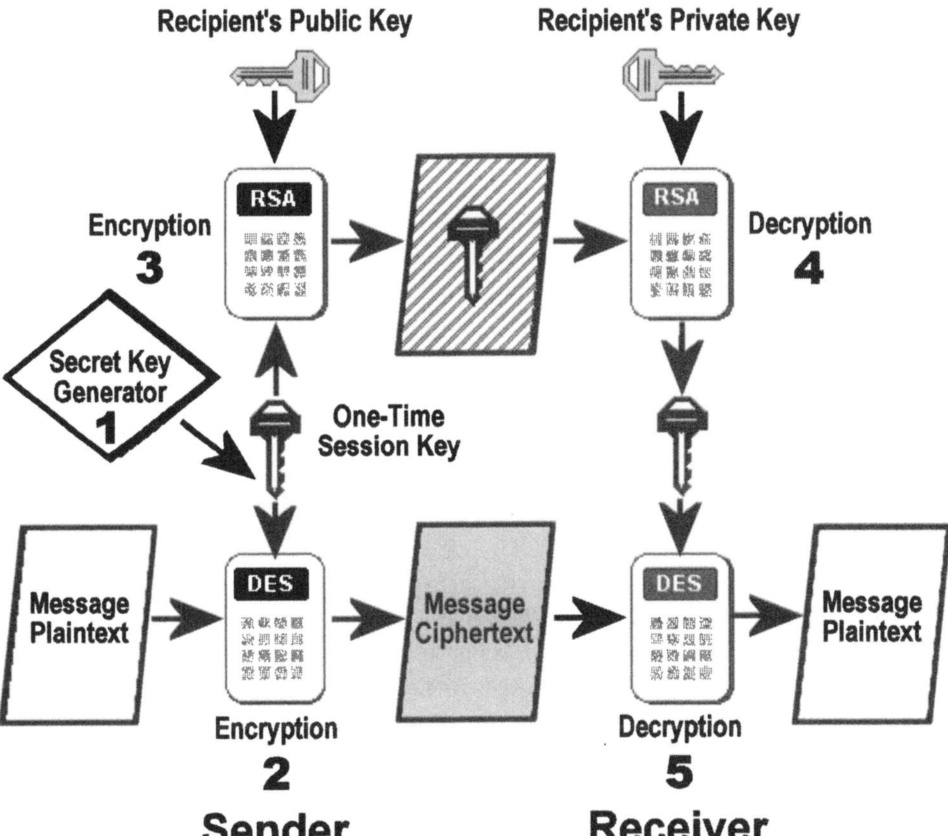

method of transmitting data. It's not.

So how do you protect your e-mail? One way is with encryption, the process in which data is scrambled so that only an intended recipient armed with the proper decryption key can decipher it.

There are two methods to encrypt data. One approach uses the same key to encrypt at the sender's end and decrypt at the receiving end. The other, more interesting approach is a two-key system that uses a public key and a private key, which are mathematically related. The public-key can be used by anyone. It can be given to your co-workers or posted on a Web site. The other key is private and is customized and retained by the person who created it. With the two-key system, anyone may encrypt a message with the public key. However, only the intended recipient holding the private key can decrypted the message.

The two most popular public-key encryption standards are DES (data encryption standard) and PGP (Pretty Good Privacy). Both offer two-key security.

Digital signature schemes are a similar technology that use public-key cryptography to authenticate e-mail. With digital signatures (not to be confused with electronic signatures), e-mail forgeries are virtually impossible. Digital signatures also use a two-key scheme: one private and one public. When someone sends an e-mail with a private key, the recipient who has the public key can be sure of the identity of the sender. While straight encryption turns an e-mail into an indecipherable piece of data, a digital signature strictly provides an authenticating key without altering the e-mail itself.

PGP is a 128-bit encryption program. The number of bits in an encryption scheme increases the level of security exponentially. For example, a 2-bit algorithm takes twice as long to crack as a 1-bit algorithm, and a 3-bit algorithm takes twice as long as the 2-bit scheme. Rest assured, 128-bit encryption is virtually impossible to crack.

As in many areas of computing, standards are lacking when it comes to encryption. There is no industry standard that lets different encryption programs communicate with one another. Two popular protocols used with e-mail to encrypt data are S/MIME from RSA Data Security and Open PGP from Pretty Good Privacy. If you use either protocol to encrypt a message, the recipient must use the same protocol in order to receive it. In short, you have to choose the encryption scheme that supports your e-mail software.

PGP is the dominant encryption software of the two protocols mentioned above, and it is compatible with all e-mail programs. However, other encryption packages get the job done. Some e-mail packages, such as Outlook Express, have S/MIME built in, so you might want to choose this standard rather than applying PGP.

Of course, encryption isn't just restricted to e-mail. Any file on your system can be encrypted.

Symantec's (Cupertino, CA – Cupertino, CA – 408-517-8000, www.symantec.com) Norton Your

➤ Electronic Signatures and Digital Signatures

Electronic signature software will be cornerstone of the upcoming wave of new e-business applications. An electronic signature is basically a scan of a written signature gathered with a stylus and tablet. To validate the authenticity of the signature, a user creates an electronic signature card and repeatedly signs their name. The software uses a simple algorithm to learn how the signature is created. It senses the speed of the signature, what it looks like and how much pressure was applied to the pen. Users then share the signature card with the people they do business with. Signatures applied to new documents with the software are verified against the signature card to assure their authenticity.

Electronic signatures will see broad application throughout the computing world. They will minimize paperwork and cut snail mail costs. An electronic document can be received, edited and returned quickly and easily. Applications might include bank loans and approvals, purchases, contract negotiations, human resource and accounting documents and all matters in which a signature is required either by law or policy.

Eyes Only for various flavors of Windows keeps your private data private on the desktop, laptop, network or over the Internet. It offers automatic encryption and decryption, ensuring that no one can read your files without your permission.

Secure Session and Transaction Systems

Once people and servers have been authenticated, secure sessions can commence over the network. This involves software that incorporates such protocols as:

- **S-HTTP (Secure Hypertext Transfer Protocol)** which is a version of the HTTP protocol that encrypts HTTP message exchanges between client and server (it's used by Mastercard and Visa).
- **SSL (Secure Sockets Layer)** is like S-HTTP but goes further in that it can encrypt all data that is exchanged, not just HTTP. It was developed by Netscape and offers RSA and DES Encryption. It's a Presentation Layer (Layer 6) protocol but it can support any Application Layer (7) protocol.
- **Private Communication Technology (PCT)** protocol, a security protocol that provides privacy over the Internet. Like Secure Sockets Layer (SSL), the PCT protocol is intended to prevent eavesdropping on communications in client/server applications, with servers always being authenticated and clients being authenticated at the server's option. However, this protocol is said to correct or improve upon several weaknesses of SSL.

Additional Network Security Risks

Hackers, eavesdroppers and other malicious individuals aren't the only risk to an individual or corporate system. Computer viruses get the most media attention, but there are also other risks that need to be guarded against, such as those related to e-mail, FTP and TCP/IP.

Viruses

A virus is a piece of self-replicating program code, just a few tens or hundreds of bytes in size. Like a real virus, it uses something (your PC) to copy itself and mess up you PC and network in the process: Viruses can corrupt or erase files, display amusing messages, graphics or sounds, or just crash your system regularly. Viruses are stealthy and try to conceal themselves or even attack anti-virus programs, such as those from Norton and McAfee. Some viruses contain a "payload" and a "trigger," which causes the virus to deliver its payload. The trigger may be a particular system date, the number of re-boots or the number of floppy disks infected.

Boot sector viruses infect the first sector (boot sector) of floppy disks and the partition sector or the boot sector of hard disks when the PC is booted. Executable file viruses infect program files on local drives or network drives. Macro viruses infect the program-like macros within files such as Word documents and Excel spreadsheets.

A polymorphic virus is a virus that cannot be detected by searching for a simple, single sequence of bytes, since it mutates with every replication. It's similiar to a Stealth virus, which are viruses that actively try to conceal their presence.

A Trojan Horse is a program intended to perform some covert and usually malicious act but unlike a regular virus, however, it doesn't reproduce itself.

A worm is a virus-like program that propagates itself from system to system, using up resources and slowing down the network and/or user's workstations. Unlike a virus, it does not attach itself to a host program. The "I love you" virus is an example of a worm.

➤ The "Seven Deadly Sins" of Computer Virus Protection

As a service to executives and managers, United Messaging (West Chester, PA – 888-993-5088. www.unitedmessaging.com) a messaging and e-mail outsourcing provider, has created this list of what it has found to be "the Seven Deadly Sins of computer virus protection." Avoiding them will help keep your company virus free and forward moving. Committing just one could spell disaster. United Messaging updates this list and keeps it posted on www.unitedmessaging.com:

Sin #1 Letting down your guard. Don't assume "it can't happen to you." Implement a layered defense and ensure that it is updated regularly – and that someone monitors that defense 24x7. It's vital that your mail server be located behind a firewall. Make sure antiviral filtering and virus scanning is up and running on your mail server(s) and ensure they are constantly updated in real time. Frequent, timely updates are essential – a system that was updated last week is typically out-of-date when it comes to the latest bug. Make sure antivirus software is installed on every desktop. Even if software is updated, a virus may slip through and be dormant until activated by opening the file, so scan all saved files regularly.

Sin #2 Security Shortcuts. Don't cut corners when it comes to security. Many organizations justify low expenses for security by pointing to their blemish-free record of security incidents. Ask what they use to detect break-ins or infections and the real blemish becomes obvious. All network components, applications, and servers should be monitored. Use secure logging and make sure you audit it daily.

Sin #3 Communication Failure. Don't assume all IT staff knows your system and processes for combating viruses. Clearly defined roles and responsibilities will maximize coordination and minimize duplication. Since speed is critical, you don't want people guessing about what the next step should be. Make sure your team knows what tools and resources are at their disposal. For a widespread event, make sure end-users are able to communicate with the incident response team in a pre-defined automated manner. Communicate the team members, processes, and procedures to all employees.

Sin #4 No Backup Plan. Expect the Unexpected.. Don't wait until tomorrow. If a virus does destroy your files, you'll need recent backups of vital files to recover. Viruses can strike at an inconvenient moment, despite your best efforts. It's imperative that you consistently backup your files and store them offline.

Sin #5 End-user Dependence. Don't assume your employees, as skilled as they may be, can keep you out of trouble. Employees may be empowered, trustworthy, dependable and knowledgeable, but it only takes one error or omission to open the door to a serious virus. The best defense is to make sure that it never becomes an end-user problem. Push the antivirus software out to users and log them. Don't depend on users to be diligent about security. Users who access free mail accounts from inside the corporate network can introduce many infections. Prohibit access to unsecured systems through policy and tools – and don't forget communication and training.

Sin #6 Maintaining a Closed Door Policy. Don't ask, don't tell, is no way to deal with the issue. Don't keep your security procedures a secret. In many organizations, security is someone else's problem. Make it everyone's responsibility. Clearly define the rules and make sure everyone follows them. Spell out the consequences. Help users view security from a personal perspective with the example driven from the top. When it's part of the corporate culture, the mission becomes possible.

Sin #7 Taking on More Than You Can Handle. Don't overburden your already frenetic IT staff with frustrating e-mail virus issues. Consider outsourcing your corporate e-mail to a messaging specialist for better budget control, reliability, and tighter security. Even large companies now find it a daunting task to keep up with the necessary technologies, measures, updates, and audits. Outsourcing makes good financial sense and often leads to improved messaging strategies that, in the end, may lead to a more rewarding business afterlife.

SMTP E-Mail Risks

Turn Commands. A "turn command" is a command sent to an SMTP/SendMail server that tells the server to forward your domain's e-mail (e.g. all mail to xyz.company.com) to an address. Note that the address specified in the turn command is not necessarily a friendly address, so a hacker could have your e-mail forwarded to an unfriendly domain where it could be read.

No "Mail From" Verification. Since SMTP e-mail is essentially an ASCII text message with SMTP commands, hackers can falsify "from" information and bypass your existing security.

E-Mail Bombs. E-mail bombs are programs similar to batch routines that send multiple e-mails to the same address. E-mail bombs generally use up lots of valuable disk space and may overload your SMTP e-mail or Web server with the incoming data.

Viruses. Viruses most often enter networks and PCs when they are sent to users in the form of SMTP e-mail attachments.

FTP Risks

Anonymous FTP. Anonymous FTP, similar to a "guest" account on a network file server, allows any user to login to your FTP server to download data. Anonymous FTP could provide front door access to your FTP site and beyond, and most administrators would be cautioned against its use.

Inherent TCP/IP Risks

Denial of Service. DoS (denial of service) attacks cause the loss of access to a resource rather than allow the attacker to gain unauthorized access to the resource. They involve overloading a resource such as disk space, network bandwidth, internal tables of memory or input buffers (buffer overflow). The overload causes the host or particular service to become unavailable for legitimate use. This could be blocking access to a resource all the way up to causing a host to crash. There are numerous denial of DoS attacks and the solutions to them are not always easy. Some attacks can be blocked with filters. Others can be as simple as turning the particular service off if it is not needed. But others cannot be easily blocked while still allowing normal use of the service. For example, consider the case of the Firebox II from WatchGuard (Seattle, WA – 206-521-8340, www.watchguard.com), a firewall security applicance for business networks. When it first appeared in 2000 it was discovered that it was vulnerable to a DoS attack if the feature "external access to FTP Proxy" was enabled, which is not the default setting. A remote attacker could, therefore, cause a denial of service by connecting to the FTP proxy port and flood the Firebox's port with a large number of FTP and SMTP requests, which could cause the port and the proxy handling service to crash. This kind of attack could also crash all other services on the device, and consume up to 100% of the CPU resources. The firewall would have to be shut down and restarted to regain normal functionality (fortunately, the problem was quickly dealt with and a software patch for the router soon became available at the WatchGuard Software Patches Web page).

IP Sniffer. A program that steals network IP addresses by reading IP packets. Hackers can trick routers by sending harmful data masked with internal IP addresses.

IP Spoofing. A process by which unfriendly IP packets are sent into a network with a "friendly" IP address found with an IP Sniffer (similar to a Trojan Horse process).

Scanner. A program which scans IP networks and maps IP addresses to domain names using a DNS server. Scanners let hackers determine valid internal IP address ranges and specific addresses for key network servers, gateways, and DNS.

World Wide Web Risks

Web Links. Hot Links expose Web and IP addresses which can be used as a back door into a network.

Web Resources. There are literally hundreds, if not thousands, of Web and FTP sites with hacker programs, RFC's (requests for comment), and explanations of how to hack network operating systems, UNIX servers, passwords, etc.

Firewalls

A firewall is a barrier between your network and the Internet, through which only authorized traffic can pass. As traffic passes between your network and the Internet it's examined by the firewall which follows the strict guideline of "whatever is not expressly permitted is denied".

Firewalls can protect your network from unprotected networks, including the Internet. A firewall uses rules that determine what traffic is allowed or disallowed, supports a variety of IP applications and different Internet services, and defines which IP addresses and hosts you wish to permit or deny.

Most firewalls are used to screen traffic between a company's internal network and the Internet, but firewalls can also secure one part of a network from another. For instance you could secure your corporate accounting department from your subsidiary's network. Some security experts think that a firewall is an effective way to protect your network from malicious Internet hackers.

Hardware Firewalls

Firewalls can be implemented as relatively standalone hardware devices, particularly if a high-band width connection to the Internet is needed. Examples of hardware firewalls are the Cisco PIX 515 and the WatchGuard Firebox II.

Software Firewalls

Firewalls can also exist as PC software. Personal firewalls typically run below the normal PC network drivers, or as a replacement for the network driver. They're quite flexible, but older CPUs will take a bit of a hit in performance.

The earliest firewalls were network level firewalls, or "packet filters" which tracked the source and destination addresses of IP packets on the network – the simplest data to verify when receiving data from the Internet. While packet filtering is implemented on most routers and is transparent to users, packet filters can be easily bypassed with IP Sniffers/Spoofers which allow hackers to trick routers into permitting unfriendly data onto the network. Routers working alone have no means of detecting spoofed packets – allowing unauthorized access your network.

By focusing only on address information, packet filters looks exclusively at the lowest layers of the OSI model – those dealing only with physical device connection to a circuit, and the routing of packets to those devices. Other firewall technologies look at other layers. Ultimately the best firewall examines all layers of the OSI model simultaneously to provide the best security, doing so without slowing down applications or users on the network.

In the next generation of technology, "proxy servers" or "application gateways" scrutinize the communication between IP applications by examining the actual data being transmitted. This thwarts hackers who spoof IP packets to gain unauthorized access to your network. In short, such application gateway firewalls check traffic at the application level – for example, FTP, e-mail or Telenet. An application-level firewall also often readdresses outgoing traffic so it appears to have originated from the

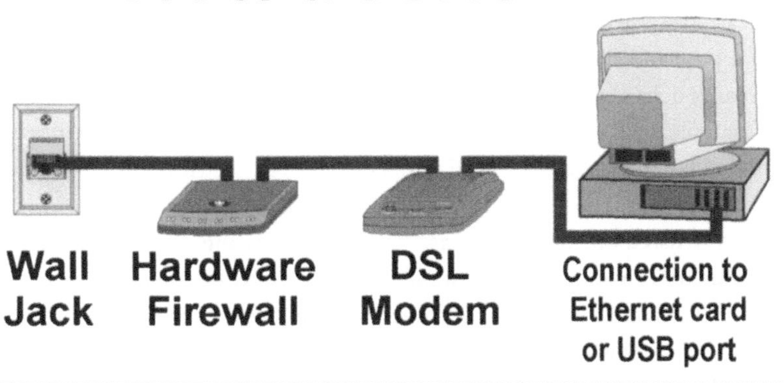

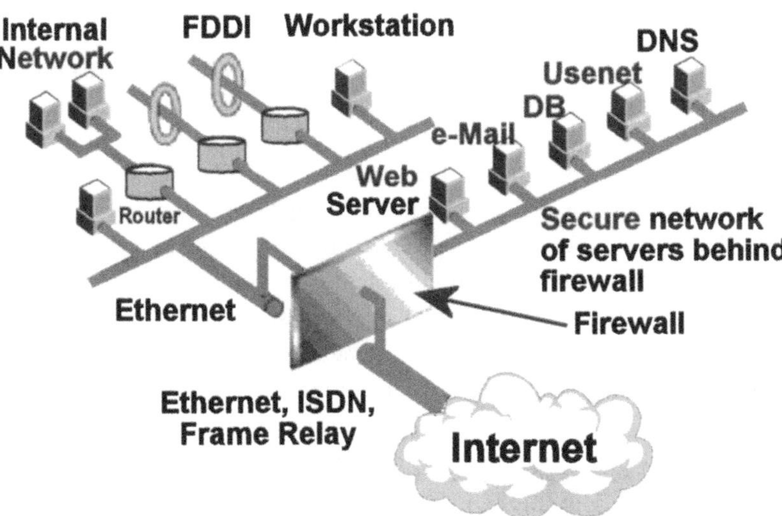

firewall rather than the internal host.

Thus, while a firewall based on packet filtering must allow some level of direct packet traffic between the Internet and the hosts on the protected networks, a firewall based on proxy technology doesn't have this characteristic and can, therefore, provide a higher level of security. Application Gateways can also validate other security keys, like user passwords and service requests, that only appear within the application layer of the OSI model. Since application gateways operate as one-to-one proxies for specific applications, there must be a proxy agent installed for every IP service (i.e.

RealAudio, Web/ HTML, FTP, etc.) to which users need access.

These proxies understand what valid data looks like, and can tell the gateway to pass that data on to the proxy's counterpart on the local net. Although application gateways offer a high level of security, they have several drawbacks. For example, there is a lag between the release of new IP services and the availability of corresponding proxy agents, meaning users might need to wait for mission critical applications to be available to them. Also the installation, maintenance, and upgrade of the proxy agents impose additional administrative and maintenance burdens. Furthermore, proxies introduce performance delays since incoming data must be processed twice – by the gateway and the proxy agent. For example, FTP sends a request to the proxy FTP agent, which in-turn speaks with the internal FTP server to complete the request. FTP data is also verified through the proxy.

Another kind of firewall is the "trusted gateway." This kind of firewall relies on very secure operating systems, called "trusted operating systems." These are operating systems that satisfy a number of stringent governmental or international standards security requirements. The trusted gateway firewall system itself is divided into three software compartments: that which interacts with the Internet, that which interacts with the enterprise, and a trusted gateway that mediates communications between the other two compartments. The operating system prevents applications that run in one compartment from accessing resources outside that compartment. Any application that runs on the Internet compartment (e.g. a Web server), can only have access to resources in the Internet compartment (e.g. public HTML pages), or else it must use the trusted gateway to ask for information from the enterprise compartment.

A firewall is not always a single computer or a standalone device. A firewall may be composed of two filtering routers and one or more proxy servers running on one or more "bastion" hosts, all of which are interconnected via a small, dedicated LAN stretching between the two routers. The external router blocks attacks that use IP to break security (IP address spoofing, source routing, packet fragments), while proxy servers block attacks that would exploit a vulnerability in a higher layer protocol or service. The internal router blocks traffic from leaving the protected network except through the proxy servers. The trick is to formulate the specific criteria by which the packets of an intruder can be distinguished from those of legitimate and unauthorized users.

In the ancient days when Unix was the only real operating system and Windows was just a gleam in Bill Gate's eye, what we now consider traditional firewall technology just wasn't a viable option for most organizations because they were Unix based, requiring Unix gurus and Unix training. Today, however, even residential PCs should have firewalls attached to their broadband modems. Such firewalls include the Sygate Personal Firewall Pro for small businesses and the McAfee Firewall, which is a software firewall designed specifically for home users. These are very easy to use and configure, and has almost any feature a home user could want. At less than $30 with free updates, they are also an excellent value.

The Three Myths of Firewalls

Not everyone is enthralled with firewalls. Bob Blakley, a Security Architect at IBM, coined the following "Three Myths of Firewalls." His "Three Myths" explains why cryptographic techniques such as those used by Kerberos are better when compared to stopgap attempts that some firewalls represent.

1. **We've got the place surrounded.** Firewalls make the assumption that the only way in or out

of a corporate network is through the firewalls; that there are no "back doors" to your network. In practice, this is rarely the case, especially for a network which spans a large enterprise. Users may setup their own backdoors, using modems, terminal servers, or use such programs as "PC Anywhere" so that they can work from home. The more inconvenient a firewall is to your user community, the more likely someone will set up their own "back door" channel to their machine, thus bypassing your firewall.

2. **Nobody here but us chickens.** Firewalls make the assumption that all of the bad guys are on the outside of the firewall, and everyone on the inside of the can be considered trustworthy. This neglects the large number of corporate computer crimes which are committed by insiders.
3. **Sticks and Stones may break my bones, but words will never hurt me.** This myth may also be restated as "Sticks and Stones may break my bones, but Word will never hurt me." Newly evolving systems are more and more blurring the lines between data and executables. With the advent of Word macros, Javascript, Java, and other forms executable fragments which can be embedded inside data, a security model which neglects this will leave you wide open to a wide range of attacks.

Intrusion Detection Systems

Firewalls impose a barrier at the point of connection between the Internet and the protected network to keep out hackers. Although firewalls are a necessary component to a security strategy, they are commonly known not to be foolproof. They are difficult to install and configure correctly, even by experts. Minor errors in configuration can leave your systems wide-open to hackers. An additional difficulty with firewalls is that they need continuous updating and monitoring to make sure the configuration matches the needs of the organization deploying this type of technology.

Even when firewalls are working at their full ability, such perimeter defenses, while essential, are not capable of telling you what happened when system misuse has occurred. Firewalls can tell you that there was an attempt attack may have occurred and it was stopped. But if the attack is really successful, there won't be an alarm. Any experienced security professional will tell you that no matter how much effort is given to perimeter defenses, security breaches will happen, especially when you consider experts' estimates that up to 70% of all system misuse is caused by people within the affected organization. These are people who already have access to the network and the appropriate passwords.

Real-time network intrusion technologies are an effective second line of defense behind firewalls. Indeed, a sophisticated intrusion detection system (IDS) may consist of three key components: firewalls, some kind of network intrusion detection, and host-based data and network integrity (DNI) tools.

Roughly speaking, an intrusion detection system can assume three forms:

- **Network Intrusion Detection System (NIDS).** This monitors packets moving across the network and determines if a hacker or cracker is breaking into a system, or is causing something disruptive such as a denial of service attack. Such a system might find that many TCP connection requests are occurring to different ports on a target device, thus uncovering a TCP port scan. A NIDS can run on the target machine, watching its own traffic, or it can run on an independent machine and watch ***all network traffic***, including traffic from a hub, router or probe. A "network" IDS can monitor many devices, while other, lesser IDSs monitor only the machine on which they are installed.

- **System Integrity Verifier (SIV).** This looks for backdoors left by hackers, the tell-tale sign of which is usually one or more changed files. The most popular SIV is Tripwire from Tripwire, Inc. (Portland, OR – 503-223-0280, www.tripwire.com). Since SIVs are watching files, they can also scrutinize file-like operating system components, such as the Windows registry and "chron" configuration. They may also detect when a legitimate user somehow acquires root/administrator level privileges. In the case of Tripwire for Servers, it's a bit more sophisticated in that it utilizes known facts about the system it is protecting, such as which system calls are being made by the software, not just which files have been recently altered. Tripwire collects and monitors all of this data. Tripwire is an "anomaly-based" SIV in that it operates under the premise that any attack no matter how complex will alter the system in some way. When the system detects any deviation from normal, it alerts the system administrator.
- **Log File Monitor (LFM).** This keeps tabs on network services-generated log files. Like an NIDS, these systems are also looking for suspicious patterns, but in this case in the log files. A tampered log file almost certainly suggests that the system is being hacked. A typical example would be a parser for HTTP server log files that are looking for intruders who try well-known security holes, such as the "phf" attack. An example of an LFM is "Swatch" which is a very simple, freebie program written in the Perl programming language that monitors log files. Swatch lets you to automatically scan log files for particular entries and then take appropriate action, such as sending you e-mail, printing a message on your screen, or running a program.

Wireless LAN Security

The most popular wireless LAN is based on the 802.11b standard which broadcasts over the 2.4 GHz frequency band. Of course, these systems have their own security issues. And because wireless networking broadcasts information that can be intercepted more easily than wired communications, a number of security concerns should be carefully considered before deciding on deployment of this technology. Hackers and other malicious parties now regularly drive around office parks and neighborhoods with laptops equipped with wireless network cards attempting to connect to any discovered wireless networks (this is practice is called "war-driving").

There are now Web sites that publish the locations of discovered wireless networks (e.g., www.netstumbler.com). The range for many wireless 802.11b devices for home use is 300 feet and this is growing as manufacturers introduce new products. Hackers often enlarge their antennas to their wireless network cards to increase the reception range of their cards.

Unfortunately, there are a number of security vulnerabilities associated with the 802.11b networking protocol:

- Sever Set ID (SSID) usually sent "in the clear" (e.g., unencrypted). SSID is a configurable identification that allows clients to communicate to the appropriate base station. With proper configuration, only clients that are configured with the same SSID can communicate with base stations having the same SSID. Unfortunately the Wireless Equivalent Privacy (WEP) encryption standard employed by 802.11b does not encrypt the SSID so if an attacker can receive your wireless network signal it is only a matter of time before they intercept your SSID. Once they have the SSID they can generally connect to your network unless other steps are taken (see section on mitigation techniques below).

- WEP encryption scheme is flawed. WEP can be configured in three possible modes: no encryption, 40 bit encryption and 128 bit encryption. WEP encryption has a number of vulnerabilities that allow attackers to eventually compromise data encrypted only with WEP. Today, there are readily available tools to automate the process of cracking WEP encryption. These tools take a lot of network traffic (millions of packets) to get the WEP key. On most home networks this would take longer than most people are willing to wait. Still, if the network is very busy, the WEP key can be cracked and obtained in as little as 15 minutes. A fix for the WEP encryption scheme is not slated to be addressed until sometime in 2002.
- Many wireless base stations have Simple Network Management Protocol (SNMP) enabled. If the community string (essentially a password) for this service is not properly configured, an intruder can potentially read and write sensitive data on the base station.
- All wireless technologies are subject to denial of services attacks. An attacker with the proper equipment can easily flood the 2.4 GHz frequency with spurious transmission so that the wireless network ceases to function.

Compounding the security risks inherent in wireless networks is that most base stations are configured to the least secure mode when they come out of the box. This makes installation easier, but puts the onus of security on the user installing the wireless network. For example:

- Most wireless base stations have the SSID set to a default value. These are well known and if not changed allows anyone who knows the default SSID value to connect to your wireless network.
- Many wireless base stations come configured in a "non-secure access mode" which allows any computer to connect to a base station with or without the appropriate SSID.
- Most wireless base stations come with WEP turned off, i.e. no encryption is being used. This allows even casual attackers to monitor your wireless network traffic. Even though WEP is weak it should still be used as it can impede attackers.
- Most wireless base stations come configured with well-known default SNMP community strings. These vulnerabilities allow a number of different attacks to be perpetrated, such as:
  - Connection of unauthorized hosts, typically a laptop or PDA, to your base station without authorization (depends on configuration – see an upcoming section on mitigation techniques).
  - Interception and monitoring of wireless traffic on your network (often even if encrypted).
  - Hijacking existing sessions (e.g. it is possible for an expert attacker to take over your secure encrypted Web-based session with your online banking institution).
  - Denial of service attacks (overwhelm the radio frequencies employed by the wireless network with spurious traffic).
  - Attack another wireless client directly, by bypassing base station. If a wireless client, like a laptop or desktop, is running TCP/IP services like a Web server or file sharing, an attacker can exploit any misconfigurations or vulnerabilities with another client.

Although a wireless network can never be made as secure and robust as a wired network, there are several steps that a user can take to better secure their wireless network:

- Use additional encryption beyond WEP. For example VPN, Secure Socket Layer (SSL) and Secure Shell (SSH) traffic are all encrypted before transmission and, therefore, are far less susceptible to compromise even if WEP encryption is not enabled or has been compromised by an attacker.

- Enable 128 bit WEP encryption (see vendor documentation).
- Change SSID to a hard guess password (include letters, numbers and characters).
- Enable any additional authentication schemes supported by your base station. Two common examples are authentication-based Media Access Control (MAC) address or WEP authentication keys. If your wireless base station supports either of these protocols, configure them according to your vendor's documentation.
- Disable broadcasts of SSID in the wireless base station beacon message (see vendor documentation). In most default configurations the base station regularly broadcasts the SSID making it much easier for an attacker to intercept the SSID. Even when disabled, the wireless clients will transmit the SSID (albeit much less frequently), so a patient attacker will eventually get the SSID.
- Disable SNMP on wireless base station and wireless client(s) (see vendor documentation). SNMP allows for remote administration across the network, but for home use, it is safer to manage controls using a direct connection to the base station. (See vendor documentation for where to connect to USB or other port.)
- Ensure that the administrative password used to configure the wireless network base station is changed and difficult to guess (i.e., not a dictionary word and includes letters, numbers and characters).
- All wireless client computers should be treated as if they were directly exposed to the Internet. That means additional steps must be taken to secure these hosts. Ensure all clients with a wireless network card have:
    - A personal firewall installed (even if your network also has a firewall installed at its Internet connection);
    - File and printer sharing disabled;
    - SNMP disabled;
    - NetBIOS protocol disabled over TCP/IP (see vendor documentation); and
    - All TCP services that are unnecessary disabled..

While there are risks associated with using wireless networks many can be mitigated with careful planning and configuration. For many users the benefits of wireless networks will outweigh the risks. However given the weaknesses of WEP encryption, any sensitive or proprietary data transmitted should be encrypted prior to transmission by other means (e.g., VPN, Pretty Good Privacy, SSL, SSH). Generally, with the proper precautions, users can safely use home wireless networks.

EVALUATING THE SECURITY RISKS

The company should already have an on-going program of security, maintenance and monitoring to take care of potential security breaches and security holes. If not, do so IMMEDIATELY – perform a security analysis on the entire IT system and then document potential weak spots. Then perform a security audit with outside auditors to check for potential security holes that might have been overlooked. There are numerous security consultants who can help in this endeavor.

I strongly suggest that your company retain a security expert to perform a detailed review of the business's internal procedures, network topology and permissions, access controls, hardware, software and utilities that could possibly compromise the company's IT infrastructure.

Today's average IT ecosystem is made up of a complex combination of hardware, software and other components (hosted, custom-developed, off-the-shelf) that are provided by a multitude of different vendors. It's virtually impossible to find and resolve all security risks. The IT staff can't really evaluate the security design of all the software components – all aren't under its charge. Therefore, as a practical matter, the IT department must depend on each component within the ecosystem being well designed from a security standpoint, and then strive to minimize exposure resulting from integration within the ecosystem.

Network systems are continuously being modified and with each change comes new security risks. This means it's almost impossible to prevent new security holes from popping up – ***diligence is the only defense.***

IN SUMMARY

All in all, midsize companies with some distributed systems, linked WANs, integrated enterprise systems, and interactive Web sites are in something of a pickle. They don't have the financial resources of large companies, but are just large enough that they may require the full spectrum of security measures to be in place and ready for action. Smaller companies and SOHOs are a bit better off, security-wise, because they usually don't have many, if any, distributed systems, making their networks and data easier to protect.

Remember: Identifying security threats is essential to ongoing security maintenance. Everything from e-commerce to e-mail demands consistent and effective security protection. Anything else leads to the disaster that you're trying to avoid in the first place.

CHAPTER 8

VPNs Put Employees Virtually Anywhere

THE NATURE AND SEVERITY OF A DISASTER, be it a terrorist attack, earthquake, fire, flood, an accident at a chemicals plant, or some city-wide damage caused by rioting Enron stockholders, is what determines how long your business will be disrupted. Location-specific businesses are at a disadvantage, for even a well-secured building equipped with the latest fault tolerant technology and top-notch disaster recovery and security policies will ultimately succumb (or the roads, bridges and public transportation leading to them will collapse) if a "disaster" actually lives up to its dictionary definition: "An occurrence causing widespread destruction and distress; a catastrophe. A grave misfortune."

Hence: A business that "can't get out of harm's way" is a business that is a sitting duck.

We've seen that the primary means of minimizing the impact of disaster on a business is to make it as "distributed" as possible. We've looked at how data and applications can be distributed, cleverly placed on networks and "remote vaulted" to distant, secure buildings either owned by your company or supplied by a hosted backup service.

But if you can distribute data and computing power to make it safe from disasters of the natural or man-made variety, what about your employees? Can't you distribute them too? Can't you extend your business environment, the applications and the data, to wherever they are, so that if disaster wipes out your headquarters the remaining network can shoulder the burden as best it can until some temporary, larger telework centers are brought on line?

Yes, it's possible – and even vital, since many companies that have moved to such "teleworking" and "virtual teaming" operational environments are now positioned, via remote access, to continue functioning and "roll with the punches" in the event of a real disaster.

The traditional, primitive way that a company "distributed" itself was through expansion. A company that started at one location would grow up to a certain point and then, like some great amoeba, would shoot out some pseudopod-like branch offices. People would have to commute to these clones of headquarters, work at their PCs and send and receive files piecemeal between the main office via some kind of Wide Area Network (WAN), such as 56 Kbps modem, an ISDN line or perhaps even a T1. Buildings on the same campus could even have the same actual physical connections in a closed network: twisted copper LAN cable, point-to-point microwave links, optical fiber or other lines of communication strung between the buildings.

Granted, a closed network ***appears*** to have some advantages. Data and network traffic should not be accessible from outside the network, for example. But it's expensive to run your own network, and, unless some kind of ungainly encryption technology is retrofitted to all of the applications, such a network could easily be physically breached: wiretaps, eavesdropping on microwave transmissions, etc.

This kind of network can also be easily attacked. And, as we've seen, under classic business models, if you lose your headquarters, you've literally lost the center of your business.

Until now.

There exists an underlying, enabling technology which ties together the other items found in this book. It gives data, conferencing and other communications a hackproof and high-speed channel. It can extend your business' network beyond the firewall to your employees' homes, or to laptops in hotel rooms, making the "digital, virtual office" possible. It can give you a "private channel" on the Internet that can be shared by employees or used in conjunction with your strategic partners anywhere in the world like an extranet. It's the Virtual Private Network.

Virtual Private Network (VPN) technology now provides the cheapest, fastest way to build readily remote-accessible voice/data WANs. Capable of yielding higher bandwidth connections than wireless / mobile connections, VPNs can give employees who are teleworking or at small branch offices more of the performance and capabilities they'd enjoy if they were right at headquarters. Connections between multiple servers can also be established. Employees can be kept at whatever safe remote location is desired.

Paul Ferguson, a consulting engineer at Cisco, elegantly defines a Virtual Private Network as follows: "A [VPN is a] communications environment which is closed to access from outside a defined community of interest, constructed though some form of partitioning of a common underlying communications medium, where this underlying communications medium provides services to the VPN on a non-exclusive basis."

Put more simply, a VPN is a ***private*** voice/data network, run over ***public*** network facilities. It's a beautiful concept: transparently harness the enormous resources of the public network to overcome geography and time zones, render infrastructure uniformity, centralize management, and enjoy true economies of scale. A VPN is generally less expensive to build and operate than a dedicated real network, because the virtual network shares the cost of system resources with other users of the real network. Best of all, it can serve as one of the key enablers for teleworking, putting your employees out of harm's way, scattering them about the world, if you like. In theory, a VPN lets you build out as you grow, and (to some extent) pay as you go.

VPN EVOLUTION

It's a pretty venerable concept, too. In one form or another, VPNs have been marketed for about 15 years, over several generations of technology. Just as the term "media server" has changed over the past few years – from a storage server for streaming multimedia to a media processing server accessed via an IP network – so too has "virtual private network" undergone a metamorphosis, thanks to IP.

VPN versus Direct Dial

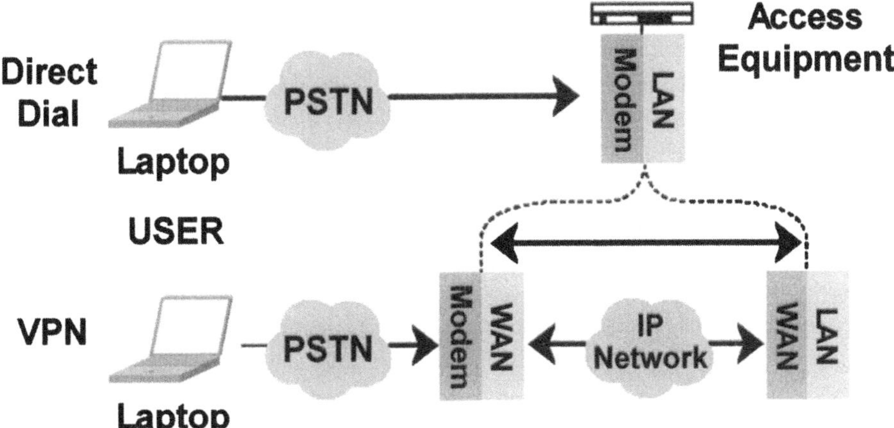

Circa 1985, AT&T began offering "software defined networks" (SDNs) – long distance WANs based first on dedicated i.e., T1 direct to an AT&T POP, then later, circa 1989, also on switched connection through Local Exchange Carrier access. A SDN was implemented by building a database at a Network Control Point (NCP) to record phone numbers in use at a customer's various facilities. When someone (or some machine) dialed out, network resources queried the database to determine if the call was on- or off-net; and routed and billed accordingly.

Customers liked SDN because it gave them lower prices on interoffice long distance, a single bill for standard LD and "leased" connections, and rapid, flexible provisioning at new sites. AT&T liked SDN because it let them move customers away from real leased lines (which wall off expensive CO real estate, switch, cross-connect, copper and fiber facilities, then use the walled-off bandwidth inefficiently) onto "virtual" leased lines – connections traveling over shared public-switched infrastructure, but billed at insider rates under long-term contracts. The result: more predictable traffic, more efficient use of all network facilities, more long-term contracts, and reduced churn.

On the data side, the development decades ago of the X.25 protocol provided a means for sending private packet-streams across the public network; but early-1980s X.25 infrastructure offered transmission rates more appropriate to low-bit rate CPU/dumb terminal connections than true client/server, high-rate bulk file transfer or real-time (e.g., voice) transport. For a while, it looked like X.25-over-ISDN might evolve into a flexible VPN protocol under the SDN aegis. But the slow delivery of ISDN by carriers stalled this trend, and the market changed focus.

SMDS (Switched, Multi-Megabit Data Service) enjoyed a brief vogue as a means for constructing metropolitan-area VPNs, but was swiftly supplanted by frame relay and the somewhat-similar ATM (at least re cell-sizes: both SMDS and ATM employ 53-byte cells). These packet/cell-based data-transport protocols offer "virtual circuits" (VCs) – a switching methodology, expressed at Layer 2 of the Open Systems Interconnection (OSI) stack, that aligns nicely with certain VPN requirements. Packets/cells traveling across a Frame Relay (FR) or ATM network don't contain source and destination addresses; instead, they carry a pointer to a virtual circuit, belonging to the customer. The virtual circuit – which must be provisioned on each network resource in the route from source to destination – is normally characterized by a firm, committed information-rate (minimum bandwidth guarantee) and other QoS parameters.

Virtual circuit-switching is very fast, compared with address-based routing; and remote provisioning makes it fairly easy for carriers to build VPNs on a FR and ATM infrastructure (or, more likely, a combination of the two). But there are drawbacks to FR and ATM, especially when used – as is increasingly the case – to carry IP traffic. Inter-protocol (i.e., IP address to FR Data-Link Circuit Identifier or DLCI to ATM virtual circuit) conversion takes some time. It's not especially efficient to stuff IP packets into FR packets, then slice them into ATM cells – this generates lots of header overhead. And because FR and ATM are virtual circuit-oriented, networks built on these protocols have only a limited ability to "self-heal" and route around congestion.

Complicating the picture, is the fact that nobody's quite sure how secure FR and ATM networks really are. Certainly, neither FR nor ATM provides the kind of packet-level, end-to-end authentication and encoding required by high-end customers; especially when FR and ATM VPNs (and of course, the LANs they bring together) are interconnected at many points with the public 'Net.

The ideal – and experience with FR and ATM VPNs is making this more clear – is a system that brings the "virtual" aspect of the VPN acronym all the way home. You want your public network to comprise discrete, functionally uniform, scalable, generic components; capable of cooperative self-provisioning; supportive of QoS; easy to install, scale up, and administer. You want that network to be ubiquitously available (ergo, you have to be prepared to leverage the Internet). And you want equally scalable, equally flexible, remote-administrable endpoints and "on-ramps" (routers, gateways) capable of requesting services from the network and enforcing airtight, end-to-end security through deep encryption.

What you want is what the current crop of IP VPN solutions can deliver, especially when coupled with pure IP and Multi-Protocol Label Switching (MPLS) in SP/Internet "midbone" and (eventually) backbone.

IP VPNs operate at Layer 3 ("the network layer") of the OSI model. The differences between Layer 2 and Layer 3 VPNs are really striking. Layer 3 tunneling is more scalable and is application and network independent. It can be applied to any form of routable communications (voice, video, data), thus providing an effective scalability pathway.

Today's IP VPNs generally exploit "tunneling" (a methodology that guarantees the safe passage of a packet of data through the Internet using encryption to protect the data payload, as well as the source and destination addresses), encryption, authentication, and access control technologies and services. They carry traffic over the Internet, a managed IP network or a provider's backbone.

IP traffic reaches these backbones by traveling over any combination of access technologies, including T1, Frame Relay, ISDN, xDSL, ATM or simple dial-up access. At the endpoint, access router, or other "onramp," IP (and other kind of packets, such as IPX packets) are encapsulated in a tunneling protocol such as IPSec (see below). These packets are, in turn, packaged in an IP packet containing the address of the corporate network, the packet's ultimate destination. When you start the session the local network access server at the POP can be used to give you your own IP addresses, allowing you to retain your unregistered address for working with the LAN.

The Virtual Private Network Consortium (www.vpnc.org) believes that there are three major tunneling protocols that can be used to create VPNs by encapsulating IP packets in them:

- **Internet Security Protocol (IPSec)** is from the Internet Engineering Task Force (IETF). Many expect IPSec to become the dominant way to provide security services for the Internet Protocol (IP), both IP version 4 (IPv4) as well as version 6 (IPv6). IPSec involves encryption algorithms, authentication algorithms, and management of encryption key exchange. Whereas customer premise equipment-based VPNs use IPsec tunnels to protect all data entering or leaving the enterprise network, Network-based VPNs usually rely on private layer 2 links – usually DSL or Frame Relay connections – between the customer site and the POP or CO. From that point, IPsec tunnels encrypt data sent over shared infrastructure: the provider's backbone network and the public Internet.
- **The Point-to-Point Tunneling Protocol (PPTP)** is a proprietary protocol originally designed as an encapsulation mechanism to allow the transport of non-TCP/IP protocols (such as IPX) over the Internet using the Generic Routing Encapsulation (GRE) protocol. PPTP was developed by and has been proposed as a standard protocol by Microsoft, Ascend, 3Com, and others. Microsoft has built PPTP into its full line of Windows server software. PPTP is usually installed on a company's firewall server. Novell's BorderManager also supports this protocol, as can other operating systems using PPTP software.
- **Layer 2 Tunneling Protocol (L2TP)** is an IETF standard tunneling protocol that evolved from a combination of PPTP and the proprietary Layer 2 Forwarding (L2F) protocol found in Cisco's

Avaya's Hardware-based VPN Service Units (VSUs)

Top: The VSU-7500 is a high-availability high-performance gateway with hardware redundancy that supports up to 7,500 simultaneous IPsec tunnels at 100 Mbps. Typical users include large enterprises, application service providers (ASPs) and managed VPN data service providers.

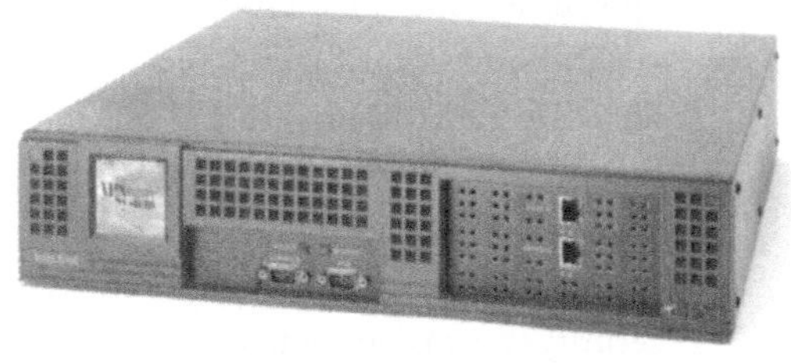

Middle: The VSU-5000 is a high-capacity high-performance VPN gateway that supports up to 5,000 simultaneous IPsec tunnels at 90 Mbps. Typical users include medium to large enterprise corporate headquarters for broadband site-to-site and remote access VPN deployments.

Bottom: The VSU-2000 is a high-capacity high-performance VPN gateway that supports up to 1,000 simultaneous IPsec tunnels at 45 Mbps. Typical users include medium to large enterprise corporate headquarters for broadband site-to-site and remote access VPN deployments.

Internetwork Operating System (IOS). PPTP is geared toward ISPs and has provisions for call origination and flow control, while L2F has less overhead and is suited for managed networks. L2TP is said to have the best features of both protocols. L2TP is used by ISPs for secure, node-to-node communications in support of multiple, simultaneous tunnels in the core of the Internet or other IP-based network.

TYPES OF VPN PLATFORMS

The VPN products themselves can be grouped into three categories: Hardware-based systems, stand-alone software packages, and firewall-based systems with encryption patches. These days all such products support both LAN-to-LAN and remote dial-up connections.

- **Hardware VPN products** are typically encrypting routers that store encryption keys on an ASIC chip, and are more difficult to crack than their software-based systems. Encrypting routers also have greater throughput than software-based systems, and should be used whenever a large "pipe" enters the enterprise. Some hardware VPNs demand that you install a special card in each PC on the network.

Premier examples of hardware VPN technology are the VPN Service Units (VSUs) sold by Avaya (Basking Ridge, NJ – 908-953-6000, www.avaya.com). These are dedicated, hardware-based VPN gateways that provide overlay remote access and site-to-site VPN and firewall services on existing enterprise networks. The VSU series of VPN gateways consist of six models: the VSU-5 for small and home offices, VSU-100 for small and medium businesses, VSU-2000 for branch offices, and the VSU-5000, VSU-7500 and VSU-10000 for large enterprises and managed data service providers. The VSU gateways are centrally administered and configured using Avaya's "VPNmanager" policy-based management application, and support remote access services with the "VPNremote" desktop VPN client software.

Unlike firewall or router-based VPN solutions, the VSU gateways feature a dedicated high performance IPSec packet-processing engine and real-time data compression designed specifically to handle high-bandwidth VPN traffic. The VSUs offer wire-speed performance ranging from 16 Mbps to 100 Mbps for "Triple" or 3DES-encrypted IPSec traffic, and bridging of non-VPN traffic at even higher speeds can be achieved.

Smaller standalone VPN systems are quite inexpensive. You can buy a SOHO VPN purpose-built appliance from SonicWALL (Sunnyvale, CA – 408-745-9600, www.sonicwall.com) and get a full stateful-inspection firewall plus antivirus software for about $600 (All SonicWALL firewalls are aimed at the SME market and provide cost effective protection from Internet security threats). Or you could buy a FireBox SOHO from WatchGuard (Seattle, WA - 206-521-8340, www.watchguard.com) for about the same price, ***and*** get a firewall and a year's worth of service for free. WatchGuard's innovative LiveSecurity Service enables organizations and users to keep their security systems up-to-date through WatchGuard's broadcasts of threat responses, software updates, information alerts, expert advisories, support flashes and virus alerts over the Internet.

Software-based VPNs are generally more flexible than hardware-based VPNs, and are used in situations where remote sites encounter a mixture of VPN and non-VPN traffic, such as Web surfing. Software systems have less performance than hardware-based systems and tend to be more difficult to manage. Still, it is rather tempting to just download some client software, run it on some PC, and

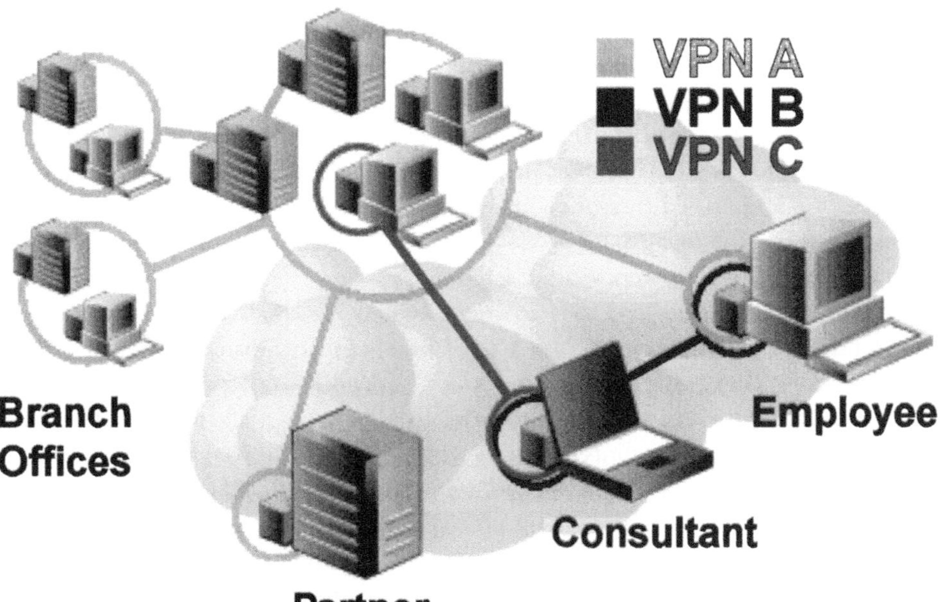

have a VPN up and running.

Firewall-based VPNs, on the other hand, can leverage the firewall's existing security mechanisms, including restriction of access to the internal network. They also address translation and can execute powerful authentication procedures. Most commercial firewalls "strip out" services that may pose security problems, thus hardening the host operating system kernel. Firewall-based VPNs are inexpensive and are great for intranet-only implementations.

Hosted VPN Services allow you to dump most of your VPN headaches on a friendly service provider. These providers can be of prime importance if you're unable to ensure that your VPN can be regenerated from another location if your headquarters gets wiped out by the next visitor from the asteroid belt or due to some event from a "home-grown" source, i.e. earthquake, flood. Such services have spent fortunes on developing incredibly resilient and secure VPNs.

For example, SafeNet, Inc. (Baltimore, MD – 410-931-7500, www.safenet-inc.com) offers managed security services, called SafeNet Trusted Services, at its Baltimore location. SafeNet Trusted Services provides VPN system availability in excess of 99.9%, yet saves you money by eliminating the investment in equipment, facilities and security expertise necessary to manage VPN security on the Internet and other TCP/IP networks.

SafeNet Trusted Services begin with defining your VPN security policy. The definition process includes creating essential administrative functions, such as designating a Customer Service Manager and Network Administrator. Emergency notification and escalation procedures are set. A list of authorized users and an access control policy is established for each VPN user and secure site.

These policies define frequency of PIN and key changes, actions to be taken during security alerts and reports to be issued.

However, if necessary, SafeNet Trusted Services can create the infrastructure needed to run a VPN for you immediately, thereby allowing your company to re-start operations without having to have established its own team of security experts first; and secondly, it can act as a backup in the event of a disaster such as a fire or flood, thus enabling the continuity of service.

SafeNet Trusted Services delivers standards-based VPN security compliant with national and international standards endorsed by IETF, ISO, ANSI and FIPS. SafeNet's hosted VPNs are based on a potent combination of secure technologies, from IPSec and Public Key Infrastructure (PKI) standards, including 3DES-based encryption, user authentication, message authentication, role-based policy management, and fine-grained access control. SafeNet assures its customers of compatible, scalable and interoperable network security solutions.

They also offer enterprise-level VPN security solutions, which are a favorite with such major financial institutions including JP Morgan & Co., Chase Manhattan Bank, Interbanking Argentina, Bank of New York, The Central Bank of Sweden, and Citibank.

SafeNet's Trusted Services management centers are built in physically secure facilities with double doors controlled with smart cards and monitored with surveillance cameras and are designed for continuous operations. They are equipped with redundant power supplies and communications, are backed up to remote sites and meet various official audit requirements, such as the Statement on Auditing Standards No. 70 (SAS70), a widely recognized auditing standard developed by the American Institute of Certified Public Accountants that signifies that a service organization has had its control objectives and control activities examined by an independent accounting or auditing firm.

In the case of OpenReach (Woburn, MA – 888-783-0383, www.openreach.com), you can purchase one of their pre-configured systems, or you can simply download their software and use it on your PCs. OpenReach technology is based on a packet-filtering firewall.

Also, the Hosted Business VPN Service (HBVPN) from eTunnels (Seattle, WA – 206-239-9800, www.etunnels.com) does deals with service-oriented channel partners to resell a variety of low-cost, end-user-managed Internet-based VPN services to get a business's VPN up and running without too much fuss.

End-users of the eTunnels system can securely extend corporate LAN or WAN resources to mobile and/or teleworkers, build and easily manage branch-to-branch VPNs between satellite offices, and quickly set up secure business partner and/or customer extranets.

With eTunnels, there's no hardware to buy, configure, or maintain. The eTunnels system replaces hardware VPN implementations with a hosted, pay-as-you-go application- and user-oriented system that makes it easy for VPN users to securely connect to VPN resources such as file shares, printers and applications.

Setup and administration for the channel partner is kept simple and straightforward via bundled professional services and service templates, customizable GUI client and server software and pre-built Web-based administration tools. Billing and customer technical support is transparently handled by eTunnels.

The eTunnels Business VPN Client is a Windows 95/98 and Windows NT/2000 GUI software client with an integrated mini-browser that's installed on the end user's workstation. This client software manages the secure connection between the end-user clients and servers using IPSec tunneling.

The eTunnels Business VPN Server is a Windows NT/2000 or Linux 2.x software service or daemon that's installed on corporate application, file and print servers. This software manages the secure connection between the server's resources and each end-user workstation, using IPSec.

The eTunnels Network Server co-ordinates the setup and teardown of the IPSec tunnels between eTunnels network elements (clients, servers and gateways) by handling Internet Key generation and exchange through a secure SSL conversation with the network elements.

The eTunnels Web-based Company Management Tool is a secure Web site through which the end-user can configure and view the company-specific aspects of the eTunnels VPN-On-Demand service.

Finally, eTunnels' "VPN-on-Demand," is a new approach to creating security relationships that makes dynamic provisioning of VPNs faster than ever. Called a "zero configuration" technology, eTunnels' VPN-On-Demand is easy to deploy and manage, and won't put a strain on IT resources. The VPN-On-Demand platform makes it easy to rapidly build and manage multiple, overlapping, and secure networks that can scale to any number of people. The platform delivers versatile solutions for a variety of secure networking challenges including branch offices, mobile workers, partner networks, and wireless LANs.

VPN-On-Demand offers dramatically increased efficiencies over traditional VPN technology. Features include remote configuration capabilities, central administration across any number or type of VPN element, granular manageability down to individual users or applications, independence from network / topology / devices, transparency to intervening address translation / firewalling / proxy services, end-to-end security, and support for multiple, concurrent and overlapping VPNs. The system offers independent manageability among multiple IT authorities within a partner extranet

The VPN-On-Demand platform is built upon two technological innovations, but regardless of what platform you use, other options include the types of algorithms used for encryption – 1) DES/3XDES, RSA, RC4, RC5, IDEA, and 2) key exchange (IKE, SKIP).

Enterprise customers with branches having lots of users and a T3 connection will probably buy separate boxes and custom hardware to support all the sites on their VPN, while smaller companies or companies with small branch offices may not need them, particularly if those sites have dedicated Internet connections that are T1 or smaller. This is because, given the processing power of newer PCs, some PC-based VPN gateways can easily handle a T1-size link to the Internet, which is equal to 1.5Mbps. That means that you don't necessarily have to buy "designed single-purpose boxes" that act as the gateway between the VPN, multiple PCs at individual sites and their connections to the Internet.

This PC versus special-purpose VPN box argument is reminiscent of a similar controversy that stormed in the early 1990s – whether to use a dedicated external router or an internal serial interface card, all of which went by the misnomer of "router cards." Computer makers stated that external routers are largely unnecessary devices for most scenarios if you are running one or more UNIX-like hosts already, and are less functional choices even if you're not. Router companies countered by adding more features.

While the performance of the VPN-specific functions running on PCs may be adequate for small offices and their T1 connections, such smaller organizations might still want more features than they can get on a PC platform, and they may not want to dedicate a non-fault tolerant PC or server to serve as a firewall / VPN, so they may actually prefer a standalone box. While there is no "one-size-fits-all device," some companies do offer a wide range of products. NetScreen (Sunnyvale, CA – 408-730-6000, www.netscreen.com) has one of the broadest security product ranges in the industry, from the first gigabit-speed security systems for Internet data centers and service providers (the NetScreen-1000) all the way down to their single teleworker solution (the NetScreen-5XP).

Another established Internet Security market leader is CheckPoint (Redwood City, CA – 650-628-2000, www.checkpoint.com), which offers VPN and firewall solutions to meet the needs of everything from SOHO users to the largest of enterprise applications.

For example, CheckPoint's VPN-1 / FireWall-1 SmallOffice delivers market leading VPN-1 / FireWall-1 technology to small and medium businesses (SMBs), enterprise branch offices and MSPs that provide managed security services to small office environments. VPN-1 SmallOffice is available pre-installed on a variety of low-cost, high-performance security appliances and can also be deployed on standard server platforms.

For PC-based systems, CheckPoint even offers the VPN-1 Accelerator Card II, an ingenious PCI add-in card that improves VPN-1 gateway performance by off-loading processor intensive cryptographic operations from the host CPU at the gateway to a dedicated processor on the card itself.

BUT ARE VPNS REALLY THAT GOOD?

Yes, they are. VPN applications now include: Remote access to a corporate site wherein secure connectivity is provided to remote and mobile users, secure site-to-site connectivity for multiple branch offices, extranets where secure connectivity is provided to suppliers and key business partners, and internal corporate usage on intranets.

When it comes to remote access VPNs, the basic concept is to give teleworkers and mobile workers a secure way to get back to a corporate network over the Internet or a service provider's backbone.

In a typical remote access VPN, teleworkers and mobile workers can connect back to a corporate network over the Internet or a service provider's backbone after the user's special client software recognizes the specified destination and negotiates a secure tunnel between the network access server and a tunnel-terminating device on the LAN.

To do it, the user just dials into a Service Provider's (SP's) or Internet Service Provider's (ISP's) Point of Presence (POP), where the SP or ISP authenticates the user and establishes a tunnel (either over the SP's private network or the Internet "cloud") back to the corporate LAN's edge, where the user may once again have to authenticate him- or herself to gain access to the corporate network. The encrypted packets traveling back and forth are wrapped in IP packets to tunnel their way through the Internet during the course of the encrypted VPN session. The VPN server negotiates the VPN session and decrypts the packets. The unencrypted traffic then flows as it would normally to other servers and resources.

Contrast the above procedure to the traditional "dial-in" technique whereby a user makes an expensive long distance call into a bank of modems, a remote access server or concentrator located at headquarters with a VPN.

Some companies can cut telecom charges from $1,000 per month per person with a dial access account to an ISP's flat monthly rate of less than $20 per month.

Prior to adopting a VPN, a company may have relied upon a dedicated link to an ISP for Internet access and a 24-channel T1 line into a remote access server to support dial-in users. Using a VPN means that you don't need the costly T1 line for dial-in access – the traffic from these users now travels over your existing Internet access line. In one fell swoop, a single VPN gateway and one "pipe" can eliminate the need for multiple network edge devices and at the same time support many different kinds of applications.

In the case of site-to-site connectivity, as in the remote access scenario, branch offices, satellite offices, telework centers and teleworkers' operating out of a home office connect to corporate headquarters through "tunnels" that transport traffic over the Internet or via a provider's backbone. Thus, a company can reduce communications costs by paying only for the access line from a remote location to the nearest service provider's POP, rather than paying for a long distance link all the way to corporate headquarters. In this way offices in foreign countries can circumvent international tariffs.

Also, many sites have multiple access lines for data: one to trade data with headquarters and a second for Internet access. Industry studies have found that more than 70% of sites are overspending on access lines, in this fashion. Using VPN technology for site-to-site connectivity allows a branch office with multiple links to remove any extra lines used to communicate data to and from headquarters, moving traffic instead over the existing Internet access connection.

A site-to-site VPN can be installed as quickly you can secure a high speed Internet connection (a few weeks) as opposed to installing a leased high-speed data services line (several months in the US, even longer in other countries).

When used internally, VPNs' encryption, authentication and access control services can segment populations on an intranet. For instance, a Human Resources (HR) department could let employees check on their accrued vacation time, but not be able to see their performance reviews – managers, meanwhile, can see both. When used in this way, VPNs are a more flexible successor to the old "Virtual LANs" (VLANs), which were all proprietary and didn't work in IT environments where there was a melange of different vendors' hubs and switches. VPNs can ignore the different kinds of equipment by relying upon IP-based tunnels nailed up between a user's workstation and a server, thus creating a milieu that's similar to physically segmenting users on discrete LAN segments.

All of these various applications for VPNs are not mutually exclusive. A company could deploy a VPN to reach its branch offices, then let single remote users access the network too, and ultimately allow the network to be accessed by privileged outsiders, all done by just changing some settings in software.

Large-Scale, Multi-Owner VPNs

Extrapolating on the idea that a VPN can be an extranet, linking partners together over a "private channel" on the Internet, one could envision really large, secure, dispersed quasi-public high bandwidth VPNs. After all, from the 1960s through the 1980s, US universities and the military used the Internet to share information and ideas "freely" among themselves (meaning that they essentially had a private network, since the average person didn't own a mainframe or minicomputer to join in the fun). VPNs bring back the secure "private club" concept. Commencing on June 30, 2000, the world's first optical VPN became the platform for perhaps the world's most advanced scientific communication network. Called G-WIN, it was handed over by Deutsche Telekom to the German Research Network Association (DFN). Alcatel, a leader in optical networking, contributed the essential technical elements to the network, which connects all universities and research laboratories in Germany.

At the world premiere, G-WIN deployed Alcatel's (Paris, France – +33-01-40-76-1010, www.alcatel.com) Web-based Alcatel 1355 Virtual Private Network (VPN) Manager. It's designed to administer optical SDH VPNs (the next model will serve SONET and WDM networks). It allows enterprises and new carriers to buy bandwidth from operators instead of building their own networks, VPN owners (including medium and large enterprises) to allocate and manage their VPN resources from their premises, via an Internet browser installed on an ordinary PC, which is connected to the Operator's network management system and provides simultaneous access to a set of pre-defined services.

The functionality offered (i.e., VPN resources definition, connectivity provisioning, fault reporting and monitoring of performance parameters) lets end users manage their VPNs in real time. It also allows end users to drastically decrease service set-up time and quickly verify that operators are supplying the necessary quality of service.

Based on standard Internet technology, the system offers considerable functionality, such as end-to-end bandwidth-on-demand capability and the freedom of managing VPN networks in real time while closely monitoring their quality down to single bit errors detected on the transmission line.

VPN Telephony

What one would really like to see from VPNs is not just creating private data connections, but voice traffic with a high Quality of Service (QoS) as well – "VPN telephony." Once VPNs are in wide use, they'll provide the opportunity to integrate other types of communication such as Voice-over-IP (VoIP) and video conferencing. Some major telecom players are interested in IP-based VPNs. Up until late 1997, the Regional Bell Operating Companies (RBOCs) appeared to have little interest in IP VPNs. But, following a wave of IP VPN efforts by AT&T, GTE, MCI WorldCom, Concentric Networks and others, the RBOCs suddenly started playing catch-up, buying up companies with VPN expertise. VPN telephony products for the customer premise are also starting to appear.

For example, the Clarent VPN Telephony Suite from Clarent (Redwood City, CA – 650-306-7511, www.clarent.com) can run on a Celeron-based PC with 64 MB of SDRAM that's equipped with telephony connections such as a digital T1 / E1 or analog POTS lines (RJ-11 jacks). The system gives service providers a novel way to add value and flexibility to the corporate communications services they offer by enabling voice and fax to travel over VPNs. The product is essentially a voice and fax-over-IP gateway for connecting telephone voice calls and fax-over-VPN and public IP networks.

Clarent's product doesn't generate the VPN itself, but acts as a front-end that handles multiple gateways and does VoIP processing by efficiently structuring packets and getting them out to the right gateways.

Each call is routed to a gateway within the VPN if one is designated. If the VPN doesn't cover certain cities or countries, corporate traffic can still be routed over public IP voice networks. Calls can access the public network from within the VPN, but public calls cannot access the VPN from outside. This allows low-cost voice and fax communications anywhere in the world, least-cost routing, and the security of a closed, private network.

In terms of voice quality, Clarent's patent pending technology improves packet loss recovery, reduces latency and delivers the optimum sound quality within the lowest bandwidth even while other traffic consisting of fax and data also travels along the single, secure network. For VPN service providers, adding voice and fax to the VPN leads to a closer bond with the customer, allowing the provider to move beyond simple bandwidth leasing to become a complete one-stop shop for communications service on a global basis. Also, service providers can now offer more competitive rates to customers because the cost of administration, billing, and account maintenance of multiple services are now consolidated into a single system.

The Clarent Command Center provides billing and management for networks of multiple VPN domains, providing hundreds of thousands of simultaneous telephone calls through a single point of administration. Different inbound and outbound charges can be set for each VPN domain. Free Calls or Call Blocking options can be set for individual corporate customers. In terms of call detail recording, the system records both inbound and outbound call information including customer account, time and duration of call, destination and call termination code.

The Clarent Command Center consists of a software program attached to an ODBC-compliant relational database. The Clarent Command Center package can run on a variety of high-availability Windows NT platforms connected to the Internet, from simple servers to high-speed systems with RAID drives, tape backup and hot-swappable power supplies. Multiple Clarent Command Centers running on separate machines and attached to a single billing database can increase uptime through redundancy – keeping the system running even when one Clarent Command Center becomes unavailable.

Aside from NT servers, the database can be housed on any high-availability server, even a mainframe that's running a standard RDBMS such as Oracle or Microsoft SQL Server. By putting the database on a more powerful system, Clarent Command Center can support additional gateways within a single network.

The Clarent Service Editor and the IVR toolbox support customization of voice prompts either on-premise or remotely via a service provider office.

The system can scale from four to thousands of simultaneous calls in each corporate site, with easy upgrades as branch offices grow in size and traffic.

As VPNs are made to handle more and more voice and data traffic, performance becomes a key issue with VPNs and so the advantages of a hardware-based VPN system become glaringly evident. Using the software approach or just plugging accelerator cards into a router will probably always be part of the VPN picture, but when it comes to doing things like voice and integrated voice and video, there's really no choice but to move into a hardware environment. Fortunately, the encryption chip market has elevated the level of performance across most of the hardware-based VPN vendors.

Trying to implement IPSec packet handling in software or just as an add-on to a router not only takes processing power away from the core application of the equipment, but to add insult to injury, it's not even that effective at processing VPN traffic. Hardware-based VPNs such as Avaya's, however, can handle IPSec traffic basically at line speed. It can handle Triple DES-encoded traffic at 80 or 90 Mbps and does so without introducing any packet latency or queuing into the system, which is very important for doing VoIP over VPNs.

Since packet latency and QoS concerns are crucial in VoIP-based "VPN telephony," the ability to process small packets is very important, because VoIP applications generally default packets to very small sizes.

The performance differential for small and large packets can be measured, especially in VPN equipment, because the equipment does all of the IPSec encapsulation around IP packets, so more layers of overhead are introduced there. Thus, as the number of packets needed to move some specific amount of data increases, the amount of overhead that has to be dealt with increases also. At some point it isn't feasible for a software VPN application to try to handle VoIP. Significant additional latency will be added into the system because of the less efficient software-based processing.

Small IP packets introduce a lot of overhead, lowering the performance of routers, switches and VPN devices. So with an application such as VoIP, VPN systems that normally transport 80 or 90 Mbps when moving large packets suddenly find that they're moving only about 20 Mbps when smaller, 64-bit traffic is flooding the network.

Additional techniques to enhance the QoS for VPN VoIP applications can involve "marking" packets so they can be prioritized, allowing for differential service. "Important packets" (e.g., for voice and video) can be given special treatment, provided that all of the routing equipment between the endpoints can read the markings. IP version 4 already provides an eight-bit Type of Service (TOS) field in the IP header (IP version 6 has an equivalent Traffic Class byte). The first three bits are used to indicate one of eight possible levels of packet precedence, which allows an IP node to designate specific packets as having higher priority than others.

Avaya's VSU firmware includes an embedded operating system, which adds additional functionality. They can mark the Type of Service bits in the packet headers with differential service markings and propagate those through the "clouds" if it's a DiffServ backbone or even if it's both a DiffServ and MPLS-enabled backbone.

But with VoIP, of course, the real challenge has always been the latency, delay, jitter and other anomalies picked up by packets as they traverse the vast Internet cloud and how to work with your service provider to get meaningful service level agreements to deal with these packet delivery phenomena. To achieve QoS will require things such as differential service classifications so that when the packets are being routed onto the backbone, they can be treated appropriately.

MANAGING VPNS

Performance is only one of the key VPN issues. Manageability has now become a crucial issue. Certainly the gateways have to be fast, and have advanced functionality in terms of things such as QoS, dynamic routing, and network address translation. These types of features need to be on par for a vendor to be a first tier player in the VPN space. But given that those functions are in place, the next question is: How do you deploy, scale and manage a VPN?

The real focus today is on VPNs moving away from the gateway and up to the system level – where policy, authentication, and related information is stored and administered. Where is that information being configured and how is it being delivered out to the gateways?

Several VPN companies have introduced a directory server-based Lightweight Directory Access Protocol (LDAP) management platform that uses a very efficient communications protocol between the manager and VPN gateway devices. The upshot: this kind of architecture lets you configure all VPN parameters centrally at a directory server, and then deliver them out to the devices using the LDAP protocol. Contrast that with routers, especially when you've got network administrators scattered about on IOS consoles setting routing tables to make a VPN work. The directory approach is clearly better.

The differentiation seen today among the different VPN players centers on the management platform. When Microsoft's Active Directory first appeared and the company had high hopes for the product, one almost got the impression that Microsoft thought VPNs were just an extension of their OS, since they promoted the idea that when Active Directory became part of managing an NT network, you could also use it as a platform to distribute out policy information for VPN usage. Initially, however, Microsoft was at a slight disadvantage in the VPN marketspace, since it offered a software-based gateway, essentially an add-on to their 2000 Server.

One big challenge for managing people and VPNs is ease of use for the end user. In the case of Avaya's VPN equipment, it has a central point of configuring policy information at the directory server, so it not only pushes the information out to the gateways – every time the user logs in or authenticates himself or herself to a gateway, the VPN device then immediately pushes out real-time policy information to the desktop which requires no intervention from the user at all. The user simply logs on using his or her secure ID and a password or a user ID password or a digital certificate. Once authenticated, LDAP capabilities within the Avaya system automatically configure the software on the user's desktop for the life of that session.

Directory servers are having a big impact on the most recent generation of VPN products in terms of allowing systems to be developed that can support many users. Systems now exist in the field incorporating thousands of gateways. If those systems needed personnel to be stationed locally to go out and put their hands on a console on each one of those gateways, such a cumbersome, ungainly situation would break quickly from a management perspective.

Another issue is whether anyone will use customer premise VPN equipment at all, since, like much of computer telephony functionality, VPNs can be delivered as an enhanced service. This is Nortel Network's approach with their Shasta 5000 product.

Nortel Networks (Saint John, NB, Canada – 1-800-4-NORTEL, www.nortelnetworks.com) is interesting in that it produces VPN equipment for both the corporate and the carrier / service provider market.

Keerti Melkote, Director of Product Management for Nortel Networks, tells the author that "Nortel looks at VPNs and sees a trend – VPNs as tools for outsourcing at a very high level. The price points suggest that in the future much of VPN service will be purchased by companies from carriers and service providers."

Initially, Nortel acquired from Bay Networks the Contivity Extranet Switch, which combines remote access protocols, security, authentication, authorization and encryption technologies into a single solution. The switch can support a huge number of simultaneous user sessions. The architecture for the switch is user-centric, where an individual user or group of users can be associated with a set of attributes that provide custom access to the Extranet. In effect, you can create a personal Extranet based on the special needs of the user or group.

Nortel commissioned The Tolly Group (Manasquan, NJ, 732-528-3300, www.tolly.com) to evaluate its Contivity 2600, which is designed to serve large branch offices or data centers that support up to 1,000 VPN tunnels. Tolly engineers subjected the Contivity 2600 to a battery of tests to determine the switch's single-rule firewall and IPSec gateway bidirectional zero-loss performance, as well as to benchmark switch performance when both services are vying for bandwidth.

Test results show that the Contivity 2600 delivers up to 316 Mbps of bidirectional zero-loss throughput when handling 1,518-byte frames across two port-pairs in a firewall configuration. Results also demonstrate that the Contivity 2600 achieves up to 112 Mbps of bidirectional zero-loss throughput when handling 1,518-byte frames in a VPN IPSec gateway scenario with 3DES and SHA-l.

Furthermore, testing proves that the Contivity 2600 provides near line-rate firewall throughput (190 Mbps) while simultaneously passing in excess of T3 rates (80 Mbps) of 3DES, SHA-1 and IPSec traffic.

Nortel improved upon this when they teamed up with RADWARE/RND Networks (Mahwah, NJ – 201-512-9771, www.radware.com) to announce a joint product offering that combines the Nortel Contivity Extranet Switch product family with RND Network's Web Server Director (WSD), an Internet Protocol (IP) load balancer. Both products will be reference-sold by their respective companies.

The RND WSD optimizes the use of multiple Contivity Switches by intelligently load balancing VPN traffic between units. It continually monitors the status of each switch and throttles traffic to the units based on predefined, real-time performance and health parameters. WSDs can also be installed in pairs, eliminating any single point of failure in the network.

Thus, the Contivity Extranet Switch provides "Personal Extranets," while the WSD provides automatic VPN server failure detection and failover capabilities. It's a combination that ensures that end users always access the optimal VPN location. The Contivity Extranet Switch can inform the IPSec client of a fail-over destination to be used in future connections. If the Primary Switch fails to respond, the Contivity Extranet Access client will automatically try the fail-over destination.

"The other thing that Nortel is seeing is actually a much bigger trend," says Melkote. "Which is, what's really the most efficient way to provide VPN services? One of the issues of the customer premise (CP) based systems is that they provide services but you need a technically skilled staff to pull it off because you're managing the whole thing yourself."

"And if you ever use a server that you buy from the VPN service provider, then the expenses are quite high because they have to come to your premises and manage it from there," says Melkote, "So the price points associated with managed CP services are fairly high and don't scale down to the small and medium sized enterprises that want access to these types of services too."

"The new trend relates to how you can provide these services without having to send dedicated personnel to the customer premises. Basically, this comes back to network based services, or providing these types of services using a device in the cloud, in the network itself. And that's where the new platforms like the Nortel's Shasta 5000 come into play," says Melkote.

The Shasta 5000, what Nortel calls the first true Broadband Service Node (BSN), offers seamless migration from conventional broadband aggregation to high-tech subscriber services. With the Shasta 5000, you take all of the functions that you had enjoyed previously in CP VPN technology and integrate them and virtualize them into a single device at a point in the network that Nortel calls the "subscriber edge" of the network. It's where the typical subscriber enters the network.

"Some of the services that the Shasta 5000 can provide are the very same things that can be found at the CP level," says Melkote. "Access to the VPN and intranet, access to the Internet, access to an extranet – be it a shared extranet or a private extranet. And, of course, it provides a network-based firewall that actually protects the VPN traffic from the Internet."

This model allows a service provider to lower the cost of operations because one can now deliver services almost dynamically since you're now provisioning them in a centralized manner.

The Nortel Networks' Shasta 5000 is powerful enough to allow service providers to aggregate tens of thousands of subscribers onto one platform and apply highly customized IP services. Its scalability allows for up to 32,000 subscribers and 112 processors. The Shasta 5000 BSN integrates with other access and core products in the Nortel Networks family. Services supported include individual firewalls, VPNs, and traffic management, provisioning subscriber self-provisioning, and back-office integration.

The Shasta 5000 BSN enables service providers to aggregate different types of access technologies, such as DSL, dial-up, Frame Relay, ATM, cable, and wireless onto a single platform. Funneling all subscribers through a common platform improves subscriber management and integration with back-office systems.

The Shasta 5000 BSN combats the vulnerabilities of broadband "always on" access by providing many security services designed to support single-subscriber applications, and boosts security on a site-by-site basis by supporting multiple users per location. These security services include IPSec encryption, firewalling, anti-spoofing, and Network Address Translation (NAT).

The Shasta 5000 allows for the creation of "Personal Content Tunnels." These subscriber specific tunnels are marked with a high level of priority (QoS) to ensure the fastest possible content delivery.

Working in conjunction with the Shasta 5000 BSN, the Nortel Service Creation System (SCS) provides a service definition system to allow for the rapid creation of customized, mass market IP service offerings. With it, customers can now sign up for their network services through any standard Web browser.

Nortel's Contivity switch family also serves as the basis of the VPN Remote Access Suite offered by Genuity (Burlington, MA – 781-262-8730, www.genuity.com), formerly GTE Internetworking. Genuity's Secure Remote Access Suite for Nortel Networks' Contivity switch is a service that's designed to make secure communications and connectivity cost-effective and convenient for large enterprises. The new integrated service allows a customer to achieve a full eBusiness network solution by combining its managed infrastructure services rather than having them delivered only as individual offerings.

The VPN Remote Access Suite should give enterprise customers a sort of "virtual IT department" to monitor availability and performance, and maintain aggressive service level guarantees, all in one. The result is a one-stop-shop for provisioning, support and billing.

The service combines VPN Advantage for Nortel Networks' Contivity switch managed VPN service with Genuity's own DiaLinx remote access service and a choice of its dedicated Internet connectivity services – Internet Advantage or BizConnect – based on bandwidth and service level requirements.

By combining managed VPN, remote access and dedicated Internet connectivity in a single service, companies can focus on their core business objectives without worrying whether their Intranet communications are secure and reliable. Companies also benefit from having their entire Internet communications infrastructure managed by a single service provider, ensuring timely customer support and comprehensive service level agreements (SLAs).

Genuity backs the solution with SLAs for busy free dial availability of 97% and initial modem connect speed of 45.6 Kbps, 99% of the time. And since Genuity owns all of the hardware and software, the burden of capital investment and technology obsolescence is now on Genuity, not you.

Genuity's VPN Remote Access Suite for Nortel Networks Contivity is available now and starts at less than $20 per user per month.

Not to be outdone, The CryptoCluster VPN Gateway from Nokia (Mountain View, CA – 415-883-2967, www.nokia.com) is a scalable and high reliability VPN product for large network providers and corporations. It's said to be the only VPN solution based upon a true clustering technology (derived from Nokia's acquisition of Net Alchemy) that allows individual VPN gateways to work together as a single unit.

Nokia's cryptographic clustering technology ("CryptoCluster") offers two features not normally found in VPN solutions:

- **Active Session Failover.** If any member of a cluster becomes unavailable, the cluster will automatically reassign the active sessions among the remaining members with no disruption in service. In the case of the Nokia VPN gateway solutions, security associations (SAs) are securely shared across the cluster, avoiding costly re-keying in the event a member becomes unavailable.
- **Dynamic Load Balancing.** Cryptographic clustering enables multiple Nokia VPN gateways, collaborating via high-speed network connections, to function as one extremely powerful gateway. As more gateways are added "live" to the cluster, the load is automatically re-balanced to include the new gateway, with no impact on current cluster operations.

The CryptoCluster product family includes three VPN gateway platforms, management software and remote client software, starting with the CC0500 CryptoCluster VPN gateway for SOHOs (4.5 Mbps, 500 IPsec Tunnels), the CC2500 CryptoCluster VPN gateway for regional or mid-size offices (45 Mbps, 5000 IPsec Tunnels), and the high performance CC5200 CryptoCluster VPN gateway for enterprise headquarters, carriers or large service providers (150 Mbps 50,000 IPsec Tunnels)

Each of these units can be clustered together with any combination of additional Nokia CryptoCluster VPN gateways, sharing VPN session load in a manner proportional to their capacities.

Nokia's CryptoConsole management software is a Java-based management tool that provides a single interface to manage, define and deploy security for all Nokia VPN resources even if they are in different security domains, ensuring consistent security policies across all systems. For remote clients, Nokia also provides an IPSec client for various flavors of Windows. The client allows you to establish secure tunnels from your Windows system to CryptoCluster gateways or other IPSec-compliant devices.

And of course the router companies have jumped on the VPN bandwagon. Just one example is Cisco Systems' (San Jose, CA – 408-526-4000, www.cisco.com) Integrated Communications System (ICS) 7750, an advanced IP telephony solution that delivers a managed Web-based communications applications that can transform branch-office and midmarket business environments into eBusinesses. It features an expandable platform that enables businesses to take advantage of all existing Cisco IOS services such as VPNs.

The 7750 serves as an inexpensive platform for quick deployment of advanced applications for eBusiness such as unified messaging, integrated Web call centers, data/voice collaboration and networked video.

The ICS 7750 has everything that's needed to deliver converged data, voice and video – multiservice router/voice gateways based on well-known Cisco IOS technology, application servers running core voice applications, call processing software, integrated Web-based system management, and a data switching interface for seamless connectivity to recommended Cisco Catalyst QoS enabled switches.

The ICS 7750 is a six-universal-slot, industry-grade chassis system. Each universal slot accepts a card – the multiservice route processor (MRP) or the system processing engine (SPE) – to address a range of data and voice connectivity and application needs.

The MRP is the multiservice router/voice gateway running Cisco hardware and Cisco IOS software to support both digital and analog voice trunk gateways and WAN interfaces – all on a single modular card. It offers a wide selection of data and voice interfaces, allowing your business to match bandwidth to the needs of each site as they grow. Businesses can leverage all existing Cisco IOS services such as VPN, firewall, IP Security (IPSec), and QoS to ensure quality voice and data transmission.

THE FUTURE

Looking toward the future, VPNs will be everywhere, making possible a new generation of distributed data and voice applications. VoIP VPNs will happen, and large corporations will be able to shuttle both voice and data out on these virtual networks, with performance and security guarantees backed by service level agreements. Future plans for VoIP VPN vendors include letting companies terminate IP calls on the public network. Imagine: some day, all phone calls leaving your desk, both internal and external, will end up as IP packets on these VPNs, and the good old carriers will be needed just for the last stretch of copper – and not at all if the call ends up in your organization. Employees will be able to be placed at home or at whatever secure locations deemed necessary, secure in the knowledge that even a tremendous disaster will not completely obliterate their working environment.

CHAPTER 9

Telework

THE SEPTEMBER 11 ATTACKS ON THE UNITED STATES displaced many corporate offices and their workers. To stay in business, companies temporarily located their corporate offices in other buildings and/or had their employees set up shop in their homes via cell phones, laptops, high-speed access lines, an Internet account and e-mail.

If a company was lucky, it had a teleworking program in place prior to the 9/11 attacks, which served to help everyone in the organization to make the transition – logistically, technically, and emotionally – to a new environment. Thanks to their forward-thinking approach, businesses found their crisis management chores a little less difficult.

While workers who have had some familiarity with teleworking prior to September 11th will find it relatively easy to make the transition to full-time teleworking in crisis situations, teleworking novices may find it more difficult. Still, there are almost always a sizable number of employees who travel frequently, and are thus accustomed to remote-access working, which definitely helps to gracefully ease them into a teleworking setting.

Though there are a growing number within the business community who have already instituted telework programs, the September tragedy was a wake-up call for everyone to recheck their business-contingency and disaster-recovery plans and re-examine how they could quickly shift, if necessary, dozens or even thousands of employees to a distributed office environment. And since the September terrorist attacks forced hundreds of companies to set up remote offices for their employees just to stay in business, as companies rethink their business-continuity plans, they're often including teleworking as one of the crucial elements.

Teleworking offers businesses an efficient way to get displaced employees back to work. John W. Loofbourrow Associates had offices based in the World Trade Center. It had an active teleworking program. When the disaster occurred the company's systems collapsed along with the buildings, but because there was an existing infrastructure in place, it took only two weeks to restore all systems to working order. According to Gartner analyst John Girard, companies that have established "any kind of work-at-home or mobile work schemes stand the best chance to get their employees back to work quickly and safely."

DISASTERS COME IN MANY SHAPES AND SIZES

That's something California businesses have known for some time. They already had earthquakes, flooding, mudslides, gridlock and an energy crisis to contend with, now terrorist attacks must be added to their growing list of worries. It's not unusual for, say, a California law firm to have a remote-access program in place (even if they don't actually sanction teleworking, *per se*) so that their staff can access the firm's information from virtually anywhere when a disaster strikes. If an energy crisis, mudslide or earthquake strikes, the attorneys can ensconce themselves safely at home and still deliver a productive workday.

As you can see, terrorist attacks aren't the only events that might suddenly keep large numbers of employees out of a corporate office. Add to the above-mentioned scenario Mother Nature's handiwork – hurricanes, floods and snow storms – and one soon realizes how any company, anywhere, can be brought to its knees by an unexpected disaster.

Potential technical, logistical, training, support, and even emotional challenges are plentiful when deploying a teleworking workforce on sudden notice. Logistical problems range from shipping computers to employees locked out of their usual offices, to getting telecom companies that are swamped with emergency repairs to install broadband services or a second telephone line for dial-up modems in employees' homes.

Perhaps the biggest hurdle when handling new teleworkers is getting them set up with additional phone lines or high-speed services. "In the best of times, it can be a challenge coordinating schedules so that individuals can be home when the technician arrives," says Ray DuBois, VP of New York business renewal at American Express, one of the corporations displaced by the terrorist attacks on the World Trade Center.

This challenge can be magnified since a disaster can also leave cable operators, DSL providers, and local telephone providers with the job of managing very high workloads under trying conditions (some telecommunication providers had their infrastructure severely damaged by the September 11 terrorists). Verizon Communications Inc., which provides telephone services in New York, its surrounding suburbs, and nearby areas of New Jersey, says the average customer wait for installation of DSL (Digital Subscriber Line) and other high-speed services is 10 days in most parts of its service area. But a Verizon spokesperson admits that the waiting time for such services in the New York area is longer, in part because of the damage to the company's facilities and lines during the collapse of the World Trade Center buildings.

To help compensate for those waits, many employees, who don't already have second phone lines in their homes, will rely on cell phones for voice, and home phone lines for data communications until telecom companies can service their homes. For example, within days of the WTC disaster, Zurich North America, a company that had more than 600 employees located at One Liberty Plaza (which is quite near the World Trade Center site) began shipping laptops and cellphones off to its workers so they could begin the process of setting up a teleworking environment.

THE CHALLENGES

Challenges abound. Users get frustrated because applications that work so effortlessly in their office LAN environments can be painfully slow when accessed from home, especially if the company and the teleworkers don't have high-bandwidth access available. Proper training and support are key factors in the success of a teleworking workforce, whether work-at-home programs are permanent or makeshift. For instance, American Express had more than 5,000 workers displaced because of the destruction of the WTC complex, including the 800 teleworkers who already use teleworking techniques to get their jobs done. The company's IT department has assembled a team of approximately 200 IT workers who are supporting this huge staff. IT personnel can even train new teleworkers on how to use the system via telephone.

Then there are the purely technical tasks to contend with. Companies need to round up IT staff to handle the installation of communication software and applications needed on each worker's system, train employees on how to use them, and troubleshoot hardware problems remotely. In some instances, legacy applications need to be Web-enabled and secured.

Security becomes paramount, especially because home users can be particularly vulnerable to hackers and other breaches. Then there's the added cost of purchasing or renting the computers and providing high bandwidth connections – yet, ironically, all of this can actually be less expensive than setting up space for each displaced employee in a temporary corporate office environment.

Merrill Lynch & Co. is an example of the typical large corporation at the WTC complex prior to September 11th. The company had up to 1,500 employees regularly teleworking, and that number has quadrupled since 9/11 according to Bernadette Fusaro, vice president of work/life strategies. Merrill Lynch was profoundly affected by the terrorist attacks, but it already had a telework program in place, which it could further build upon, and so the company was able to quickly get its employees working at full capacity again.

For the new teleworkers, Merrill Lynch provided technical training as well as guidance on effective ways to work remotely, although the amount of instruction was reduced to just the most important basics. The IT staff provided employees with training on remote access and the human-resources staff offered guidance on how to be productive from home.

Under normal circumstances, Merrill Lynch teleworkers would receive up to a week's worth of technical training, especially if they're not comfortable working remotely, says Janice Miholics, VP of global telework strategies. Although in this instance the training is more streamlined to get the company back in business faster, the training is still thorough enough to ensure that the employees can work with confidence.

Merrill Lynch also finds it useful to keep tabs on how its teleworkers cope, particularly employees new to teleworking. The firm's HR group provides remote employees with questionnaires every 30 days to gauge how they feel about their "telework" situations. "We want to know how they've handled the relocation and if they're interested in going back into an office or continuing to work from home," Miholics says.

For teleworkers – particularly those who are unexpectedly working remotely – keeping up productivity can be a problem, but emotional hurdles present the greatest challenge. "Many home workers are surprised to find out how much more efficient they can be" when there's no water cooler to congregate around, says Fred Crandall, a partner at the Center for Workforce Effectiveness, a consulting firm that studies virtual office environments. But new teleworkers can be stricken by loneliness. Companies should discuss what it's like to be "virtual." To prevent employees from starting to feel isolated, departmental groups or teams should try to meet via phone or in person at least once a week via audio or video conferencing to brainstorm and congregate. Teleworkers should make the most of chat and Web collaboration technologies where participants can access and discuss PowerPoint presentations and other material while online together.

Quite a few corporations expect many of their displaced employees to continue to work from home occasionally or permanently even after everything is "back to normal." According to Milholics, employees and employer, alike, feel that teleworking "is a godsend for people with long commutes."

WHAT EXACTLY IS TELEWORKING?

Although a relatively new sounding concept, the basic idea has been in use for many years. People have always worked outside the corporate environment, using the fax machine, snail mail and land-line telephones as their lifeline to the corporate office.

➤ Telework Disaster Planning Checklist

Here's a suggested telework checklist for emergency planning. This checklist isn't meant to be all-inclusive, but it is a good starting point for meaningful discussions to begin within the company.

1. Distribute corporate assets – this includes employees. Use the military strategy of asset dispersal to spread the risk (specifically, data, files and staff) to distributed locations – for example to teleworkers' home offices, off site data storage locations, telecenters, etc.
2. Consider allowing part of your workforce to telework. Remote locations such as home or a local telecenter are often the best places to work during disasters, particularly when getting to the corporate workplace may be impractical or impossible. While telework is not suitable for all types of situations due to the nature of the job or the personality of the worker, it can temporarily work in many situations.
3. Make telework part of your emergency preparedness and business continuity strategy. Emergency preparedness is like insurance: you hope you will never have to use it, but it's good to have, just in case. Have telework one among many potential solutions; a telework program can mitigate the negative impact of a disaster and allow the company to balance its business objectives with its employees' needs.
4. If a telework program is in existence, expand and fine-tune it. If there is no telework program, develop one. It's easier than you think. Even an *ad hoc* program can be of benefit, although you should-formalize it later, since a complete and well-designed formal program invariably brings the biggest benefits and savings.
5. Build a "telework kit," which consists of the basic telework guidelines, a list of important telephone numbers, e-mail addresses, passwords and procedures for staying in communication and protecting key data.
6. Buy laptops for the employees and configure them to work with the corporate systems. Don't forget extra rechargeable batteries and a back-up power-supply. Laptops enable your staff to work from wherever space and facilities they can find.
7. Equip a remote location or individual employees' homes to access, deliver, and receive information – from anyone, at anytime, and from anywhere.
8. Arrange for a corporate communications infrastructure and contingency plans that lets it appear to the outside world as if everyone is physically located at one common central site.
9. Identify ahead of time which of your business tasks are suitable for teleworking and then have the staff take relevant files home. Note: even just a phone, pen and paper and the telekit can allow the work to continue in the short-term.
10. Read up on telework and emergency preparedness.

For the purposes of this book, teleworking is defined as the practice of working outside the corporate environment on a computer that's linked via communication technology to the corporate information infrastructure where messages and data can be transferred. Yet, this only defines the basic concept of teleworking. There are many factors involved in working away from the corporate environment, such as the distance from the office (10 miles means you can pop in with little advanced notice; over 200 miles means a visit to the corporate offices is a planned event); communication methods (either using "old fashioned" methods, a corporate VPN, some kind of remote access system or a digital workplace set-up); and the primary work location.

While there are different types of teleworkers, all require that the employer and employee make certain that everything is set up and all the necessary equipment is on hand for an efficient and effective work environment. The different typical teleworking categories are as follows:

- **Part-time Teleworker**. The employee's primary place of work is in the corporate office, but the employee may spend a certain amount of the normal work day at a remote location (client's office, home, telework center), carrying out functions similar to those they would do if they were physically at the corporate location. The average part-time teleworker spends only around 1.5 days per week at the remote location, so most of their duties are actually performed within the corporate environment.
- **Full-time Teleworker**. The employee's primary place of work is outside the corporate environment, while his or her counterpart spends the majority of their time in the corporate "cubbyhole."
- **Satellite Office**. While not strictly a teleworker since the employee still works in a corporate environment, the location is remote from corporate headquarters where the workers' counterparts are located. Many times a satellite office is set up in a location where a large concentration of employees critical to the operation of the company reside. This reduces the necessity for these individual employees to travel to corporate headquarters on a daily basis. In this teleworking situation, the employer ensures that everything is set up correctly and all the necessary hardware, software and communications are available.
- **Corporate Telework Center**. Buildings chock full of technology – fully loaded and configured computers (hardware and software), high-speed connection, printers, copiers and fax machines. Corporate telework centers are constructed in a central location close to a concentration of corporate employees' homes. For example, News Corp. (a large news and entertainment corporation with corporate offices situated in mid-town Manhattan and reporters and writers scattered throughout the New York metropolitan area) if so inclined, might construct one corporate telework center somewhere in the middle of Long Island, New York; another in Rockland County, New York; and a third in northern New Jersey. The employees would obtain an office assignment in the telework center closest to their home, regardless of job function. This enables employees to avoid long commutes, corporations to have a distributed workforce, and virtual teams to work via a secured environment and state-of-the-art technology links.
- **Third-party Telework Center**. Employees of several organizations usually use such a center. Hotels, airports and other places frequented by business traveler often offer this type of center. These telework centers provide a higher speed connection; state-of-the-art computers loaded with every imaginable computer program; printers and copiers (including color) and fax

machines. The teleworker can schedule time as needed to perform their computer-and communications-related work.

- **Mobile or Nomadic Teleworker**. This is often the bailiwick of the sales representative or "road warrior." Employees in this group often use some form of teleworking plus digital workspace since they are constantly on the move and normally don't have a base office.
- **SOHO Teleworker**. When someone talks about teleworking, they are almost always referring to corporate employees working at home or elsewhere away from the corporate office environment. However, many companies depend on long-term outside contractors to perform various crucial corporate tasks. The same technologies and worker values that cultivate interest in corporate teleworking also fosters enthusiasm in the SOHO (Small Office Home Office) environment. Give a smart person a cell phone, a laptop with a modem and a fax machine, and he or she is in business.

Before September 11th, teleworking was seen as a stagnating non-trend. Many managers had decided that it was too difficult to manage a network of dispersed remote workers; and employees felt stigmatized if they weren't seen around the office. But after the terrorist actions, many workers and their employers have re-evaluated their respective priorities. When an event that ends life so swiftly and on such a vast scale occurs, it causes people to wake up and more carefully reconsider previous decisions.

Various Considerations

This does not mean that every employee should work at home or in a satellite office or telework center. Use teleworking selectively and appropriately to decentralize the corporate office.

When a corporation takes on the telework concept it needs to consider the role teleworking will play within its corporate culture (the set of habits and patterns of how staff works together and how things get done), i.e. the effects teleworking has on the corporation as a whole. When teleworking is instituted in an organization – whether it has five employees, 500 or 5000 employees – it will affect the flow of work, the way people communicate with each other, and much more.

A team effort is required for teleworking to succeed. But in many businesses, teleworking has been planned and implemented by only one segment of the corporate staff: the Information Technology (IT) staff, or the Facilities and Real Estate staff, or the Human Resources staff. While it's commendable that one group took the initiative, no one department or group can manage teleworking on their own.

The best teleworking programs are the ones that involve a planned structure formulated with consideration of cultural changes, and anticipates what changes may be necessary for teleworking to

➤ Telework or Telecommute?

What's the difference between the terms "telecommute" and "telework"? Very little. Telework tends to be the term more commonly used outside the US, while "telecommute" is used more domestically. I prefer the word "telework" because it's a more accurate description of the concept: "tele" means "distance", *ergo* "telework" means "work at a distance." Telework advocates believe that "telecommute" has too strong a connotation with commuting, and that "telework" is a broader and more inclusive term. But, whatever you choose to call the concept of an employee working outside the formal corporate milieu, the underlying concept is the same – the decentralization and distribution of the corporate workforce by applying various methods to bring work to the workers.

succeed in the long term. Yet, many in the business community fail to recognize the integrated, connected aspect of teleworking, and the corporate culture changes wrought by teleworking.

This inconsistency causes the success rate of telework programs to vary widely. Companies that find teleworking to be a success understand and, thus, properly implement a telework program. Others complain that teleworking is more "trouble than it's worth." These companies have either half-heartedly adopted teleworking or improperly instituted the program. Then there are the businesses that are still fighting tooth and claw against teleworking of any sort. Hopefully, this chapter will help the remaining stalwarts in the business community to see the error of their ways.

A Historical Perspective

Many people think that teleworking or telecommuting is an outgrowth of the wholesale adoption of the personal computer during the 1980s and the invention of the 9600 baud modem in 1987; thus, the phenomenon must have begun by 1990. That's wrong. Many cottage industries – knitting, garment sewing – tried it, and even IBM, surely the most corporate of all companies, used teleworking in the late 1960s by installing IBM keypunch machines at employee's homes to produce data-entry cards. What about the sales person who travels to visit customers? This employee group fits the definition of a teleworker. And traveling sales people have been around for years, the vast collection of traveling salesmen jokes attest to that fact.

Admittedly, however, prior to the late 1980s, teleworking was considered something bizarre or unusual. Going to the office every day was the norm, so almost anyone who performed their corporate office duties away from the office was looked on as different and even a little strange (sales professionals being the exception).

The fundamental idea behind telework is the decentralization of the corporate environment. But in today's information-based economy, all employees don't need to be in one single location or to work together during the same time period to be productive.

For teleworking to become a concept recognized within the business community at large, the more forward thinking employers had to take a chance (as is necessary for any innovation to catch on). It meant that their teleworkers had to figure out how to carry home a heavy computer (the "transportable" computers of the time were a burden to carry around, much heavier than modern laptops). These workers likewise struggled with modem connections that were too slow to be of much value. They also had to be very careful about their work schedule and to be sure they made it into the office for scheduled meetings. While many believed that it would just be a matter of time before every employer started using teleworking, and that it would be offered to everyone – just as employers offered paid vacation time, insurance, and other benefits to everyone – it hasn't happened. In fact, it ***shouldn't*** happen – there will always be a need for a corporate headquarters, populated by real, office-bound people.

The Current TeleWorking Climate

That brings us up-to-date. Today many corporate meetings are held via audio or video conference calls. And as the reader will see, a laptop, PDA or handheld computer, a cell phone, modem and Internet access is all that's needed for a virtual office setup (sometimes referred to as a "digital workplace").

It's very easy to look at all of the exciting technology available today and to think about how effortless teleworking can be, thanks to broadband and wireless access, VPNs, the Internet, intranets, corporate portals, state-of-the-art laptops and PCs, handheld computers, PDAs, cell phones, software and more.

There is a plethora of technology for teleworking – hardware, software, and telecommunications – enabling corporations to institute effective teleworking infrastructures. Most of this technology is very good, yet it takes a consortium of complementary technologies to help implement a good teleworking infrastructure. With so many vendors touting their wares, companies have great difficulty selecting the necessary components: Laptops, PCs, Handhelds? Cell phones, a RIM-like Device, PDAs, perhaps wait for a three-in-one device (cell phone, pager, PDA)? VPN, Corporate Portal, Intranet, Virtual Office via Internet, a Remote Access System? The list goes on and on. The problem isn't the technology, it's how to gather the right technology together to form a viable cohesive whole.

What the Future Holds

Now, let's take a glimpse into the future (perhaps Charles Dickens and H.G. Wells have been too big of an influence on the author?) where work will be something you do, not someplace you go. Where the employee who toils daily within the corporate office environment is considered "bizarre" and the employee who teleworks at least a portion of the workweek is considered "normal."

In such an environment, the employee's lifestyle will improve. For many employees the worst part of their day is the grueling commute to and from their place of work. Teleworking will enable employees to spend more time with their children and their aging parents since working close to or within their home makes it easier for the employee to be in the home and its surrounding environment a few more hours each day.

I must emphasize that everyone will not be a teleworker, and even for those who do telework, most won't do it full-time. Yet, I also must emphasize that the concept of telework will become universally accepted as a normal occurrence in the workplace.

As the reader now understands, during the information economy of the last 20 years, teleworking has progressed from a concept to questioning if teleworking was possible, to cautious optimism about teleworking. Although the corporate community hasn't progressed to the point where employees can assume some form of teleworking is naturally available, I don't think such a scenario is too far in the future because, if distributed computing, distributed storage makes good sense, why not also a distributed workforce? Smart executives will implement telework programs because it makes good business sense to do so and their current and future employees will simply assume that it is available.

WHY TELEWORK?

Other than decentralization of the corporate office complex, a teleworking program is good for the corporate bottom line. Research indicates that a defined telework program can improve hiring metrics, stem employee turnover (long tiring journeys are no longer a daily ordeal, employee's out-of-pocket travel expense is reduced, child care is less onerous) and cut facilities costs (less desks, chairs, space is needed).

Teleworking is a concept that is here to stay and the demand for teleworking technology will continue to grow. A recent survey by Techies.com indicated that more than 90% of tech professionals would like to telework at least a few hours per week. And, according to a Yankee Group report, the potential benefits for a distributed workforce have resulted in conservatively projected teleworking forecasts of 18% annual growth. The Washington-based International Telework Association and Council (ITAC) report supports the Yankee Group's conclusions. The ITAC states that in 2001 28 million US employees participated in some form of teleworking, whether it's in their home, on the road, in telecenters or in satellite offices. This is a 17% increase from the 2000 ITAC numbers. The same holds true in Europe. A Telework Association's August 2001 report shows that employed home teleworkers grew by an average of 20% per annum whereas the overall teleworking figures during this period increased by an average of 15% per annum.

But it's not all good news. While teleworking is definitely on the increase, according to a survey by ***Industrial Society***, of the 91% of European companies who practice flexible working, only 69% do so as a formal activity. The survey suggests that problems (although relatively minor) can result; such as communication difficulties, problems with managing varying working arrangements and resentment from staff bound to their corporate office.

The Possibilities

Another key question is what sort of network, equipment and support is required to provide the interface to the teleworkers. There are three basic teleworking options: work at home, at a telework center or in a satellite office. For home-based teleworking, the company should provide enough phone lines to satisfy reasonable peak demands, enough bandwidth per phone line to satisfy user needs and some kind of remote access capability.

For some teleworkers, a 56 Kbps modem will do. For most teleworkers, especially for those using imaging systems, requiring video conferencing or needing access to large graphics files, a high-speed connection may be in order.

As for computers for teleworkers, many organizations have adopted the strategy of providing laptops with docking stations instead of desktop PCs.

On the other hand, some organizations ask workers to pay for the equipment they use outside the corporate office environment. For example, because of budget constraints, few departments in the City of Los Angeles can afford to provision their teleworkers. Most require that prospective teleworkers provide their own PCs if they need them; although the city made arrangements with local retail stores to provide corporate discounts to city employees. Still, according to the city's telecommuting coordinator, lack of city-provided equipment has not been a major deterrent to its telework initiatives.

One way to reduce costs without eliminating the traditional office environment is to establish a telework center as discussed previously. This can vary in size from a few to hundreds of workstations. If the center is wholly owned and operated by the company, it can be treated just like a "nor-

➤ Nortel Networks

Nortel Networks and other visionary companies recognize the advantages of a distributed workforce and have implemented and even expanded the telework programs in their various divisions.

Studies regarding Nortel's programs report a 24% improvement in employee productivity; the employees participating in the program report an increase of 10% in their overall satisfaction with their job. That, in part, has contributed to a 24% reduction in turnover, which, in turn, has generated a hard dollar savings to Nortel by reducing costs associated with recruiting and new employee training.

mal" remote site, with local networking, a high-speed data link to the central facility, video conferencing and strict security features.

Costs and Benefits

There are two primary cost areas for a company implementing a telework program.

Startup Costs

The first are the startup costs, which include planning, selection, training, and technology implementation. Additional computers, docking stations, telecommunications hardware and software (and their installation) tend to be the dominant up-front costs for large telework programs. But there are also primary additional costs incurred for training and project management during the implementation phase.

The Computer Conundrum

The computer question is an important one. Foremost is the decision as to who's buying. Most government agencies, in this era of tight budgets, restrict teleworking to those who can provide their own PCs at home. Most private companies provide the computers for their employees.

Usually it is not necessary to duplicate the entire computer setup. Full-time teleworkers only have one computer dedicated for their use – the computer at the teleworking location. The majority of teleworkers, though, split time between a teleworking location and the corporate office, so they typically use a laptop and docking station setup, so the duplication tends to be only for the monitors and docking stations, rather than an entire system. Where LAN-like performance is a requirement, you need to factor in the cost of a high-speed connection between the telework location and the corporate office.

In answer to the security problems presented by telework programs and to gain some control over the maintenance and management of teleworkers' equipment, some companies are restricting teleworkers to using only company-owned PCs for accessing company-provided broadband services. Often these computers have their configurations locked down so that the PC will be little different from the ones in the company offices. This methodology works to some extent when you have a well-defined and funded telework program; but, unfortunately, many companies only have an *ad hoc* telework system, where teleworking is decided on a case-by-case basis.

The situation can become a bit muddled when the teleworker, rather than the business, makes all the arrangements. He or she now has a sense of ownership, especially if the IT department presents the teleworker with an attitude of "you're on your own." Although it's understandable and advisable to expect a teleworker to be able to handle basic systems maintenance, other details, such as security, should remain the province of the corporate IT department.

Operating Costs

The second consideration centers on operating costs, which consist of telecommunications charges, IT support and management and maintenance fees. Most of the long-term costs of teleworking occur in the domain of the IT department.

The Benefits

Industry research indicates that companies see a net gain from their teleworking investment through:

- **Increased productivity**. Managers find the productivity of workers who telework one or two days per week, averages 5 to 20 percent higher than their non-teleworking colleagues.

- **Decreased facility costs**. Teleworkers need less or even no formal corporate office space, depending on how often they work at home and/or at a telework center. The same holds true for parking space. Many corporations, such as IBM, Nortel, and the State of California, have adopted telework programs on the basis of space savings alone. David Mead, president of Telecommuting Success, a consulting and outsourcing firm that specializes in the design, implementation and management of telwork programs, reports that companies can expect annual savings of up to $8000 per teleworker by reducing the need for corporate facilities and associated costs.
- **Human resource savings**. Teleworkers tend to take less sick leave than their in-office counterparts.
- **Employee loyalty**. Teleworkers are more loyal to the company. Turnover rates are reduced – often dramatically.

Let's not forget the benefits accrued by the teleworkers. These include reduced operating costs for their cars, fewer trips to the dry cleaners and less expensive lunches. But the most important benefit is stress reduction since teleworkers are able to fit their work into the rest of their lives, instead of *vice versa*. Tmanage Inc. (Austin, TX – 512-794-6000, www.telsuccess.com), a company that develops and implements sustainable teleworking programs for Fortune 1000 companies, reports that its clients have shown that teleworking equates to an indirect pay raise averaging $4000 per year through savings in transportation, meals and clothing.

Even the community benefits from teleworking. Home-based teleworkers use their automobile less causing traffic congestion to be reduced, energy to be saved and air quality to be improved.

While many firms initially embark down the teleworking path for the dollar savings and decentralization, in the longer term the increase in employee morale, lower turnover numbers, and the ability to attract higher caliber workers, are the sustaining reasons that businesses will continue to expand their telework programs.

MAKING IT HAPPEN

Successful teleworking programs have eight main phases. Each has important implications.

Planning. A primary function of the planning phase is to establish success criteria, performance measures and preliminary policies for teleworking. This is also the phase where you ensure that there is upper management support, preferably among several upper managers, before you institute a telework program.

Workflow. The company needs to take a close look at the existing workflow and communication habits of the potential teleworkers. Workflow will be affected by teleworking, such as signing off on paperwork, transferring files and information, and other processes that could require special consideration when the employee is not at the corporate office. For example, if the company relies heavily on face-to-face meetings, the teleworker will need to be equipped with teleconferencing tools.

Communications. The policies that the company has in place for communication (telephone, fax, Internet access) should be extended to encompass the needs of the teleworker. For instance, if a client is billed for long distance calls or faxes, the teleworker needs to continue this practice even though there is no formal cost capture method.

Pilot Project. This is vital. The key objective, as far as the IT department is concerned, is to establish the networking functional and technology performance requirements, scheduling and costs. If your company is thinking of introducing teleworking on a broad scale, it is important to include a representative sample of all departments in the pilot project. For instance, some departments will need extensive technological support; others may need just a little help. A pilot project is where you want the shortcomings of an otherwise well thought out plan to raise their ugly head. For example, teleworking doesn't come naturally to all employees and the corporate system design may need some changes before everything works smoothly. This phase is also the time for assembling the core of the rollout plan. In essence, repeat the processes of the planning phase, but refine the numbers with the data derived during this phase.

Implementation. Don't forget the small details, such as transparent or at least glitch-free access to the department network by laptop-equipped teleworkers. If you're using a groupware product, make sure that it will support teleworkers. Try not to implement new forms or scripts at the same time the pilot project participants begin teleworking; you want to be able to separate teleworking-related problems from technology adaptation problems. A key cost factor is the contention ratio – in this case, the number of teleworkers divided by the number of telecommunications ports required to serve them. A conservative approach would have one port per teleworker, but ratios much higher than that might suffice.

Training and Support. For any project, including teleworking, to be a success there must be proper training and ongoing support. Central to this task is the help desk. Make sure that the help desk personnel realize that this is an important project and thus gives it full support. But, detailed online support also needs to be implemented and systems should be put in place so the information is kept current.

Telework Policy. Draft and finalize a set of telework policies and procedures. Review these periodically to determine if they truly meet the needs of the teleworker, the company and the workflow.

Rollout. Once the plan has been made as bulletproof as possible, the technology is in place and all the participants have been trained in both technology use and management practices, you're ready for rollout. The pilot project phase should have served to fine-tune the technology and training needs and to provide critical data for the rollout.

Don't be hasty; remember all successful teleworking programs require some adaptation time. Too many things are still in flux in the first several months after implementation for reliable follow-on decisions to be made.

Teleworking works and works well when properly implemented. Conversely, poor planning and failure to get management's backing can be dangerous to a teleworking advocate's career. But, if everything goes smoothly and the teleworker has the ability to work from anywhere in a manner identical to the employee sitting in the corporate offices and the company can document a net benefit to its bottom line, you may actually get recognition for being a stellar innovator.

The Technical Infrastructure of Teleworking

Not all telework is technology-intensive. Some teleworkers do nicely with just a phone line, day planner, paper and a few writing devices (pen, typewriter). In some organizations these low-tech users constitute as much as 30% of teleworkers. Thus, they are essentially not a problem for the IT department.

At the other end of the spectrum are the full-time workstation users who need lots of access to graphical material, databases and/or full-motion video. For them, modem connections won't cut it and an ISDN line is only marginally adequate; DSL, cable or some other form of high-speed access will be the only acceptable solution. These cases require planning and expenditures.

Most teleworkers fit somewhere between the extremes. They use PCs but not all the time. They need access to their office PCs, but "sneakernet" may suffice for file transfer. If they do need LAN or server access, the traffic is most likely to be bursty, i.e. short transmissions at irregular intervals. If you routinely monitor traffic on your system now, it's fairly simple to anticipate such usage patterns. In fact, teleworking may smooth out the peak loads because, for example, the teleworkers may check on their e-mail earlier in the day than the employees who must wait until they get to the corporate offices.

Whatever the needs of teleworkers are today, you can be sure that the technology demand – and network access needs – will increase in the future. But, of course, that would be the situation whether or not there are teleworkers. The primary difference between teleworkers and corporate office personnel is the data link location between their computers. All in all, the only necessities needed to enable teleworking are whatever the employees' current needs are, plus a telecommunication connection.

Technological advances in software, hardware and telecommunications have enabled efficient and effective teleworking. Yet, there are other system issues, such as how to provide remote access and ensure systems security, and what access should be given. For example, which teleworkers need absolutely current company data for all they do? How many of them can make do with slightly older data, perhaps on a CD-ROM? How many can wait until they take a trip into the corporate office for that access?

Remote Access

Supporting teleworkers can be a complex proposition, given the wide variety of access technology and access methods – ISDN, DSL, cable modems, satellite and even wireless. Each requires unique terminal equipment at the teleworker's location.

The best way to handle such a tangle is to place limits on both access methods and equipment; this serves to give the IT staff a manageable list of products and systems they must understand, which, in turn, enables the IT department to capture various providers' configuration information into a central repository for reference. This repository should also include useful information for

➤ Employee-related Disadvantages

The benefits employees reap from teleworking are offset by a few disadvantages, which must be taken into consideration when implementing a telework program, such as the following.

- **The lack of socialization**. One of the best places for socialization is in the office where you meet new people and make friends. When working at home the process of socialization can be stunted.
- **Establishing the right work environment**. One of the essential facilitators of working at home is to establish an area where you can work efficiently, an area that is quiet and can be set up for teleworking needs, i.e. desk and other relevant furniture. Some teleworkers have extensions built to their homes so that they can telework in their own little environment away from all the hassles at home.
- **Distractions**. Even when home-based teleworkers find just the right work space, there will still be distractions – whether it's people outside the house mowing the lawn, workmen digging to lay new sewer lines, or the chatter of children when they return from school. It's important to adjust to the environment and make sure that these and other distractions don't cause disruption.

troubleshooting connections, such as settings for SMTP servers, DNS, and IP addresses.

Still, the company must recognize that while it's preferable to limit the choices, some teleworkers will always require a unique configuration. Also, few geographic areas can boast universal availability of service providers. Nor can every telephone line support DSL or ISDN signaling.

Or perhaps the company will enact a policy of providing analog dial-up accounts by default while reserving more costly (and more difficult to service) high-speed access for high-bandwidth users, such as graphic artists and database users. Yet, even analog modems from different manufacturers can behave differently.

Remote access to corporate networks is always a worrisome issue. While some companies host their own dial-in remote access servers (due to concerns about the security of public communication networks), many businesses rely strictly on service providers for dial-up and VPN access.

Before deploying a remote access solution (RAS) to a large teleworking contingent, run a pilot test to work out all of the bugs and to find out if the RAS's service levels meet the teleworking program's needs. Reliable, efficient remote access is critical for companies with teleworkers.

Finally, whatever remote access option chosen, make sure there is easy to access and easy to understand documentation that can adequately address the teleworkers' inevitable questions.

There are numerous products and methods available to provide remote access to a corporate network outside the corporate environment. Since no two corporate networks are alike and the needs of no two businesses are the same, I'm not going to attempt to discuss these options in this chapter. There are many good books available for both the teleworker and the IT staff – one of the best is the ***Network Tutorial***, authored by Steve Steinke and the editors of ***Network Magazine***. Also visit the "resources" listed in this chapter and make use of a good search engine to obtain information on remote access methods.

Software

There are a host of vendors offering a wide variety of tools to aid the teleworker and the enabling corporation. With technology in an ever-quicker evolutionary cycle, vendors who are constantly changing names, logos and acquiring other vendors or merging with each other, it's difficult to give a appropriate list of helpful software, other than what's already set out in other chapters of this book. I suggest the reader visit the resources listed in this chapter to obtain updated information about software for teleworkers.

Instant Messaging

Keeping the aforementioned caveat in mind, I am now going to discuss a software option which many corporations don't yet offer through their corporate systems and which most teleworkers will find crucial to an effective telework environment – instant messaging.

First ICQ software, then American Online's Instant Messenger (AIM), popularized the idea (not to mention the use) of instant Messaging (IM). Then Microsoft and Yahoo! entered the picture with their own "Messenger" products and the race was on. For example, Novell, the Lotus IBM subsidiary and other software vendors are building new IM systems tailored for business use. Today, both Novell and Lotus offer wireless text messaging and have licensed AOL's IM protocol, so users of the software can communicate with AIM users. It's a vital tool for communication, much like pagers, cell phones and voice mail. Yet, you get to see that the other person is there. You don't see that with e-mail, telephone or pager.

IM allows workers to locate and establish independent sessions with others, including a group of workers, clients, suppliers, etc. for conferencing via IM. Once connected, the group or individuals can exchange comments in real-time and share files, including graphics, audio and video.

But, IM products are built upon proprietary technology, e.g. different IM products can't communicate with each other – an AOL IM user can't talk to an ICQ or MSN IM user and vise versa. Yet, this situation is changing (and there may be intercommunication by the time the reader buys this book). AOL is under a government mandate to communicate with other IM applications, and several vendors support the emerging session initiation protocol for Instant Messaging and

➤ Teleworking Tools

Successful teleworking involves the use of certain tools and supplies, depending upon the nature of the job. Not all items listed herein are applicable in every situation. Use this list as a guide only.

Technology
Computer – desktop or laptop (if laptop, a docketing station may be needed).
Monitor
Glare filter
Mouse or trackball
High-speed modem
Adapters (for car lighter or cell phone)
Document holder
Cell phone or pager
Fax machine (dedicated or multi-purpose)
Keyboard
Printer(s)
Scanner
Zip drive
External disk or CD-ROM drive
Surge protection/power strips
Copier
Personal digital assistant
Calculator

Equipment – General
Telephones & accessories
Lamps and lights
Hole punch
Pencil sharpener
Diskette & CD holders
Flashlight
White board or paper easel
Rolodex
Utility knife
Paper shredder
Headset
Microcassette or small recorder
Transcription machine
Label maker
Typewriter
Paper cutter
Letter opener
Postage scale or meter
Clocks
Rulers

Furniture
Desk and/or computer work station
Shelves
Storage cabinet
Carpeting
Trash cans
Chair
File cabinets
Bookcases
Printer/fax/copier stands
Chair floor mats
Bulletin board

Office Supplies
Toner and ink cartridges
File folders and labels
Labels (shipping, filing)
Pens, pencils, markers
Tape, various types with dispensers
Rubber bands
Paper weight
Trash cans
Paper towels/tissue
Spare batteries
Stamps
Paper
Corporate letterhead
Corporate envelopes, varied sizes
Plain envelopes, varied sizes
Scissors
Staplers, staples & staple remover
Paper & binder clips
Self-stick notes, various sizes & colors
Rubber stamps
Desk accessories & organizers
Calendars
Binders/folders

List courtesy of ALLearnatives (www.allearnatives.com).

Presence Leveraging Extensions (SIMPLE) standard. This proposed standard allows users to use the Internet for voice exchanges and to establish conferences.

Soon it will be possible to find instant messaging products that offer the following (note: some of these innovations may already by available):

- Application sharing (one user opens the application and multiple users can run that application at the same time) via IM, which represents an important function as IM systems extend into collaborative applications.
- The transmission of real-time video to enable virtual meetings.
- A virtual workspace where workers can post documents, spreadsheets, and other data so that a group of workers can review and comment on the items simultaneously.
- Whiteboarding, which allows the simultaneous sharing of a graphic space among a group of workers. Each worker can access the graphic space and add comments or information. The whiteboarding concept can include collaborative document access through word processing programs.

Ultimately, IM will be the cornerstone of the teleworker's tool kit.

Resources

There are numerous associations and resources to mine for teleworking news, products and services, research, resources, etc. The following list should give the reader a starting point.

Associations

American Telecommuting Association
www.knowledgetree.com/ata
Canadian Telework Association
www.ivc.ca
European Telework Online
www.eto.org.uk
Home Office Association of America
www.hoaa.com
International Telework Association and Council
www.telecommute.org
The Telework Association
(www.tca.org.uk)

On-line Resources

ALLearnatives
www.allearnatives.com
eGrindstone
(www.egrindstone.co.uk)
Gil Gordon Associates
www.gilgordon.com
JALA International, Inc.
www.jala.com

June Langhoff's Telecommuting Resource Center (www.langhoff.com)
Mobile Computing
www.mobilecomputing.com
Mobilis
www.mobileinsights.com
TELEWORKanalytics international, Inc
www.teleworker.com

UPS

Teleworkers need power protection. But they have different power protection needs than their corporate office counterparts. See Chapter 5 for discussion of this important issue.

Security

While September 11th was a wake-up call, companies have been examining the issue of security in the teleworking world for years and were establishing small remote office environments as a partial answer, long before the terrorists struck.

Companies realize that they need secure access so no one can snoop on their systems. Corporate IT departments, therefore, aren't opening the door to teleworkers and other remote access workers until they get intrusion-detection or malicious-code-blocking software on their desktops.

Teleworking is gradually becoming a part of many workers' everyday life. Outside sales staff get up-to-date sales data just before meeting with a client; parents work from home when the children have the day off from school; and formal teleworkers who save hours each workweek by skipping their commutes.

IT departments have typically been less enthusiastic about employees who work outside the office. A remote workforce makes it tough to centralize and control IT systems. And having hundreds or thousands of remote-access points into the company's network is akin to trying to secure a store with the doors left permanently unlocked.

Some businesses see home offices as a way to cut back on costly leases for satellite offices. Others see teleworking as a means to distribute their workforce so a disaster doesn't bring the business to a grinding halt. So, it appears that the number of teleworkers isn't going to decrease. This forces the IT department to face an overwhelming security challenge that results from enabling remote workers.

Security must be a basic element of the teleworking network design. Security precautions start with restricting access to the company's data. A variety of technology solutions are available, ranging from a packet filtering router at the network interface to elaborate systems with multiple firewalls and more.

As the number of teleworkers rise, so will security concerns. You want employees to have full access to corporate e-mail and applications from anywhere they're working, but openness invites invasion. Traditional remote access solutions are mostly secure, but long distance fees can add up, and users with cable modems and DSL connections to the Internet don't get to use their high-speed connections to get into the corporate network.

Teleworkers Raise Security Alarms

Most IT professions know that remote and mobile users can be the weakest link when it comes to network security. The growing array of security threats to corporate systems means that IT managers need to install personal firewalls on every desktop and keep antivirus software running at all times. If you're not protecting remote users with a personal firewall along with antivirus software, you're setting yourself up for disaster. But the IT department should also continuously check these systems, perhaps run a vulnerability assessment each time a remote desktop authenticates to the network in order to ensure the remote system hasn't already been compromised.

Although the business community has a number of technologies to help protect remote access systems, security ultimately rests with the remote user – the teleworker. The vast majority of successful attacks come down to the end user – they click on things they shouldn't – an enticing attachment to an e-mail, such as a jingle or an animated card. Teleworkers must be educated to not execute files they've never seen before. The most critical component of a good security system is not technological; it's the teleworkers themselves. The admonitions you normally give to in-office employees may need some added emphasis for this group, in particular those time-honored practices such as: don't write passwords down where your teenage hacker can find them; change passwords regularly; use a password-keyed screen saver; and so on.

Yet, lest the fears of rampaging teleworkers spreading chaos keep you up at night, remember that teleworkers tend to be more loyal to the company and less likely to give you trouble than some of the non-teleworkers.

VPNs

A partial answer to the security conundrum is the virtual private network (VPN). It can provide secure and flexible access for traveling and work-at-home employees. It also works for secure office-to-office connectivity.

The phrase virtual private network seems imposing, but a VPN boils down to special software in client PCs connecting across a corporate intranet or the Internet to special software in a in-house dedicated box or server. Encryption at both ends keeps data safe, and you can choose from several options for user authentication. Installing a VPN often requires a substantial up-front effort for configuration and software deployment. But once everything is running, a VPN offers much lower connection costs than traditional remote access servers, since long distance users can usually connect to a local ISP POP, and provides excellent management control and information. See Chapter 8 for a thorough discussion of VPNs.

Companies need to take seriously the security of their remote access client. It becomes a corporate network access point, so they need control of the entire client. For more information on security read Chapter 7.

Outsourced Services

Telecommunications service providers such as AT&T, BellSouth Corp. Verizon, and U.S. West Inc., as well as an assortment of other vendors, are offering outsourced services to help companies deal

➤ Telework Centers

Some organizations with teleworkers, such as the California Franchise Tax Board (which collects state income taxes in California), allow some teleworkers to work only from telework centers, where physical security can be maintained more readily than in the employees' homes.

with teleworking issues ranging from installing DSL to managing the teleworker.

For instance, BellSouth's TeamTelework Connections service is aimed at helping large companies implement formal programs for employees working from remote locations. It includes telework program development, installation of voice and data services, remote office setup and management, and technical support for workers.

U.S. West's teleworking package, Extended Workplace Solutions, includes training and support services to help new teleworkers learn to use their communications tools and manage their time. Extended Workplace Solutions evaluates a company's teleworking needs based on the people and processes involved, as well as the technology, telecommunications, and support that will be required.

Developing a telework program can be a frustrating exercise for an IT department, which must consider issues such as security, capacity, and user support. But these issues are better managed if IT gets involved early in the process. Always keep in mind that a successful telework program involves the IT department as a key player in the development of the formal program.

A formal program requires a commitment from the top down and team effort by representatives from human resources, IT, facilities management and others. U.S. West recommends that the task force developing the telework program provide the IT department with complete profiles of the teleworkers, so the IT staff can understand teleworkers' needs for voice, bandwidth, and mobility, and standardize technology choices and contain implementation costs.

TManage, Inc. (Austin, TX – 512-794-6000, www.telsuccess.com) is the nationwide leader of remote networking solutions. The company can deliver an end-to-end solution set that enables a company to securely and efficiently connect, manage and support a distributed workforce. TManage does this by utilizing proprietary tools, along with partnerships with industry leaders to deliver robust managed solutions for teleworkers, telework centers, satellite offices and mobile workers. The TManage Web site has a comprehensive list of resources on telework, and also provides industry research, a glossary and access to its financial and environmental impact calculators.

The Digital Workplace

With technology reshaping the business landscape, it becomes easier for teleworkers to function as if they were in the corporate environment. With Web-based tools, virtual teams can work more efficiently. Web/video-conferencing and collaboration Web sites that enable the sharing of files and posting of schedules abound.

The digital workplace is a concept that offers Internet-based work space where team members can store files, share documents, track issues, check team calendars, conduct ongoing discussions, collaborate and more.

Hosted applications are not only convenient for the teleworker, but can serve as an outsourced tool for businesses looking to eliminate overhead costs associated with hardware, software, networks, training and IT staff.

At a basic level, these sites provide the employee with e-mail, chat forums, document management functions, workflow and the ability to share calendars and contact lists. Most of these digital workplaces charge in the range of $10 to $15 per month per user; this includes disk storage of around 20MB per user (but the cost can range upward to $500 or so per month, per user). Note that some Web sites offer their digital workplace features for free, but the user must deal with advertisements.

Let's take a look at some of the digital workplaces available for the typical teleworker.

Centra (Lexington, MA – 781-861-7000, www.centra.com) offers real-time Web collaboration products, content creation tools, and knowledge delivery systems for real-time business meetings and events through a simple Web browser interface.

Creative Solutions (Dexter, MI – 734-426-3750, www.creativesolutions.com) offers Virtual Office, which is an ASP for practicing accountants. Virtual Office provides an anytime-anywhere, online access to all of the vendor's tools, which it refers to as Creative Solution Accounting Productions.

E-Room Technology (Cambridge, MA, 617-497-6300, www.eroom.com) provides Internet-based communication tools for collaboration with customizable databases that can be used for customized project management. It even offers its own set of industry specific templates and methodologies which it calls e-Practices. These tools allow for customization to accommodate workflows and scalability of projects. E-Room also features a project calendar, which can synchronize with the team's Microsoft Outlook calendars.

LiveOffice (Torrance, CA – 800-251-3863, www.liveoffice.com) is an all-inclusive ASP for the teleworker. It offers an extensive suite of tools including e-mail, personal and group calendar, address book, document storage, discussion board, chat, task list and news feeds. It's designed to support teamwork as well as to provide a good set of tools for individual productivity. For example, the LiveDesktop feature can integrate with Microsoft Office, ACT and Goldmine.

OpenAir (Boston, MA – 617-351-0230, www.openair.com) provides Internet-based systems for client billing and management reporting.

Sitescape (Wilmington, NC – 910-256-5038, www.sitescape.com) enables the management of projects through collaboration technology that provides software, which allows groups of people to easily share information, work together on documents, and facilitate meeting online.

StoragePoint Corporation (San Diego, CA – 858-650-3100, www.storagepoint.com) offers WebDrive, which allows subscribers a secure place to save, get, share and organize folders and files using only a Web browser. WebDrive can be visualized as a virtual disk with automatic encryption and compression that provides a fast upload and download to and from any computer with Internet access. StoragePoint services are compatible with corporate firewalls and proxy servers, and accessible through Windows Explorer and Internet Explorer on any PC running Windows.

TeamOn Systems Inc.'s (Issaquah, WA – 425-369-5700, www.teamon.com) corporate headquarters solutions provide office-wide e-mail, shared and personal workspaces, document sharing and storage, calendars, contact management tools and more. Check out their Website – it provides a comprehensive, functional and exceptionally well-designed set of tools, especially tuned to team use.

Even internet messaging (IM) services have adopted the digital workplace concept. You can now get IM services over the Web from any computer, anywhere, without the need to install the proprietary software. This allows the teleworker to just hop on the Internet, go to a special Web site, type in the correct user name and passwork, which calls up his or her IM account and presto! their family, co-worker, buddy lists are ready and waiting. AOL gave their IM users this option when it launched the "AIM Quick Buddy" (recently renamed the AIM Express). ICQ offers a similar alternative at http://lite.icq.com. The same is true for Yahoo! Messenger, although it offers a slightly less direct route.

This is just a sampling of the digital workplaces available. There are many more. Go to your favorite search engine and type in "digital workplace" to help you locate a Web-based product to fit your specific needs. But, as a caveat, the ASP model is still evolving and the Internet can be a risky place to do business, so consistently backup any data you store on a digital workplace site, for it may be "here today and gone tomorrow."

An Exercise in Recursion: Managing Management

By far the highest barrier to acceptance of teleworking is departmental managers. Effective teleworking management requires that managers focus first on clearly specifying the tasks to be performed, then concentrate on the results produced by their employees, rather than the work processes used to produce those results. Yet many managers feel dependent on eyeball contact with their employees as a prerequisite for effective supervision since they assume that their workers won't work well if they aren't there to keep them in line.

While this is a valid assumption for some employees, there's a question that begs to be asked: "Why did you hire such problematic employees in the first place?" The best employees are people who know what they're doing, can communicate effectively and are disciplined enough to complete an assignment without much direct supervision. This is also the ideal group for teleworking.

On the other hand, not everyone (not even the "best" employee) is suited to home-based teleworking, not even part-time. Some employees need to be near fixed files or equipment. Others don't work well without daily social contact. Still, in many cases, with the proper preparation and management, these people can thrive in a telework center environment.

Training and Support

For a telework program to be a success, teleworkers and their managers need training. First, there's the need for training in the technology that aids the teleworker and manager in communicating with each other, others in and outside the company, and the company's customers, especially if the telework program introduces new technology, such as chat or audio/video conferencing. If new technology is being introduced, it should be emphasized during the training period that the teleworkers are more likely to be on their own and, consequently, should ***themselves*** take pains to learn the most possible from the training sessions.

The Teleworker

Employees not only need to learn how to use new technology tools, but also how to maintain them, and where to go for help when they need it. For instance:

At AT&T, teleworking simulations were created that let prospective teleworkers work on-site as though they were working remotely. By practicing using only e-mail, the telephone, and voice mail to communicate, they are better prepared to forgo face-to-face communications when the time comes.

At BellSouth, teleworkers must be able to disassemble and reassemble their computers, modems, printers, and fax machines before they can enter the telework program. Some companies test workers' competency with software before they are sent off on their own. Such training has many benefits, including reduction in long-term technical support.

The Manager

Second, but equally important is management training. Both teleworkers and their managers need to know the basics of working together apart. Topics to be covered include establishing goals, recognizing results, setting schedules, communicating effectively, getting organized and managing time. This training can be critical to the success of a telework program.

The Help Desk
An effective help desk is more important to teleworkers than to people in the office. Repeated surveys of teleworkers have shown that there are two prime sources of help in most corporations: The help desk personnel and the departmental techie. Naturally, the techie is no longer easy for the teleworker to exploit, so the help desk becomes the only source of support.

Making it Work
As part of any telework policy, the company should draw up a document detailing specific issues, such as responsibility for home-office equipment and expenses, liabilities for furniture and equipment, the extent of company insurance, the number of days the teleworker is required to be in attendance at the main office complex, and the frequency of contact from the remote location. Both the company and the teleworker should sign this document.

Although workers' compensation regulations in most states hold the employer responsible for an employee when he or she is working, regardless of location, a signed agreement can mitigate a company's risk in this area.

The University of Texas' Houston Health Science Center is among a small group of employers who have developed a detailed telework program. The school's Work/Life Program produced an extensive guideline for telework, which recommends that employees prepare thorough written proposals before approaching their supervisors with requests to telework. For example, it suggests that employees begin by preparing a log recording what they do on their job during a two-week period.

These tasks should be divided into three categories to determine:

- What can be done remotely?
- What must be done within the confines of the corporate office?
- What can be done in either location?

If a substantial amount of work can be done remotely, then that job can probably be done by a

➤ A Tempest in a Tea Pot

The question of corporate risk and responsibility with regard to teleworking managed to set off something of an uproar in the past when the United State's Occupational Safety and Health Administration (OSHA) released a five-page document on the subject.

In response to a query sent by a company asking for help in defining its responsibilities to its teleworking employees, OSHA said that an employer would be responsible for preventing or correcting hazards in a home office. According to the OSHA document, "Employers must take steps to reduce or eliminate any work-related safety or health problems they become aware of through on-site visits or other means." A statement which both jolted employers into pondering liability issues, and caused teleworkers to conjure up paranoid visions of company inspectors repeatedly visiting their homes! AUGH!! Although OSHA quickly assured the public that it wasn't demanding home-office inspections, this short-lived tempest underscores the fact that important issues regarding teleworking does exist.

teleworker.

IN SUMMARY

Properly designed and implemented, teleworking works for a great variety of employees – anyone whose job involves solo thinking or communication that does not require face-to-face interaction – about half of the industrial countries' work force. Employers like teleworking because it produces net benefits to them, which is also why teleworkers and the communities they work in like it. Everybody seems to win, except possibly the downtown building owners and shopkeepers and the transportation industry.

Teleworking is an idea whose time has arrived. Each employee location is an extension of the corporate office, the should be able to respond to co-workers, customers and partners in the same manner as someone located in one of the corporate cubicles, because technology can deliver comparable voice, data, and video communication tools used by corporate office workers. After all, you've read to the end of this chapter because you understand that teleworking is a viable option for the decentralization of a company's most valuable asset – it's workers.

CHAPTER 10

Devising a Master Plan

BACK IN MY COLLEGE DAYS OF LONG AGO (the mid-1970s), I once had the special privilege – and it was considered a real privilege in those days – of seeing my first computer, an IBM 370 architecture system.

Or rather I should say I ***didn't*** see it. All I actually saw was a CRT terminal, one of many provided under the auspices of the State of New Jersey, which had peppered them around various academic institutions residing within the borders of the Garden State. The actual computer itself was "somewhere." Perhaps it was at the state capital at Trenton, perhaps at Rutgers University or the Newark College of Engineering (now NJIT). The point is, my terminal could have gone up in flames or been the victim of a Mississippi-like flood and the computer itself would still be humming and computing merrily along, no worse for wear – indeed, with one less foolish college student stealing CPU cycles, the system would now be running a bit more efficiently!

Unfortunately, the high priest mentality of the old days that kept the average person from experiencing the high technology in all its glories also did a darned good job of protecting its assets. Computers were hidden away in "secure locations" in places ranging from distant office buildings to underground bunkers, protected by nasty-looking people stationed in front of bland-colored metal doors, behind which were rumored to be rooms built with raised floors, computer-safe fire-suppression systems, and enough battery and diesel generator power to run a system clear through the pending nuclear war that somehow never did arrive to put the Internet through its ultimate test.

In our modern democratic society (the one certain miscreants seem to be intent on destroying these days) computers are everywhere. Many people can afford one, and most businesses are literally based on the concept of distributed PCs and servers connected by communications networks, both of which are now vulnerable. That's why in this chapter we'll explore the basic elements of risk, and to introduce you to the concept of a security risk assessment methodology.

SEMANTIC SUBTLETIES

Now, the activity of protecting the integrity and uptime of your business or other organization can be refined into several subtly-related approaches: business continuity, disaster recovery, contingency planning, business resumption planning or continuity of operations. Sometimes the differences between these are semantic, though a chief difference between business continuity planning (BCP) and disaster recovery planning (DRP) is that BCP applies to the continuation of business functions following an event. BCP events can impact communications and IT capabilities directly, such as a hurricane or one of those contrived California power shortages, or it may involve something completely non-technical, such as a strike by non-IT personnel or Enron-like financial malfeasance that puts a company's business workings in jeopardy. DRP, on the other hand, is more precisely defined, as it applies to "incident response" or the recovery from a more concrete incident (usually a sudden, unforeseen disaster) on an otherwise stable organization, and the restoration of communications and IT support of business functions. In both fields of endeavor, the goal is to keep an organization's critical functions operating, be it a mighty multinational enterprise or the humblest SOHO.

BUSINESS CONTINUITY LIFE CYCLE

A company is a dynamic entity and the research, development and deployment of continuity / disaster / security plans and policies are also a dynamic affair. Since no plan or policy, once formulated, can remain in place forever untouched, businesses tend to repeat the process at whatever intervals are deemed necessary, giving it somewhat the flavor of a "life cycle."

Risk Analysis

First of course, risks must be identified. Security risk analysis, popularly known for decades as risk assessment, is fundamental to the security and even the very integrity of any organization. It ensures that controls and financial expenditure are efficiently formulated and deployed to counter the risks to which the organization is exposed.

Identification of Critical Functions

The first step in the analysis is to identify a business's more critical functions. This can be uncovered through onsite observation and research or can be described by an organization's senior management. These functions must be ranked or prioritized in some way, since few businesses will actually spend whatever is necessary to provide maximum uptime and security. A company cannot make every system redundant with automatic failover – just imagine if every worker had a human counterpart sitting in an office on the other side of the world, patiently waiting to spring into action in case the first worker was incapacitated! Once prioritization has taken place, decisions can then be made specifying the level of redundancy for each.

Determination of Resources Supporting the Critical Functions

All of an organization's resources, from computers and data to people and paper clips, support the most critical company functions and "make them happen." What you are ultimately protecting or duplicating are not really the functions, but the resources that drive the functions. The second step is, therefore, to determine what resources support the critical functions. Staff tends to fall into hierarchical subsystems such as teams, departments and divisions, while physical resources tend to cut across these and are clumped together by related functionality (computing infrastructure, communications infrastructure, physical infrastructure, applications and data). As with any network of relationships, resources existing in one functional area may be a critical component for a function in another area.

Identification of Possible Disastrous Events.

The third step is to take each resource and come up with a listing of the phenomena for each that make a contingency plan necessary in the first place. This can take the form of a simple maxtrix of events against resources. The description of how the resources can be affected by events is also important, and all of this information can be used as a sort of database or decision matrix when formulating plans.

For example, threats to the computer infrastructure include power surges, hackers, water and fire (sometimes euphemistically referred to as "thermal degradation" of people and materials). Fire can not only burn or melt computer or telecom platforms - burning polyurethane or polyacrylonitrile or many items composed in part of nitrogen will also generate the potent gas toxin hydrogen cyanide (HCN). Even if you put out a fire quickly, there are enough smoke particles generated to get into sensitive electronic equipment and wreak havoc.

Defining Plans

At some point you'll have to stop adding to your matrix and actually start defining plans. Each resource has one or more effects on a function's availability and its need to keep continuing. This step deals with determining the effect on a given resource by a particular event, finding a list of redundant or alternative resources to minimize the functional effect, and selecting the appropriate alternative that makes sense from a business point of view (e.g. cost or downtime). Service level agreements (SLAs) should be reviewed at this point and adjusted if necessary, appropriate back-up facilities should be specified, as well as alternative staffing arrangements.

In the not so recent past, when corporate honchos sat down and discussed telecom and datacom in the context of disaster management planning, they uttered such spine-tingling questions as: "Do we really need redundant phone services?" or "Do we have an extra router locked up in the basement storage cabinet."

Contemporary disaster management planners, on the other hand, need to move communications out into the limelight under the rubric of remote access, particularly with consideration to the VPN "glue" holding together a vast, ghostly network of secure data backup locations, teleworkers, telecenters and video conferencing units, all of which will be carrying on company tradition long after the worst has occurred.

During "peacetime" teleworking's chief attributes are economic and social. They lower costs once a secure and flexible network that extend out to employees is set up enabling the employees to spend more "quality time" with their families and less time commuting to work under variable weather and social conditions. They also increase productivity and attract and retain personnel who might not have ever considered working for your company. But when fortune ceases to smile and shrieks instead, teleworking becomes a primary disaster recovery strategy.

After all, in the wake of the 1993 World Trade center bombing, the California earthquake of 1994, the Federal Building bombing in 1995, and during the mid-west floods of 1997, it was the organizations already in the teleworking groove that stayed up and running while many others were never heard from again.

On September 20, 2001, Gartner, Inc., (Stamford, CT – 203-316-1111, www.gartner.com) the world's leading information technology research and advisory firm, released a fascinating Research Note having the staid title, "Disaster Management Plan for Remote Access." This particular Gartner

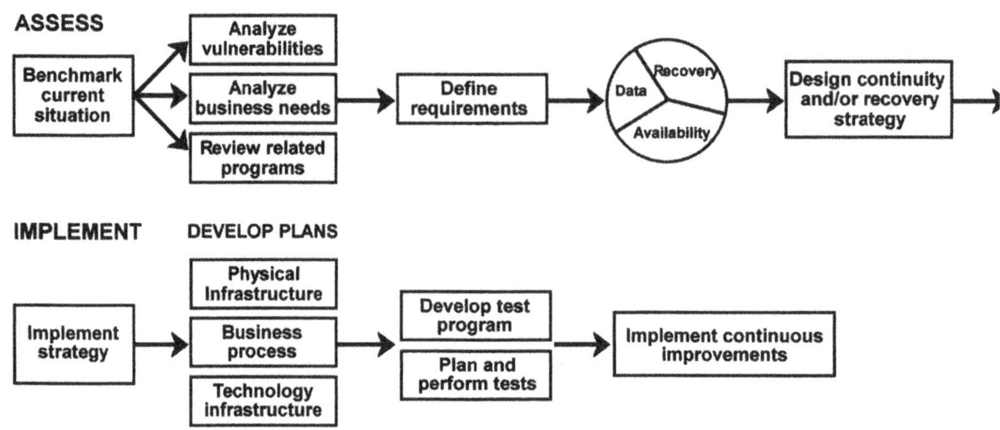

For effective Business Continuity Planning (BCP) and Disaster Recovery Planning (DRP), extremely thorough risk assessment must take place, with critical functions identified along with what resources support these functions, what kind of phenomena would disrupt them and what a planned response for each of these would be. Your research may indicate that a full-blown business continuity or disaster recovery plan for your organization is unfeasible or simply impractical from an internal business perspective and should be outsourced.

Research Note found that, "The moment that disaster strikes, whether a terrorist or military attack, earthquake or other dangerous event, marks only the beginning of disruptions that can last for weeks to years. Enterprises that rely on location-specific business operations can be severely disabled. Those that have implemented any kind of work-at-home or mobile work schemes stand the best chance to get their employees back to work as safely and quickly as possible. Those enterprises that have moved to virtual teaming as a primary work format will be ideally positioned, via remote access, to move quickly in the event of a disastrous interruption to operations."

Gartner's Research Note lists various steps for managing remote employees to help cope with the aftermath of a disaster. Essentially, enterprises should have pre-selected staff set up as disaster management teams (DMTs), with each consisting of representatives from various departments such as Human Resources, Information Technology/Services, and other Divisions and/or Business Units – all of them well-indoctrinated in the ways and benefits of remote access and teleworking. It is this first line of defense that will engender the greatest restitution of the business process by immediately coordinating employee activities with the aid of the remaining communications infrastructure. The efforts of these teams will, of course, be concentrated in geographical areas where their company has a major presence.

The teams should be armed with lists of who is at what site, what kind of site it is (teleworker home, telecenter, temporary hotel space) how many employees are working per site, access to home addresses and contact numbers, and a description of remote-access capabilities at each site. Lists, like any other corporate data, should be distributed among members of the team, so that all employees can be contacted quickly by each team member (by any way possible) in the event that normal business communications channels are kaput.

As in the case of the World Trade disaster, these lists could also be used to verify casualties and missing persons.

When disaster actually strikes, Gartner says that the first members of the staff to be contacted should be a core group of essential personnel (this makes sense since these will be most of the people who do all the work anyway!).

In emergencies, DMTs act almost like little military command and control units – or even better, like Internet routers attempting to route their high priority "packets"(personnel) – those company employees essential to the business at that point in time – around dangerous biohazards, or wrecked transportation systems, to work where they can do the most good under the cirumstances. Even remote access might be unattainable or unreliable because of network service outages.

Using the enterprise's network management tools, according to Gartner, DMTs should maintain an up-to-date inventory of which network services are functioning and which employees are in areas that are able to gain access. With this information, DMTs can direct employees to alternate remote work sites and set them up in virtual teams formed around critical business processes such as customer service and logistics. All virtual teams during the emergency are monitored and coordinated by DMTs.

As the size and ramifications of the disaster crystalize, DMTs can ask that essential employees volunteer to team up at each others' homes or at hotels or telework centers to create small workgroups, using broadband access, for example. DMTs should have access to a current inventory of the remote access capability of each employee (dial, cable, DSL, etc.) and access to lists of hotels and telework centers in every nearby city and region.

Under Gartner's suggested scenario, DMTs must have a way to rank the relative importance of remote access to each workgroup and department and to request that nonessential user IDs be blocked and redirected to a company information site. Part and parcel with this, employees ranked as nonessential to emergency operations must be stopped from logging in to active systems in order to avoid communications congestion such as clogging VPNs. (On the other hand, Gartner suggests that employees with broadband access and home networks should volunteer to share their connection to set up a temporary workgroup location.)

DMTs must direct employees to monitor dial-up recorded messages and Web portals to obtain company information. Gartner suggests that a special secure standby Web site must be developed and activated as a virtual command center: "The site can serve as an up-to-the-minute status-reporting medium, complete with names and contact information on accountable managers," theorizes Gartner. "The site can also point to external resources such as construction management firms, engineering, architects, office supply firms, communication firms, and other third-party providers that have been preselected as part of the disaster recovery operations plan. Because employees need to retain their sense of community in an emergency, DMTs should set up private newsgroups and chat systems offloaded from the main e-mail systems."

DMTs will also have to ensure that there is no Act Two on the disaster front and that all of the data centers, security and intrusion systems continue to run. No doubt the locations of available storage resources and the datacom services allied with them, will be revised regularly. New remote users lacking prior experience with the Internet must be reminded and warned of the absolute requirements to use personal firewalls, to avoid sending company business over public mail systems, to avoid accepting executable e-mail, and whatever other cautions are necessary to ensure the integrity of company information and systems. Ad hoc demands for VPNs to create dynamic groups of users can be made easier by supplementing the VPN architecture with software tools designed to support rapid reconfigurations.

Employees are not by any means passive under the Gartner plan. They must move heaven and earth to contact their company to signal that they're alive and breathing, and to reveal their location and to verify that they've physically able (or not) to assist in re-establishing business operations. At the same time, employees must keep their access sessions as brief as possible.

Other Gartner ideas for post-disaster management include eliminating "fat" or bandwidth-hungry applications unless absolutely necessary. Instead, thin-client front-end access is the post-disaster network's friend. As the Gartner note states: "For example, if users have a choice between running Outlook 2000 or Outlook Web Access (OWA), they should choose OWA. Likewise, if an application can be accessed by Windows Terminal Services, Citrix, or other thin-client front ends, choose the thin interfaces over a full application."

Plans like Gartner's indicate that enterprises and other businesses will probably end up changing their workplace from the traditional centralized concentration of staff and resources to moving more and more employees to multiple, secure, lower-profile teleworker locations, all thanks to technologies such as VPNs and wireless.

Indeed, Gartner's conclusion pretty much described the theme of this book:

Remote access affords enterprises an opportunity to operate anywhere, anytime. Conventional telecommuting and mobile access could leave workers in the same region and collectively subject to impacts of the disaster. However, remote access coordinated by a network of competent DMTs allows companies and their people to work from anywhere and to move out of danger zones, forming global virtual teams, ultimately creating a level of resilience that will be greater than was originally planned for traditional telecommuting. Going forward, remote access is more than a nice-to-have work style: It will become a matter of business survival.

Implementing Plans

The final step is to implement the plans, thus forming the latest company policy regarding security. Such policies anchor an organization's whole approach to contingency and disaster recovery, addressing at least the basic requirements to ensure the stability and continuity of the organization. It is essential therefore that they exist, are up to date and have comprehensive coverage. Plans and policies determine the fundamental practices and culture throughout the enterprise. In the age of the microcomputer and high speed networks they are often closely associated with information security policies.

Plans must be made flexible, and sometimes when "planning" to avert disaster, one finds oneself not just using redundant versions of old systems, but newer, qualitatively different types of systems. For example, the reader may wonder as to why a huge chapter on video and other conferencing technologies is in a "survival guide" on telecom and data communications networks. The answer is that once VPNs and broadband connections have become commonplace, these conduits of copious bandwidth will allow you to distribute and ***save*** your most precious asset, your employees.

This becomes obvious when one reflects on one's mortality. Athough the author has been writing about fault resilient and fault tolerant systems since 1994, the events of September 11 reminded me that, while I sit here typing away in New York, there is no "parallel universe" version of Yours

Truly, sitting out in replica offices in San Francsico, matching my input word for word and edit for edit. Yes, the document can be backed up, but not the person. If employees work in a landmark office building, they're in danger. If they travel via aircraft, they're in even more danger. Traveling faster and faster on old aircraft technology won't reduce their peril, but not traveling at all and using new electronic conferencing technologies will keep 'em alive and kicking. Besides, as we mentioned earlier in this book, employees themselves are not stupid, and have told companies polling them (such as Avaya) that they want video conferencing and audio conferencing as an alternative to travel. Moreover, they want seamless voice, data and video to support them while they're at geographically dispersed locations – including those working safely from home.

When drawing up a good plan, documenting everything is important, since everything will probably go through several generations of scrutiny and analysis before things can be fully validated and finally "set in stone" (at least until the next major iteration). People will have to be taught how to implement the plans and test the plans to see that they actually work. One can do simple training-event-and-tests activities such as fire drills. Once cannot test a giant mudslide and the requisite response.

Wide-ranging, all-inclusive BCP and DRP occupies a lot of time, effort and money. Indeed, many conventional methods for performing security risk analysis are becoming more and more untenable as businesses become more and more complicated. It's as if a law of diminishing returns has set in, as users expend more and more effort and money and can be less sure of the efficacy of their increasingly sophisticated risk management methodologies.

Attempts at streamlining and standardizing the concepts of risk assessment have appeared in recent years. For example, COBRA (Consultative, Objective and Bi-functional Risk Analysis), consists of a range of risk analysis, consultative and security review tools. These were developed largely in recognition of the changing nature of IT and security, and the demands placed by business upon these areas. COBRA, and its default methodology, evolved mostly to handle these issues, and was developed over the course of many years of research in co-operation with one of the world's major financial institutions.

Also, there are some internationally-accepted, security standards you can refer to as you formulate the security aspects of your policy. In particular, ISO17799, is an awesome, super-detailed security standard that seems as if it is actually 10 standards, since it covers 10 different sections or areas:

1. **Business Continuity Planning.** The objectives of this section are: To counteract interruptions to business activities and to critical business processes from the effects of major failures or disasters.
2. **System Access Control.** The objectives of this section are: 1) To control access to information. 2) To prevent unauthorised access to information systems. 3) To ensure the protection of networked services. 4) To prevent unauthorized computer access 5) To detect unauthorised activities. 6) To ensure information security when using mobile computing and tele-networking facilities.
3. **System Development and Maintenance.** The objectives of this section are: 1) To ensure security is built into operational systems. 2) To prevent loss, modification or misuse of user data in application systems. 3) To protect the confidentiality, authenticity and integrity of information. 4) To ensure IT projects and support activities are conducted in a secure manner. 5) To maintain the security of application system software and data.
4. **Physical and Environmental Security.** The objectives of this section are: 1) To prevent unauthorised access, damage and interference to business premises and information. 2) To prevent

loss, damage or compromise of assets and interruption to business activities. 3) To prevent compromise or theft of information and information processing facilities.

5. **Compliance.** The objectives of this section are: 1) To avoid breaches of any criminal or civil law, statutory, regulatory or contractual obligations and of any security requirements. 2) To ensure compliance of systems with organizational security policies and standards. 3) To maximize the effectiveness of and to minimize interference to/from the system audit process.
6. **Personnel Security.** The objectives of this section are: 1) To reduce risks of human error, theft, fraud or misuse of facilities. 2) To ensure that users are aware of information security threats and concerns, and are equipped to support the corporate security policy in the course of their normal work. 3) To minimize the damage from security incidents and malfunctions and learn from such incidents.
7. **Security Organization.** The objectives of this section are: 1) To manage information security within the Company. 2) To maintain the security of organizational information processing facilities and information assets accessed by third parties. 3) To maintain the security of information when the responsibility for information processing has been outsourced to another organization.
8. **Computer and Network Management.** The objectives of this section are: 1) To ensure the correct and secure operation of information processing facilities. 2) To minimise the risk of systems failures. 3) To protect the integrity of software and information. 4) To maintain the integrity and availability of information processing and communication. 5) To ensure the safeguarding of information in networks and the protection of the supporting infrastructure. 6) To prevent damage to assets and interruptions to business activities. 7) To prevent loss, modification or misuse of information exchanged between organizations.
9. **Asset Classification and Control.** The objectives of this section are: To maintain appropriate protection of corporate assets and to ensure that information assets receive an appropriate level of protection.
10. **Security Policy.** The objectives of this section are: To provide management direction and support for information security.

For an organization to actually comply fully with the ideal levels of security set forth by ISO17799 is an expensive proposition. But it is good to use it as a guide.

An set of pre-written, ISO17799 compliant policies, and a desktop delivery mechanism can be obtained from the ***Security Policy World*** Web site (http://www.securityauditor.net/security-policy-world/). These embrace all aspects of information security, including disaster recovery and business continuity.

To Outsource or Not to Outsource

If you've sat down and pondered the ramifications of risk management only to realize that you either don't have the time or resources to undertake such a large effort, or your SOHO or 15 person organization doesn't need high levels of redundancy or super sophisticated plans and policies, what can you do?

Your initial risk management efforts may instead indicate that you should scrap plans to implement disaster recovery in-house and instead use the services of a disaster recovery firm. Arguments for in-house disaster recovery usually center on better control and lower costs. For instance, if your business is being subjected to a hurricane, all other companies within 100 or so miles are too, which means that your friendly local disaster recovery firm is either overwhelmed with frantic calls for help, or is itself in danger of being blown away too.

Also, there are many in-house cost-saving strategies that aren't available to a disaster recovery firm. For instance, your development infrastructure can be used to back up your production infrastructure. With distributed systems, similar machines at different locations can provide mutual disaster recovery, as in the case of a large Storage Area Network (SAN). Partner companies can back each other up over a VPN or extranet.

On the other hand, crafting whole redundant environments that back each other up is not an easy process. Also, hardware, software, and network compatibility must be constantly checked so that an automatic failover event will be successful.

Maintaining compatibility can be nearly impossible if different groups have access to managing the same set of equipment and over time slowly start changing things. These groups inevitably fail to notify their disaster recovery service firms of the changes in hardware and software, so when it's time for a recovery event, it fails.

As time goes on, more and more files become associated with particular applications but somehow a few of these always miss getting placed on the backup list. Sometimes a company might have a policy that backups should be done on a regular basis and then immediately shipped offsite, but somebody might violate that policy and keep the backup media in the data center just in case the machine that's been acting quirky lately actually fails.

"Political" problems can also result from implementing disaster recovery in-house. All of the infrastructure can't be protected with the same degree of assurance. To avoid internal conflict, users way want to turn to an objective third party to determine which hardware, software, and applications must be recovered most quickly, and which can wait.

Moreover, cost savings may be illusory if maintaining disaster recovery configurations guzzles all of your employee's time and forces the company to upgrade some systems more quickly or to a higher degree than would otherwise be necessary.

Furthermore, money-saving techniques inevitably involve some compromises. You may have to over-configure the two systems of an automatic failover, fault tolerant system so that either system can take over for the other without showing objectionable performance degradation. Alternatively, of course, you may decide that you can accept some level of degradation, up to and including letting the system crash!

By outsourcing to a disaster recovery firm you can retain full operating capacity during a disaster, yet from outward appearances your staff will be using normal equipment configurations and won't be terribly taxed by the situation.

The other question mark hovering over in-house disaster recovery concerns one of your most valuable assets — the staff. If the infrastructure switches over without a hitch, what about people? Will the people at the emergency site really have the time and expertise necessary to provide all the services (running reports, staffing the help desk, providing end user technical support) to actually get the business running again? Or will you have to ship people to the other site, perhaps still further extending your recovery window?

Giant enterprises may be attracted to "big names" for business continuation and disaster recovery consultations such as those set out in Chapter One. Or it can be a service such as that offered by Comdisco (Rosemont, IL — 847-698-3000, www.comdisco.com). It offers an on-line catalog from which you can choose from a variety of business recovery services, as well as software that shows you how to plan for disasters. Among the items available on-line are rentals of Comdisco's mobile units, which are equipped with UPSs, diesel generators and computer wiring, and the rental of space within Comdisco's recovery facilities. You can also rent PBXs, computer equipment, phones and office furniture from Comdisco.

AND NOW YOU'RE ON YOUR OWN!

The author hopes that this book lives up to its name, and that it will serve you as a trusty guide through the jungle of hackers, crackers, virus writers, terrorists and other evildoers bent on your destruction. Perhaps some page in this book will be the key to warding off such Wonders of Nature as lightning, floods, mudslides, and the static electricity you pick up from petting your cat.

The world became a smaller, more fearful place on September 11, 2001. But the technologies, methodologies and skills already available from the world's best minds are strong enough to protect your business, your home, and ultimately *you* from the sometimes random machinations of both Man and the Cosmos.

As I sit here in New York writing these final words, the greatest architects the world over are vying with each other to design the structures that will rise in place of the World Trade Center. You can bet that the companies occupying those new buildings will pay heed to such formerly droll activities as security risk analysis. Fault tolerant computers, VPNs, the remote vaulting of data and mobile wireless communications, all of these will make the world a better, safer place for everybody. One could easily imagine New York's other towers becoming near-empty decorative corporate landmarks, not because of a terrorist outbreak of some malevolent microorganism, but because the world's "real work" will now be done by a new army of teleworkers who have fled the great glass and steel towers of the cities to spend their working hours doing web collaboration and video conferencing, all from the safety and comfort of their homes.

As for me, well, I'll be heading into the city to check out the new buildings. You never can tell where you'll find a good restaurant.

Good luck.

Here's a Recommended Reading list of books that you may find informative:

Active Defense: A Comprehensive Guide to Network Security, by Chris Brenton and Cameron Hunt. Sybex, 2001.

Counter Hack: A Step-by-Step Guide to Computer Attacks and Effective Defenses, by Ed Skoudis. Prentice Hall, 2001.

Fundamentals of Computer Security Technology, by Edward G. Amoroso. Prentice Hall, 1994.

Hack Proofing Your Wireless Network, by Les Owens. Publishers Group West, 2002.

Hackers Beware, by Eric Cole. New Riders Publishing, 2001.

Incident Response, by E. Eugene Schultz, Russell Shumway New Riders Publishing, 2002.

Incident Response, by Richard Forno, Kenneth R. Van Wyk. O'Reilly & Associates, 2001.

Intrusion Detection: An Introduction to Internet Surveillance, Correlation, Trace Back, Traps, and Response, by Edward G. Amoroso (Preface). Net Books, 1999.

Maximum Security, by Greg Shipley. SAMS, 2001.

Security Architecture: Design, Deployment and Operations. by Christopher King, Ertem Osmanoglu (Editor), Curtis Dalton. McGraw-Hill Professional Publishing, 2001.

Glossary

1 Wiltshire: This is one of the biggest and oldest "telecom hotels" in the US. It is located at 1 Wiltshire Boulevard in Los Angeles, but is called "1 Wiltshire" for short. It is probably the largest "meet me" point on the west coast of the US, and is the gateway to the Pacific Rim.

10-100 Base-T: The number – 10 or 100 - represent the number of millions of bits per second. Base represents baseband and T represents trunk. Put them all together and you have a local area network (LAN) that can transmit data at 10 or 100 megabits per second. See also Fast Ethernet.

1996 Telecommunications Act: This US act was passed in 1996, as an extension of the original Telecommunications Act of 1934, passed (ironically) in 1934. A lot happened between 1934 and 1996, so this Act was intended to formally establish competitive structures and to foster a free market environment in the US.

2B1Q (Two Binary, One Quaternary): A line encoding technique used in ISDN BRI and in the first generation of HDSL systems. Through the use of the Pulse Amplitude Modulation (PAM) technique, 2B1Q can map two bits of data into one quarternary symbol. Thus, each symbol comprises one or four variations in amplitude and polarity over a circuit, i.e., this 4-level PAM technique allows two bits to be sent per Hertz or sine wave, allowing more data per second to be sent using a relatively low frequency. See also ISDN, Hertz and PAM.

60 Hudson Street: This is the other big meeting point and collocation facility in the US, and is probably the largest such facility in the world. Located in the Wall Street area of Lower Manhattan, it has become the model for collocation and meet me peering around the world. It is also a haven for union activity, under the table deals, graft and fraud. BEWARE!

80/20 RULE: Generally speaking, about 20% of the disk storage gets 80% of the I/O activity.

A

AC (Alternating Current): Electrical current that continually reverses direction, with this change in direction being expressed in hertz, or cycles per second.

AC line surge protector: A device which plugs between the phone system and the commercial AC power outlet to protect the phone system from high voltage surges. When a surge occurs on the power line, the surge protector sends the overload to ground. The type of surge protector that you buy will be determined by how quickly the device "clamps" or kicks in to discharge the overload, as well as the voltage level at which it clamps. How sensitive is your equipment?

Access Network (AN): The fiber connection or "pipe" and associated electronic equipment (switches, gateways, etc.) that link a core network to Points of Presence (POPs) and Points of Interconnect (POIs). See Network Service Provider.

Accounting Rate: The amount agreed upon by two national carriers as the baseline rate per minute for charging each other for the traffic they exchange. Historically, this number was not based on anything but what the two correspondents thought they could get away with.

ACH: Automated Clearing House. A form of Electronic Funds Transfer where an amount of money is transferred from one bank account to a different bank account, usually to pay a bill.

ACLS: (Access Control List Service). Restricts access to computer resources such as files and directories.

Active Server Page: (ASP) A dynamically created Web page that employs ActiveX scripting. When a surfer requests an ASP page through their browser, the Web server generates in real-time a page with HTML code, which it then sends to the browser.

Active X Controls: The interactive objects in a Web page that provides interactive and user-controllable functions.

Active X: An architecture that lets a program (the Active X control) interact with other programs over a network such as the Internet.

Address Resolution Protocol: See ARP.

ADSL (Asymmetrical Digital Subscriber Line): The most commonly used term for "full-rate ADSL" (ITU reference code G.992.1), an evolving high-speed transmission technology that send more data downstream that upstream. It expands the useable bandwidth of the existing copper telephone lines enabling these copper wires to support up to 8 Mbps bandwidth downstream (some experts say 9 Mbps) and up to 1.5 Mbps upstream. See also G.dmt.

ADSL Lite: A simplified version of ADSL which provides a lower bandwidth alternative to "full-rate" ADSL. Thus, ADSL Lite technology eliminates the POTS splitter at the customer premises (in actuality, the line is split, but at the central office rather than at the user's premises). See also G.lite, Splitterless ADSL, Universal ADSL.

Aggregator: One of the companies formed in the late 80s that used carrier tariffs to combine the bills of unrelated customers to achieve discounts, which they partially shared with the customer. Also known as an "aggrevator".

AIN: Advanced Intelligent Network. A term promoted by Bellcore (Bell Communications Research Inc., now called Telcordia), and adopted by Bellcore's owners, the regional Bell holding companies, and by AT&T and virtually every other phone company to indicate the architecture of their networks for the 1990s and beyond.

Alternative Mark Inversion: See AMI.

American National Standards Institute: See ANSI.

AMI (Alternative Mark Inversion): A line-coding format used by T-1 systems. To quote AT&T: "A line code that employs a ternary [using "threes" as a base] signal to convey binary [using "twos" as a base] digits, in which successive binary ones are represented by signal elements that are normally of alternating, positive and negative polarity but equal in amplitude and in which binary zeros are represented by signal elements that have zero amplitude."

Amp or Ampere: Quantitative unit of measurement of electrical current. Abbreviated as A.

AN: See Access Network.

Analog: Comes from the word "analogous." In the telecommunications industry, analog refers specifically to telephone transmissions and/or switching which are not digital.

ANI: Automatic Number Identification. A service where the called party gets the telephone number of the party that is calling them. The most visible example of ANI today is

Caller ID.

ANSI (American National Standards Institute): A private non-profit membership organization founded in 1918. ANSI coordinates (within the U.S.) the voluntary standard system that brings together interests from the private and public sectors to develop voluntary standards for a wide array of U.S. industries. ANSI develops and publishes standards for transmission codes, protocols and high-level languages used in the telecommunications industry.

API (Application Programming Interface): Software that an application uses to request and carry out lower-level services performed by a computer's operation system. In short, an API is a hook into software. An API is a set of standard software interrupts, calls and data formats that applications use to initiate contact with network services, mainframe communication programs, etc. Applications use APIs to call services that transport data across a network.

Applet: Small software programs that can be downloaded quickly and used by any computer equipped with a Java-capable browser. (See Java)

Application Service Provider (ASP): A third party that manages and distributes software-based services and solutions to its customers from a centralized server base.

Application: A software program that does some type of task. MSWord, Netscape, Winzip, anti-virus programs are some examples of an application.

Arbitrage: The business of buying and selling traffic without adding value. It is a commodity trade, making profit by buying time at a price that is often only marginally lower than what was paid for it.

Arbitration: The method for deciding how peripherals or other devices communicate with the main controller.

Architecture: Refers to the overall organizational structure of a given system, i.e., processor architecture or proprietary architecture. Central to an architecture is the decision about the selection of structural elements and their behavior, as defined by collaborations of larger subsystems, therefore the architectural style is the definitive guide for the system.

ARP (Address Resolution Protocol): A protocol within the TCP/IP suite that maps an IP address(es) to the corresponding Ethernet address.

Array: Two or more hard disks that read and write duplicate data. The array is treated as a single unit by the operating system.

ASP: Application Service Provider. Ostensibly, a company that offers an application over the Internet. These applications range from simple telecom services to high end payroll and accounting. The role of an ASP is largely undefined.

ASP: See Active Server Page and Application Service Provider.

Asymmetric: Not symmetric - uneven or unbalanced. In the telecommunications industry, asymmetric refers to telecom channels that provide more bandwidth in one direction than in the other.

Asymmetrical Digital Subscriber Line: See ADSL

Asynchronous Data Transfer: As applied to disk subsystems a data transfer method needing an acknowledgement for each byte of data transferred, since the data can be transmitted at at at an arbitrary, unsynchronized point in time, without synchronization to a reference time or "clock," unlike SYNCHRONOUS transmissions. Standard SCSI uses asynchronous data transfer.

Asynchronous Transfer Mode: See ATM.

ATA: See IDE.

ATM (Asynchronous Transfer Mode): A high bandwidth, low-delay connection-oriented, packet-like switching and multiplexing technique with usable capacity segmented into cells that are allocated to services on demand. The cells are presented to the network on a "start-stop" basis, i.e., asynchronously.

ATU-C: The ADSL Transmission Unit located in the DSLAM at the telephone company's Central Office, which works in tandem with the ATU-R located at the end user's premise,

ATU-R: The ADSL Transmission Unit located at the end user's premise, which works in tandem with the ATU-C, enabling the end user to Receive (but also send) data over the telephone company's copper plant.

Autoresponder: A mail utility that automatically sends a reply to an email message. Autoresponders are used to send back boilerplate information on a topic without having the requester do anything more than email a particular address. This utility is also used to send a confirmation that an email message has been received.

B

Backbone: See Internet Backbone.

Backplane: The high-speed communications line to which individual components, especially slide-in cards, are connected. For example, all the extensions of a PBX are connected to line cards (circuit boards) which slide into the PBX's cage. At the rear of the PBX cage, there are several connectors. Each of these connectors is plugged into the PBX's backplane, also called a backplane bus. This backplane bus is typically running at a very high speed, since it carries many conversations, address information and considerable signaling. These days, the backplane bus is typically a time division multiplexed line – somewhat like a train with many cars, each representing a time slice of another conversation (data, voice, video or image). The backplane's capacity determines the overall capacity of the switch. Our favorite computers (the ones that will make PBXs obsolete) have passive backplanes. The processor is on a card that plugs into a slot.

Bandwidth: The measure of how much information can flow from one point to another in a given amount of time. In the telecommunications industry, bandwidth is the width of a communications channel. In analog-based transmissions bandwidth is measured in Hertz and in digital transmissions it's measured in bits per second (bps).

Baud: The baud unit is named after Jean Maurice Emile Baudot (1845-1903) an officer in the French Telegraph Service who around 1870 designed the first automatic telegraph (teletype) and the first uniform-length 5-bit code representing alphanumeric characters (in the 1960s, Baudot code was replaced with ASCII code). The number of times per second that the signal on the line changes state (these states can be frequencies, voltage levels, phase angles, etc.), i.e., the baud rate only equals the bit rate when one is using a two-level modulation scheme, such as "no voltage equals zero" and "a pulse at positive voltage equals one." But, if no voltage equals "00" and if a pulse at a low voltage is agreed to mean "01" and another pulse at a higher voltage equals "10" and a pulse at a still higher voltage means "11," then each baud of this "multi-level code" can represent one of four possible bit states. Thus, the baud rate is not always the same as the bit rate. Four or more bits can be sent per baud.

Bell System: The former AT&T conglomerate before it was broken up in 1983. Comprised 7 regional companies, AT&T long distance (Long Lines), Western Electric, Bell Labs and others. Also had minority interested in Southern New England Telephone and Cincinnati Bell.

Berkeley RAID: The seven RAID levels as described in 1988 and 1989 papers by University of California at Berkeley researchers. Other developers have added to and refined RAID levels, but the accepted versions are "Berkeley RAID."

BIOS: Basic Input/Output System. A program, usually stored in Read Only Memory (ROM), which provides the fundamental computer services ranging from peripheral control to updating the time of day.

Bit: The basic unit in data communications. Bits compose a byte.

Blackout: A total loss of electrical power.

BLEC: An acronym for: (1) "Building-centric Local Exchange Carrier" – a supplier of telecom services for multi-tenant buildings, such as hotels, office buildings, and apartment complexes. (2) Broadband Local Exchange Carrier." A telecom service provider that limits its services to offering broadband services locally.

BOOTP (Bootstrap Protocol): A technology used by a network node to determine the IP address of its Ethernet interface.

Bot: See Spider.

bps (bits per second): When the "b" is lower case (bps) it means bits per second. When the "B" is upper case (Bps) it means bytes per second. Be careful, many people get these two terms and their corresponding acronyms confused. One bit of advice that will help you determine which means which (since columnists, authors and reports do get the terms confused): So, if you are reading about technology in the telecommunications industry - LANs, WANs, USB, local loops, etc. - the term should always refer to "bits per second" or whether an upper case "B" or a lower case "b" is used.

Bridge: A data communications device that connects two or more network segments and forwards packets between them.

Bridged Tap: Any section of a twisted pair copper wire not on the direct electrical path between the telephone company's central office and the end user's premise causing an increase in the electrical loss on the twisted pair copper wire. The vast majority of DSL services can't run over a local loop with bridged taps.

Broadband Local Exchange Carrier: See BLEC.

Broadband: A telecom transmission facility that provides bandwidth greater than 45 Mbps.

Brownout: When your voltage drops more than 10% below what it's meant to be over an extended period of time. Not good for your equipment.

Building-centric Local Exchange Carrier: See BLEC.

Bus Mastering: A bus design that enables plug-in cards to run independently of the CPU and also to access the computer's memory and peripherals on their own.

Bus: A common pathway, or channel, between multiple devices. A bus allows for connecting multiple devices, whereas channels such as a PC's serial port can connect only one device. Buses are generally hardware, in the form of an electrical connection where all circuit cards receive the same information that is put on the bus. Only the card the information is "addressed" to will use that data. This is convenient so that a circuit card may be plugged in "anywhere on the Bus." In addition to hardware buses, software can be designed and linked via a so-called "software bus." The term bus is derived from (would you believe it) a real bus that stops at bus stops. Just like a real bus, electronic bus signals travel to all devices connected to it. See also BACKPLANE.

Business Rules: A conceptual description of an organization's policies and practices enabling them to automate their polices and practices, and to increase consistency and timeliness of their business processing.

Byte: A set of bits of a specific length that represent a value in a computer coding scheme. A byte is to a bit what a word is to a character.

C

C: A very powerful programming language which operates under Unix, MS-DOS, Windows (all flavors) and other operating systems.

Cable Binder: A wire bundle typically containing between 25 and 1,600 twisted pair copper wires each carrying a different telephony service.

Cable Modem: (1) A technology that enables high-speed data signals to be transmitted over the same cable plant that sends television programs, pay-per-view movies, and so forth into the premise. (2) A modem designed for use on a CATV cable circuit.

Cache: To store data on a disk for quick and easy retrieval instead of retrieving it each time it is requested. In order to conserve bandwidth, large ISPs will cache popular Web pages.

Cache Memory: An allocated block of fast memory used to improve the performance of a CPU or disk drive. In the case of a CPU, instructions that will soon be executed are placed in cache memory shortly before they are needed, thus speeding up CPU operation. With disk drives, algorithms exist that try to determine what blocks or sectors the drive will be seeking to and reading shortly in the future or on a periodic basis. By keeping these sectors in cache memory, disk performance – and indeed overall system performance – is enhanced considerably.

Callback: Also known as International Callback, call re-origination and others. The use of existing network facilities in an application (ASP?) that allows a service provider to offer US dialtone and prices to overseas customers. Part of the service involved an unanswered trigger call, generating a great deal of controversy.

CAP: An acronym for the carrierless amplitude and phase modulation - a bandwidth-efficient line coding technique that's a variant of QAM (quadrature amplitude modulation). With CAP, the transmit and receive signals are modulated into two wide-frequency bands. CAP modulation was one of the first transmission technologies used to implement DSL service.

Carrier: Or Common Carrier. A facilities based telecommunications service provider, which offers its services to the general public.

Carterphone Decision: A legal decision by the US Federal Communications Commission in 1968 that overturned an AT&T attempt to control the use of its equipment. The implications of this decision led to the liberalization of the US telecommunications market, which is spreading throughout the world.

CDSL (Consumer Digital Subscriber Line): A flavor of DSL tradmarked by Rockwell.

Cells: The smallest component of a table. In a table, a row contains one or more cells.

Central Office (CO): (1) A circuit switch that terminates all the local access lines in a particular geographic serving area; in Europe this is referred to as a "public exchange." (2) A physical building where the local switching equipment is found. DSL lines run from an end user's premise to their serving CO. According to the context in which the term is used, Central Office (CO) can mean: building, switch or a collection of switches.

CFM: Fans capacity is generally measured in Cubic Feet per Minute (CFM)

CGI bin: A directory on a Web server in which CGI programs (scripts) are stored.

CGI Script: A program consisting of small but highly potent bits of computer code that is usually executed on a Web server so as to provide interactivity to Web pages.

CGI: (Common Gateway Interface) A predefined way in which CGI programs or scripts communicate with a Web server.

Channel: An intelligent path through which data is transferred between host RAM and some kind of I/O controller without assistance from the host CPU. Originally a mainframe concept, adopted by the PC industry. A NetWare "channel" is any kind of disk controller or adapter, of varying intelligence.

Chargen (CHARacter GENerator): A utility that provides an approximate speed for a computer's Internet connection in characters per second (with compression taken into account).

CIDR (Classless InterDomain Routing): An Internet-working term, which sounds like "cider." CIDR is a protocol that uses the existing 32-bit Internet Address Space more efficiently. It allows the assignment of Class C IP addresses in multiple contiguous blocks with more than one block of network addresses linked together logically into a "supernet."

Circuit Switched: The routing and connection of telephone calls by actually switching the termination of the call by various means. A circuit constructed in this manner is dedicated from end to end for the duration of the call.

Circuit: A pathway over which voice and/or data is transferred between remote devices. It may refer to the entire physical medium, such as a telephone line, optical fiber, coaxial cable or twisted wire pair, or, it may refer to one of several carrier frequencies transmitted simultaneously within the line.

CISC (Complex Instruction Set Computer): A microprocessor architecture that favors robustness of the instruction set over the speed with which individual instructions are executed.

Clamping Level: The voltage level above which a surge suppression device diverts energy away from the load. A good surge protector in a 120V circuit has a clamping voltage of about 135 volts. Damage to computer equipment can occur as low as 160 volts.

Clamping Time: The response time of a surge suppression device in clamping or diverting away from the load a voltage above the claming level.

Classless Interdomain Routing: See CIDR.

CLEC (Competitive Local Exchange Carrier): A term introduced in the Telecommunications Act of 1996. CLECs compete on a selective basis with other Local Exchange Carriers (LECs). While CLECs might build their own local loops, they also lease local loops from Incumbent Local Exchange Carriers (ILECs) and then resale that service to end users. See also LEC and ILEC.

Client/server: The "client" is a computer or program "served" by another networked computing device in an integrated network which provides a single system image. The "server" can be one or more computers with one or numerous storage devices.

Clock: A device that serves as the source of timing signals for sequencing electronic events, such as synchronous data transfers.

Clustering: Creating a mini-network of computers that function in a fault resilient manner. If one computer fails, others in the cluster take up the additional load. Since each member server of a cluster works on a different task, a cluster acts as a sort of parallel computing system and exhibits better processing performance than other kinds of fault resilient systems.

CO: See Central Office.

Coaxial: A type of cabling that consists of a center wire surrounded by insulation and then a grounded shield of braided wire. It's the primary type of cabling used by the cable television industry.

Co-browsing: Using a technology that allows two or more people to co-browse the Web and share Web-based applications as if they were looking at the same Web browser. As one person clicks on a link in a Web page, the result of that click will instantly show up on the computer screen of anyone co-browsing with that person. For example, if one person fills an order form and clicks submit, the result of the order submission is displayed to all participating in a co-browsing session. See also Whiteboarding.

Codec (COmpression-DECompression): A compression / decompression algorithm.

Collection Rate: This is the charge that a national telecommunications carriers charges its end users for international calls. It has no relationship to the Accounting Rate, but is obviously influenced by it.

Comment Tag: Used to insert comments in an HTML document. Comment tags are ignored by browsers (example: <!— text —> or <Comment>text</comment>).

Common Intermediate Format (CIF or Full CIF): Refers to the combination of six ISDN B channels (bearer channels), i.e. 64 Kbps circuit-switched channels. See ISDN. It consists of 30 frames per second (fps) of a 352 x 288 pixel frame or 15 fps of CIF when only two B channels (a "2B bandwidth") are used. There is also a Super CIF (SCIF) format (704 x 576 pixels).

Common Mode Noise: Abnormal signals that appear between a current-carrying line and its associated ground.

Competitive Local Exchange Carrier (CLEC): See CLEC.

Concentrator: (1) A type of multiplexer that can combine multiple channels onto a single transmission medium allowing the individual channels to be active simultaneously. In an ISP concentrators are used to combine dial-up modem connections to feed the flow of data traffic onto faster T-1 or T-3 lines that connect to the Internet. (2) In client/server networks, which utilizes a star or spoked wheel type of layout, the switching device (also called Hub or Switch) that's located at the center of the layout from which cables spread out to all of the computers. See also Hub, Multiplexer, Switch.

Conferencing: A catch-all term that includes everything from electronic bulletin boards and chat rooms to real-time video conferencing, application sharing / data collaboration, audio conferencing and even instant messaging or "IM" (AOL, MSN Messenger, ICQ).

Connectivity: The property of a network that allows dissimilar devices to communicate with each other. It also refers to a program's or device's ability to link with other programs and devices.

Controller: A hardware or software component which translates requests from the CPU and operating system into commands for the read/write actuators for disk drives.

Controllers also can translate between between the sequential bit stream used by disk circuitry and the parallel digital data used by the host computer. IDE and SCSI drive host adapters are often mistakenly called controllers (I should know, I do it myself) – the actual controller is built into the drive itself.

Cookie: An HTTP header that contains a string that a browser stores in a small text file in the Windows/Cookies directory (for Microsoft Internet Explorer) or in the Users folder (for Netscape Navigator) on a computer's hard drive. Cookies store information supplied by a user to be accessed at a later period in time but it is important to state that a cookie can't interact with other data on a computer's hard drive.

CPE: Customer Premise Equipment. Originally this referred to equipment on the customer's premises which had been bought from a vendor who was not the local phone company. Now it simply refers to telephone equipment – key systems, PBXs, answering machines, etc. – which reside on the customer's premises. "Premises" might be anything from an office to a factory to a home.

CPU (Central Processing Unit): A programmable device that can process digital information.

Crawler: See Spider.

Cross-connect: The connection of any telecommunications line in the carrier's system with any other lines also in that system.

Crossover Cable: This type of cable allows two Ethernet devices (two computers or a computer and printer, etc.) to connect directly without the use of an Ethernet hub. A crossover cable transposes the transmit and receive wire pairs, so that pin 1 on each side connects to pin 3 on the other, and pin 2 on each connects to pin 6 on the other.

Cross-selling: Offering a product similar to the one the customer is interested in if the chosen product is unavailable.

Crosstalk: In communications, a phenomenon in which a signal in one or more channels interferes with a signal or signals on other channels. This can be because of quantum effects where electron positions become a "blur" and can jump to another line, or from corona discharge or radio frequency emissions from the line. In an analog multiplexer, crosstalk is the ratio of the output voltage to the input voltage with all channels connected in parallel and turned off.

CSU/DSU: See DSU/CSU.

CTI: Computer Telephony Integration. The first step in the converging of voice and data communications.

CTR (Click-Through-Rate): The ratio of impressions to click-throughs.

Current: The flow of electricity expressed in amperes. Current refers to the quantity or intensity of electricity flow, whereas voltage refers to the pressure or force causing the electrical flow.

Customer Premise Equipment. See CPE.

D

DACS (Digital Cross-Connect Systems): A Digital switching device in telecommunications for routing T1 lines. See Cross-connect

Data Availability: The amount of time a system needs to restore a failed network to full functionality.

Data Currency: This refers to any discrepancies in a file before a crash and after the crash.

Database Publishing: Allows businesses to leverage existing data and data management assets. Many of today's database applications can create files usable by electronic publishing software. By establishing communication the database can continue managing data, and the publishing system can be used as an information synthesis tool to gather data from a variety of sources (databases, graphics, and text) and present it in a single, cohesive document.

Datagram: A single packet of information sent over a network. Usually part of a complete data stream that has been split into pieces to expedite its transmission. It is the basic unit of information in TCP/IP.

DBMS (DataBase Management System): Software that controls the organization, storage, retrieval, security and integrity of data in a database. It accepts requests from the application and instructs the operating system to transfer the appropriate data.

DC: Direct Current, electrical current which flows in one direction.

Dedicated Array Processor: A processor on a hardware-based array that controls the execution of array-specific functions, such as rebuild parameters.

DHCP (Dynamic Host Configuration Protocol): A protocol in the TCP/IP suite that allocates IP addresses automatically to any DHCP client (any device attached to your network, such as your computer) so that addresses can be reused when the client no longer needs them.

DID: Direct Inward Dialing. A group or range of telephone numbers that share the same group of trunks to actually complete calls.

Digital Certificate: A small piece of unique data used by encryption and authentication software. A digital-based ID that contains a user's information. It accomplishes this by attaching a small file containing the certificate owner's name, the name of its issuer and a public encryption key to the information that is transmitted over the Internet.

Digital Loop Carrier (DLC): (1) A remote device often placed in newer neighborhoods to simplify the distribution of cable and wiring from the local telephone company. (2) Equipment used to concentrate many local loop pairs onto a few high-speed digital pairs or one fiber optic pair for transport back to the Central Office (CO).

Digital Signal Processor (DSP): See DSP.

Digital Subscriber Line (DSL): See DSL.

Digital: The use of a binary code (bits) to represent information. The main benefit of transmitting information digitally is that the signal can be produced precisely and although it will pick up interference or garbage along the way, the telecom industry has found ways to regenerate the signal back to crystal clarity. The signal is put through a Yes-No exercise - is this part of the signal a "one" or a "zero"? Then the signal is reconstructed to what it was at the beginning of the transmission, amplified and sent on its merry way. This all means that digital-based transmission is "cleaner" than analog transmission.

Dip: A short term voltage decrease. See also "Sag".

Directories: A directory is basically a manual entry database system for which a URL is submitted along with a descriptive title and summary for the web site.

Discrete Multitone Modulation Technique: See DMT.

Disk Array: Also called a Drive Array. Although any set of disk drives put into a common enclosure could be called an array, in terms of RAID technology those drives are subject to a hardware or software-based controller that makes them appear to be a single drive to the host CPU or operating system.

Disk Farm: A collection of independent computers whose collective disk capacity is harnessed for a particular set of interrelated activities. See Clustering.

Disk Pack: An assembly of magnetic disks that

can be removed from a disk drive along with the container from which the assembly must be separated when operating.

DLC: See Digital Loop Carrier.

DMT (Discrete MultiTone): A modulation technique developed by John Cioffi while working at Stanford University. Most of the ADSL equipment installed today uses DMT. It divides upstream and downstream data into 255 separate subchannels (sometimes called "bins"), each 4.3125 kHz wide and modulated by QAM. About 247 of these channels are used for actual data. Low-frequency subchannels typically carry more bits per hertz (cycle) than high-frequency subchannels since low-frequency subchannels are less affected by attenuation. By using many narrow-band carriers, all transmitting is done at once in parallel with each carrying a small fraction of the total data. See also CAP and QAM.

DNS : Domain Name Server, or Domain Naming Service. The service provided on the Internet to translate alphabetic names used by browsers to the actual 32 bit binary addresses used on the Internet. A DNS server is a server on the Internet which provides DNS service.

Domain name: Unique address of a Web site. The address that gets you to a Web site, and consists of a hierarchical sequence of names separated by dots (periods). Also known as a Web address. It can identify one or more IP addresses. See URL.

Downtime: Signifies a time interval when a system is not in use, either because of equipment failure (unplanned downtime) or special scheduled maintenance (planned downtime).

DS0: Digital Signal, level 0. Equal to 64 kilobits per second, the smallest basic voice grade channel (using PCM).

DS1: Digital Signal, level 1. A T-1 circuit, comprised of 24 DS0s or 1.544 MBps.

DSL (Digital Subscriber Line): A family of technologies that use sophisticated modulation schemes to send data over the telephone industry's existing twisted pair copper wire infrastructure. These technologies are sometimes referred to as "last-mile technologies" because they are used only for connections from a telephone switching station or CO to the end user's premise, but not between switching stations.

DSL (Digital Subscriber Line): A technology that dramatically increases the digital capacity of ordinary telephone lines (the local loops) into the home or office. DSL speeds are tied to the distance between the customer and the telco central office.

DSL Modem: The CPE that must be installed and in working order for the DSL service provider to send data signals to the end user. Note: The term, "modem," is a misnomer, the device is actually a DSL transceiver, although it can be a type of bridge. See Bridge, CPE, DSL.

DSLAM (DSL Access Multiplexer): A line-interface network device at the CO (but can be remote) for DSL service that intermixes voice traffic and DSL traffic onto a customer's DSL line via signals received from multiple customer DSL connections. The DSLAM then puts the signals onto a high-speed backbone line using multiplexing techniques. It also separates incoming phone and data signals and directs them onto the appropriate carrier's network.

DSP: (1) When used as a noun - Digital Signal Processor (DSP) – DSP refers to a specialized computer chip designed to perform speedy, complex operations on digitized wave forms. (2) DSP can also refer to Digital Signal Processing – the manipulation of analog information, such as sound that has been converted into a digital form.

DSU/CSU (Digital or Data Service Unit/Channel Service Unit): Communication devices that connect an in-house line to an external digital circuit, such as a T1, DDS, etc. and is similar to a modem, although they connect a digital circuit rather than an analog one.

Dual Pathing: An extension to the concept of clustering, dual pathing allows multiple hosts, hard drive controllers and subsystems to establish redundant paths in such a way that if any component on the bus fails, data can still be accessed.

Due Diligence: A comprehensive investigation and assessment of all attributes, issues and variable inherent in a target entity/person/product/service, which will impact upon the target's ability to achieve its strategic objectives.

Duplex: When used as a data communication term, Duplex refers to simultaneous two-way transmission in both directions.

Duplexed: Two hard disks that have separate disk controllers and are mirror copies of each other. Data is written simultaneously to both. See Mirroring.

DWDM (Dense Wavelength Division Multiplexing): The higher capacity version of WDM (Wavelength Division Multiplexing) - a means of increasing the capacity of fiber optic data transmission systems through the multiplexing of multiple wavelengths of light. See WDM.

Dynamic IP Address: Where customers of Internet Service Providers share a limited number of an ISP's static IP addresses. With Dynamic IP Addressing each computer is temporarily assigned an IP address each and every time the computer is used to connect to the Internet. See Static IP Address.

Dynamic Web page: The dynamic change in the contents of a Web page through the use of a separate file wherein the current contents of that file is displayed on all pages connected to the underlying database whenever a browser requests a Web page.

E

E-1: The "other" T-1 standard, used in most of the world outside of North America, Japan and Hong Kong. It has a bandwidth of 2.048 Mbps, and is divided into 32 channels. In voice service, 2 channels are used for signaling, leaving 30 available for voice channel Ds0s.

E-3: The term used for a digital facility used for transmitting data over a telephone network at 34.368 Mbps. See T-3, E-3's North American equivalent.

Echo: A command in a software program that sends data to another computer which "echoes" it back to the user's screen display, allowing the user to visually check if the other computer received the data accurately.

ECML (Electric Commerce Modeling Language): A universal format for online checkout form data fields. ECML provides a simple set of guidelines that automate the exchange of information between consumers and Web-based merchants.

E-Commerce: Buying and/or selling electronically over a telecommunications system. In doing so every facet of the business process is transformed: Pre-sales, updating the catalog and prices, billing and payment processing, supplier and inventory management, and shipment. By using e-commerce a business is able to rapidly process orders, produce and deliver a product/service at a competitive price and at the same time minimize costs.

EDI (Electronic Data Interchange): A series of standards which provide a computer-to-computer exchange of business documents between different companies' computers over the Internet (and phone lines). EDI allows for the transmission of purchase orders, shipping documents, invoices, invoice payments, etc. between a Web-based business and its trading partners. EDI standards are supported by virtually every computer company and packet switched data communications company.

EIDE (Enhanced Integrated Drive Electronics): A hard drive that allows fast transfers and large storage capacities. The computer's system RAM is used for storing the drive's firmware (software or BIOS). When the drive powers up, it reads the firmware.

EJB (Enterprise JavaBeans): A Java API developed by Sun Microsystems that defines a component architecture for multi-tier client/server systems.

Electrodomestic Networked Devices (ENDs): The intelligent, electrical processing tools used in domestic network environments, including all appliances, electronics, and computers that have both embedded intelligence and the ability to communicate with other devices.

EMI: Electro-Magnetic Interference, or electrically induced noise or transients.

Encoding Scheme: A process of converting data into code or analog voice into a digital signal.

Encryption: A system of using encoding algorithms to construct an overall mechanism for sharing sensitive data. The translation of data into a secret code.

ENUM: The word "ENUM" refers to the IETF protocol that takes a complete, international telephone number and resolves it to a series of URLs using a Domain Name System (DNS)-based architecture.

ERP (Enterprise Resource Planning): A business management system that integrates all aspects of a business, such as, product planning, manufacturing, purchasing, inventory, sales, and marketing. ERP is generally supported by multi-module application software that helps to manage the system and interact with suppliers, customer service, and shippers, etc.

ESDI: Enhanced Small Device Interface. An improvement over the original ST506 disk interface, ESDI was once considered a high-speed interface for PCs (one to three Mbps), but it was surpassed by IDE and even faster SCSI drives.

ESS: Electronic Switching System. A telephone switch that is totally electronic, with stored program control. The generation that replaced electro-mechanical switching systems, like #5 Crossbar and Step-By-Step.

Etherloop: Short for "Ethernet Local Loop." Etherloop is a proprietary technology from Nortel Networks and Elastic Networks that combines the best features of two technologies – DSL and Ethernet. The DSL technology is used for signal transmission and thus provides high transmission rates, and the Ethernet technology takes care of the protocol and control of the packets and provides reliability and flexibility. See DSL and Ethernet.

Ethernet: A local area network (LAN) protocol developed by Xerox Corporation (in cooperation with DEC and Intel), which is used by the vast majority of networks in operation today. Ethernet is a physical link and a data link protocol reflecting the two lowest layers of the DNA/OSI model. Ethernet technology listens before sending. When two devices attempt to use a data channel simultaneously (called a collision). Standard Ethernet networks use

CSMA/CD (Carrier Sense Multiple Access/Collision Detection), which enables the two devices to detect a collision and institute a system to avoid same.

ETSI: An European non profit organization whose mission is to produce telecommunication standards for today and for the future, and whose standards are recognized throughout the world.

Eurocard: A family of European-designed printed circuit boards using a a 96-pin plug rather than edge connectors. The 3U is a 4 x 6 inch board with one plug; the 6U is a 6 x 12 inch board with two plugs; the 9U is a 14 x 18 inch board with three plugs.

European Union: The fifteen countries that have banded together to form a loosely unified Europe. Currently the countries are: Austria, Belgium, Denmark, Finland, France, Germany, Greece, Ireland, Italy, Luxembourg, The Netherlands, Portugal, Spain, Sweden and the United Kingdom. The significance to telecommunications are the mandate on market liberalization, and the common currency, called the Euro.

Exchange Rate: The dynamic, ever changing ratio at which the currencies of different countries are exchanged.

Expansion Board: A plug-in circuit board that adds features or capabilities to a computer system, such as a fax board to be able to transmit documents through a public phone network as faxes.

Expansion Slots: The spaces provided in a computer for expansion boards.

Extranet: A private, TCP/IP-based network that allows qualified users from the outside to access an internal network.

F

Facility: A term that refers to the various elements of telecommunications services - network connections, switches, POPs, etc. It is a rather all-encompassing term, and not clearly defined.

Failover: With high-availability systems, failover involves manually or automatically switching processing and/or applications to another device, with no data loss or corruption during either the failure or changeover.

Fast Ethernet: Ethernet at 100 Mbps - 100 Base-T. And since it's 10 times faster than Ethernet, it is often referred to as Fast Ethernet. See also Ethernet and 10-100 Base-T.

Fault Tolerance: A system's ability to maintain functionality in case of a failure.

Fault Tolerant: The ability of a network or device to handle failure. Fault-tolerant networks and devices are configured to minimize the impact of failure by switching to other equipment or cabling, lowering the amount of downtime.

FCC: Federal Communications Commission. The United States federal telecommunications regulatory authority.

FDM (Frequency Division Multiplexing): A technique that allows multiple conversations or data transmissions on one circuit by dividing the available transmission bandwidth into narrower bands. Each band can be used as a separate voice or data transmission channel.

Ferroresonant Transformer: A transformer that regulates the output voltage by the principle of ferroresonance: When an iron-core inductor is part of an LC circuit and it is driven into saturation, causing its inductive reactance to increase to equal the capacitive reactance of the circuit.

FEXT (Far End CrossTalk): A kind of crosstalk that occurs at the far end of multi-pair cable systems. FEXT is an issue with very high-bandwidth services such as VDSL, which requires a short local loop; but ADSL and

HDSL and their brethren don't feel the same affect. These technologies work over longer loops so interference tends to become weakened (attenuated) during the transmission. See also Crosstalk, NEXT and Self-NEXT.

Fiber Optics: Also known as "fiber." Fiber Optics are made from pure glass. Digital signals, in the form of modulated light, can travel on strands of fiber for long distances. Innovators are continuously finding more and better ways of sending larger amounts of information down a single strand of fiber. See also OC and SONET.

Fibre Channel: Fiber Channel is an ANSI X3T9.3 standard high speed serial interface. It offers a scaleable data rate beginning at 133 Mbps, but ranging to over 1.06 Gbps. It is an ANSI standard developed as a high-speed interface for linking mainframes and peripherals. Fiber Channel can run on single or multimode fiber as well as coax cable and twisted pair. It is best suited as a backbone for a private network.

Finger: A standard protocol. A program implementing this protocol lists who is currently logged in on another host. It is a computer command that displays information about people using a particular computer, such as their names and their identification numbers. (Integrated finger is a common Unix network function that reports information relating to a user after entering his or her e-mail address.)

Firewall: Hardware and/or software that sit between two networks, such as an internal network and an Internet service provider. It protects the network by refusing access by unauthorized users. It can even block messages to specific recipients outside the network.

Firmware: Software which is constantly called upon by a computer so it is stored in semi-permanent memory called PROM (Programmable Read Only Memory) or EPROM (Electrical PROM) where it cannot be "forgotten" when the power is shut off. It is used in conjunction with hardware and software and shares the characteristics of both.

Flash: A proprietary bandwidth-friendly and browser-independent vector-graphic animation technology distributed by Macromedia. As long as different browsers are equipped with the necessary plug-ins, Flash animations will look the same. Due to the size of Flash files, many "Flash movies" are inappropriate for analog modem users, but are perfect for end users with high-speed Internet access.

FPU (Floating Point Unit): A formal term for the math coprocessors found in many computers. The modern computer has the FPU integrated with the CPU.

Frame Relay: A high-speed packet switching protocol used in wide area networks (WANs) that providing a granular service of up to DS3 speed (45 Mbps). The name comes from the fact that frame relay does not do any processing of the content of the packets; rather, it relays them from the input port of the switch to the output port.

Frames: A programming device that divides Web pages into multiple, scrollable regions. This is done by building each section of a Web page as a separate HTML file and having one master HTML file identify all of the sections.

Framesets: See Frames.

Frequency: The rate at which an electromagnetic waveform (electrical current) alternates, i.e., the number of complete cycles of energy (electrical current) that occurs in one second. It's usually measured in Hertz. See also Hertz.

F-Secure SSH : Provides for secure UNIX shell

logins. SSH creates encrypted connections that protect confidential information, such as passwords, from exposure to network eavesdroppers.

FTP: When the acronym stands for "File Transfer Protocol, it refers to a protocol that allows users to quickly transfer files to and from a distant or local computer; delete and rename files, etc. on a distant computer. (2) FTP also refers to "File Transfer Program," an MS-DOS program that enables transfers over the Internet between two computer.

FTTC (Fiber to the Curb): A mixed transmission system. A telecommunications provider would lay fiber cables to the curb of a premise and then use legacy cabling from the curb to the premise, such as coaxial cable or twisted pair copper wiring. Another mixed transmission method used in the race to provide fiber optics to end users is Fiber to the Neighborhood (FTTN), also known as Fiber to the Cabinet (FTTCab).

FTTH (Fiber to the Home): A fiber optics deployment architecture in which optical fiber is carried to the end user's premise (no intermingling of other cabling). FTTH is the dream system that allows true broadband connectivity, but it's very expensive to deploy.

G

"G" recommendations: This is an ITU-T recommendation that encompasses packet-based audio protocols that are codecs. "G" recommendations include G.711 (uses between 48 and 64 Kbps of bandwidth), G.722 (needs between 48 and 64 Kbps but also offers stereo sound), G.728 (uses only 16 Kbps), G.729 (a speech codec that can compress human speech to only 8 Kbps).

G.dmt: The official designation of "full-rate ADSL." See ADSL.

G.Lite: The most popular name for "ADSL Lite." See ADSL Lite and ADSL.

G.shdsl: A symmetric, multi-rate DSL combining the best of 2B1Q SDSL and HDSL2. It's designed for end users that use their DSL service for voice, data and Internet access services. G.shdsl can operate over a single pair of copper wires, but for speed versatility, it also can be deployed over dual copper pairs.

Gateway: In general terms, a gateway is an electronic repeater device that intercepts and steers electrical signals from one network to another. A gateway is an entrance and/or exit to and from a communications network. The networks could be, for example, AT&T, your ISP's network, a business's WAN or a home's LAN.

GATT: General Agreement on Tariffs and Trade. The initiative that preceded the World Trade Organization.

GB: Gigabytes.

Gbps: Gigabits per second.

GBps: Gigabytes per second.

Geosynchronous: Satellites that are placed in orbit 22,300 miles above the Equator, and are made to circle at the same rate as the Earth's rotation, giving the appearance of being stationary.

GIF (Graphics Interface Format): A format for encoding images into bits so a computer can read the file and display the image on a computer screen.

Gigabit Ethernet: Ethernet at approximately 1Gbps (or one billion bits per second). It's compatible with 10-100 Base-T Ethernet; but many end users will find that they need to upgrade their equipment to support such high transmission levels. See also 10-100 Base-T, Ethernet and Fast Ethernet.

Google: One of the best new search engines.

They don't hold auctions, sell ISP connections, offer free email or span the tar out of users. Highly recommended.

Ground Fault: An undesirable path that allows current to flow from a line to ground.

GSM: Groupe Speciale Mobile, now known as Global System for Mobile Communications (GSMC). The cellular mobile system in use in most of the world, except North America.

gTLD (generic Top Level Domain): A small set of top-level domains that do not carry a national identifier, but denote the intended function of that portion of the domain space. For example, .com was established for commercial users and .org for not-for-profit organizations.

GUI (Graphical User Interface) pronounced "gooey": A program with a graphical interface that can take advantage of a computer's graphics capabilities thereby making the program easier to use.

H

"H" Recommendations: See Series H Recommendations for Line Transmission of Non-telephone Signals.

H.261 (also referred to as p*64 - pronounced "p star 64"): Defines an ITU-T standard that provides the picture format, error correction, and a standard discrete cosine transform compression algorithm for data compression. It also acts as an intermediary reconciliation entity when different equipment and software must interoperate. See Series H Recommendations for Line Transmission of Non-telephone Signals.

H.320: Defines a ITU-T standard for simple image transfer for "narrow-band visual telephone systems and terminal equipment" (ISDN-based). See Series H Recommendations for Line Transmission of Non-telephone Signals.

H.323: The ITU standard that describes how VoIP devices can communicate.

Hacker: An unauthorized person who breaks into a computer system to steal or corrupt data.

Hardware: Objects that go with the computing environment that can be touched. For example, modems, interface cards, floppy disks, hard drives, monitors, keyboards, printers, motherboards, memory chips, etc.

Harmonic Distortion: Excessive harmonic (a frequency that is a multiple of the fundamental frequency) content that distorts the normal sinewave waveform.

HDSL (High Bit-Rate Digital Subscriber Line): A pioneering high-speed format created in the late 1980s. HDSL was the first symmetric high-speed data networking service in the DSL family. It delivers symmetric DSK service at speeds up to 2.3 Mbps in both directions. HDSL requires two twisted pair copper lines. See also DSL and HDSL2.

HDSL2 (Second-Generation High Bit-Rate Digital Subscriber Line): This DSL family member offers the same options as HDSL but uses only a single-wire pair. See also DSL and HDSL.

Hertz (Hz): A unit that measures waves in terms of cycles per second. This unit of measure was named for Heinrich Hertz, who first created and detected radio waves. The terminology "cycles per second," is the same as hertz. In general the higher the hertz, the shorter the wavelength.

HIPPI: High Performance Parallel Interface. It's ANSI X3T9.3. HIPPI was originally developed as the interface to move data between supercomputers and peripherals such as disk arrays and frame buffers. More recently HIPPI has been extended and is used to network supercomputers, high-end workstations and peripherals using crossbar-type circuit switches. HIPPI provides for transfer rates of 800 megabits a second over 32 twisted pair copper wires (single HIPPI) and 1,600 megabits a second over 64 pairs (double HIPPI). HIPPI connections are limited to 25 meters, although a serial HIPPI standard is being developed to extend this range for more than 10 kilometers. HIPPI is currently the most common interface in supercomputing environments. With a crosspoint switch it can also serve as a high-speed LAN.

Home page: The main page of a Web site, usually serving as an index or table of contents to other documents stored on the Web server.

Host Bus Adapter: Also called an HBA. A plug-in printed circuit card that acts as an interface between the host microprocessor and the disk controller or some kind of external data transfer bus. The HBA relieves the host microprocessor of data storage and retrieval tasks, usually increasing the computer's performance time. A host bus adapter (or host adapter) and its disk subsystems make up a disk channel. Most HBAs connect the host bus to SCSI peripherals.

Host: The computer responsible for the data. In timesharing systems the host is the minicomputer or mainframe. On a LAN or WAN the host is simply the "server."

Hot Plug: Unlike a hot swap, during a hot plug the system's activity is suspended so the bus is quiet, but the power remains on while the drive is replaced. Also called a Warm Swap.

Hot Set: In a RAID system, a set of replacement disks reading to go on-line instantly. Also called a Spareset. See also Hot Spare.

Hot Spare: Hot Sparing takes Hot Swapping one step further. A standby disk sits in the array, online and ready to go. When a disk crashes, the failure is detected and the array switches to the hot spare, automatically rebuilding data to that drive and using it as a replacement for the failed disk.

Hot Swap: The process of replacing a component such as a failed RAID drive with a new component while the rest of the system continues functioning without interruption.

Hot Swappable: Hot-swappable drives, power supplies, or other components can be removed and replaced with a working substitute without taking the system or device down (also called hot replaceable or hot pluggable).

HTML (HyperText Markup Language): Used to create documents on the World Wide Web by defining the structure and layout of a Web document through the use of tags and attributes thereby determining how documents are formatted and displayed.

HTML: HyperText Markup Language. Authoring software language used to produce World Wide Web pages.

HTTP (HyperText Transfer Protocol): Defines how messages are formatted and transmitted over the World Wide Web, and what actions Web servers and browsers should take in response to various commands.

Hub: A device that takes any data that comes into one connection, or port, and amplifies the signal to broadcast it out of all the other ports in the hub. Looking at it in reverse, a hub can concentrate the signals from several comptuers into a single point of entry to the network, which explains its early name, "concentrator." See also Concentrator, Multiplexer, Switch.

Hypertext: A type of system in which objects, whether they are text, graphic files, sound files, programs, etc., can be creatively linked to each other.

I

I/O (Input/Output): Transferring data between the CPU and a peripheral device. Every transfer is an output from one device and an input into another.

IAD (Integrated Access Device): A router that combines the best of DSL and ATM technologies to support multimedia applications over the existing telephone network. IADs are point-to-point CPE and CO devices that terminate customers' data, voice, and video traffic and trunk it upstream to a network backbone. In some ways it solves the problem of limited bandwidth available on the copper plant linking telephone networks to the end user's premises.

IAN (Information Appliance Network): A home data network that distributes Internet access and provides for interconnectivity for all of what could now be called electrodomestic networked devices (ENDs). See also Electrodomestic Networked Devices (ENDs).

ICS (Internet Connection Sharing): A Microsoft Windows operating system feature that greatly simplifies the sharing an Internet connection.

IDE: Integrated Drive Electronics. Originally known as ATA (AT Attachment). A drive interface emulating (and therefore backward compatible with) a standard IBM AT compatible MFM controller at the control register level. All controller electronics are embedded in the drive itself, so only a simple "paddle board" connects the drive to the host bus.

IDSL (ISDN DSL): A hybrid of DSL and ISDN technologies, IDSL bypasses the congested PSTN and travels along the data network using a Digital Loop Carrier (DLC). See also Digital Loop Carrier.

IEEE (Institute of Electrical and Electronic Engineers): A professional group that designs and defines network standards.

IETF: Internet Engineering Task Force. The primary planning and engineering body that develops and maintains TCP/IP standards for the Internet.

ILEC (Incumbent Local Exchange Carrier): A Local Exchange Carrier (LEC) which, when competition began (per the 1996 Telecommunications Act), had the dominant position in the market-in summary. The original carrier in the market prior to the entry of competition. See CLEC and LEC.

Image Map or Imagemap: Clickable images. The image is a normal Web image (usually in GIF or JPEG format). The map data set is a description of the mapped regions within the image. The host entry is HTML code that positions the image within the Web page and designates the image as having map functionality.

Imaging: Refers to the act of creating a film or electronic image of any picture or paper form.

Impedance Matching Transformers: See Load Coils.

Information Appliance Network (IAN): See IAN.

Infrastructure: The interconnecting hardware and software that supports the flow and processing of information.

Input/Output (I/O) Unit: Any operation, program, or device that transfers data to or from a computer, such as disks (floppy, hard, or writable CD-ROMs, etc.). I/O units can also consist of single function operations such as Input-only devices such as keyboards and mouses and output-only devices such as printers.

Integrated Access Device (IAD): See IAD.

International Organization for Standardization: See ISO.

Internet Access Node: The Internet access provider's facility for receiving communications from subscribers and preparing same for transmission over the Internet. An ISP access node typically consists of analog modem, ISDN, and DSLAMs to accept the public network subscriber connections; routers to packetize the communications into TCP/IP; and frame relay switches to serve as the fast-packet connection into the Internet. See DSLAM and TCP/IP.

Internet Address: A registered IP address assigned by the InterNIC Registration Service.

Internet Backbone: The worldwide structure of cables, routers and gateways that form a super-fast network. It is provided by number of ISPs that use high-speed connections (T-3s, Ocs) linked at specific interconnection points (national access points).

Internet Connection Sharing (ICS): See ICS.

Internet Service Provider (ISP): A company that provides its customers with access to the Internet and the World Wide Web.

Internet Telephony: The technology of conducting a telephone call over the Internet.

Internet: When used with a capital "I", it refers to the public Internet, the global and pervasive computer network, open to the general public. When used with a lower case "i", it refers to any interconnection of two or more distinct computer networks, and is private in nature.

Inverter: The part of a UPS that converts the battery's DC output into AC power.

IP (Internet Protocol): The IP part of the TCP/IP communications protocol. IP implements the network layer (layer 3) of the protocol, which contains a network address and is used to route a message to a different network or subnetwork. IP accepts "packets" from the layer 4 transport protocol (TCP or UDP), adds its own header to it and delivers a "datagram" to the layer 2 data link protocol. It may also break the packet into fragments to support the maximum transmission unit (MTU) of the network.

IP Address: A unique identification consisting of a series of four numbers between 0 and 255, with each number separated by a period, for a computer or network device on a TCP/IP network.

IP Address: A 32-bit numeric identifier for a computer or device on a TCP/IP network. The IP address is written as four numbers separated by periods (commonly referred to as "dotted decimal"), with each 4-number set being within the range of zero to 255; i.e., 192.168.0.10 could be an IP address. See also Most Significant Bits (MSB) and TCP/IP.

IP Masquerading: A common form of the NAT (Network Address Translation) protocol which can hide many private IP addresses behind a single global IP address. See also NAT.

IP-based Networks: A network using the TCP/IP suite of protocols. See also TCP/IP.

ISDN (Integrated Services Digital Network): An international telecommunications standard for providing a digital service from the customer's premises to the dial-up telephone network. ISDN turns one existing wire pair into two channels and four wire pairs into 23 channels for the delivery of voice, data or video. Unlike an analog modem, which converts digital signals into an equivalency in audio frequencies, ISDN deals only with digital transmission. Analog telephones and fax machines are used over ISDN lines, but their signals are converted into digital by the ISDN modem.

ISO 9000: The ISO 9000 is a series of five standards (ISO 9000 to ISO 9004) that were developed in 1987 by the International Organization for Standards (ISO) in Paris. The ISO 9000 series outlines the requirements for the quality system of an organization. It is a set of generic standards that provide quality assurance requirements and quality management guidance. It is now evolving into a mandatory requirement, especially for manufacturers of regulated products such as medical and telecommunications equipment. I

ISO: International Standards Organization. Also known as IOS, International Organization for Standardization. The Geneva based UN organization that sets technical standards. Best known in telecommunications for its 7 layer model, called OSI

Isolation: The degree to which a device like a UPS can separate the electrical environment of its input from its output while still allowing the desired transmission to pass through.

ISP (Internet Service Provider): See Internet Service Provider.

ISR: International Simple Resale. The use of international private lines to create the transmission paths for international voice traffic.

ITU (International Telecommunications Union): An international organization within which governments and the private sector coordinate global telecom networks and services. Although the ITU doesn't have the power to set standards, but if its members agree upon a standard that standard does effectively become a world standard. The ITU consists of three major sectors: The Radiocommunication Sector (ITU-R); the Telecommunication Standardization Sector (ITU-T); and the Telecommunication Development Sector (ITU-D).

IVR: Interactive Voice Response. The technology that allows telephony users to communicate with computers. Uses touchtones or voice recognition to receive and interpret user input, and speaks recorded or synthesized speech back. Voice Mail, Automated Attendant, and other applications are examples of IVR.

J

Java Applet: Small Java applications that can be downloaded from a Web server and run on a user's PC by a Java-compatible Web browser.

Java Script: An open source scripting language developed by Netscape (independent of Sun's Java) that enables interactive Web sites by interacting with HTML source code.

Java: A high-level object-oriented programming language similar to C++ from Sun Microsystems designed primarily for writing software to leave on Web sites which is often downloadable over the Internet. Java is basically a new virtual machine and interpretive dynamic language and environment.

JDBC (Java DataBase Connectivity): A Java API that enables Java programs to execute SQL statements similar to ODBC.

Joules: The amount of energy measured in watt-seconds that a surge suppression device is capable of directing away from the load in case of a surge or spike.

JPEG (Joint Photographic Experts Group) also JPG: A compression technique used in editing still images, color faxes, desktop publishing, graphic arts and medical imaging. Although it can reduce image files to approximately 5% of their normal size, some detail is lost in the compression.

K

Kbps: (KiloBits Per Second): One thousand bits per second. A standard measure of data rate and transmission capacity. Also referred to as kbits/s. Note that KBps refers to kilobytes by second - one thousand bytes per second - and is a measurement for physical data storage on some form of storage device: hard disk, RAM, etc. See also Bps..

Kerberos: A security system that authenticates users but doesn't provide authorization to services or databases, although it does establish identity at log-on.

Keyword: In database management, a keyword is an index entry that identifies a specific record or document. In programming, keywords (sometimes called reserved names) can be commands or parameters, which are reserved by a program because they have special meaning. On the World Wide Web keywords are the terms that you enter into the search field of a search engine or directory.

Kilohertz (kHz): One thousand cycles per second. A common unit used in measurement of signal bandwidth – digital and analog. See also Hertz.

KVA: 1000 VA.

L

LAN (Local Area Network): A short distance data communications network consisting of both hardware and software and typically residing inside one building or between buildings adjacent each other, thus allowing all networked devices to share each other's resources.

Layered Arrays: A large array that combines arrays instead of individual hard drives. This can improve performance and reliability but requires additional hard disks.

Least Cost Routing: Carriers or users who have two or more alternate routes may select the least expensive route for a given call.

LEC (Local Exchange Carrier): The local telephone company. A LEC can be either a RBOC (Regional Bell Operating Company) or an independent – GTE is a good example. With the advent of deregulation (in particular, the Telecommunications Act of 1996) and competition, LECs now are known as ILECs. See also ILEC and RBOC.

LEO: Low Earth Orbit. A satellite that is relatively low (400-600 miles above the Earth) in orbit. It loses the advantage of being geosynchronous, but requires much less power, and does not have the disadvantage of excessive propagation delay.

Let-through voltage: The amount of voltage that gets through to your equipment at a given incoming voltage. This is a level that's acceptable to your equipment.

Link: On the World Wide Web, a link is a reference to another Web site or Web page or document and it takes you to the other Web site, Web page or document when you click on it.

Load Coil: A device used to extend the range of a local loop for voice grade communications. This device is sometimes called an "impedance matching transformer." A load coil is an inductor added in series with the telephone line which compensate for the parallel capacitance of the line. Load coils can benefit the frequencies in the high end of the voice spectrum at the expense of the frequencies above 3.6 KHz. However, you can't have load coils and DSL service on the same local loop.

Load Sharing: Also called Load Balancing. The use of multiple power supplies within a disk array so that power usage is distributed equally across all the power supplies. The failure of one supply will not cause the entire array to fail.

Load: An electrical device connected to a power source is a "load." In reference to a UPS, the load is the amount of current that is required by the attached electronic equipment. Rated loaded described in the specifications of the electronic equipment is often higher than the actual power consumption of the equipment in real world use.

Local Area Network (LAN): See LAN.

Local Loop: "Loop" is the telephone industry term for a single twisted pair of copper suitable for voice service. The electrical connection "loops" from the central office serving your area to your premise and back again, thus the term "local loop."

Lucent Technologies: The world's biggest telecommunications manufacturer. Created by the spin off from AT&T of Bell Labs, Western Electric and a field sales and service structure.

M

Macros: Small simple programs written to automate specific tasks.

Mail List: A program that allows a discussion group based on the e-mail system.

Mainframe: A large computer. In the "ancient" mid-1960s, all computers were called mainframes, since the term referred to the main CPU cabinet. Today, it refers to a large computer system.

MAN (Metropolitan Area Network): A communications network that covers a geographic area such as a city or suburb.

MB: Megabytes.

Mbps (Megabits per second): When referring to measurement of data rate and transmission capacity in a telecommunications, networking or LAN transmission facility, Mbps means a million bits per second. Note, however, that when you see either Mbps or MBps in the context of computing, such as data storage, the reference is to a million bytes per second. While there are typically 8 bits in a byte, there could be more than 8 or fewer than 8 in a byte. Also a million bits per second can be depicted as Mb/s. See also Bps.

Mbps: Megabits per second.

MC: Multipoint Controller. An endpoint that manages conferences between 3 or more terminals or gateways.

MCI: Microwave Communications Inc. The first strategic competitor to AT&T. Through its activities, especially those of a legal nature, its founder, the late William McGowan, was able to force the first significant market opening in telecommunications.

MCU: Multipoint Control Unit. An MC and an MP combined together to form a basic switching unit for VoIP.

Megahertz (MHz): One million cycles per second. A common unit used in measurement of signal bandwidth – digital and analog. See also Hertz.

Meta Tags: HTML code between the <HEAD> and </HEAD> section of a web page. Meta Tags are accessed by a search engine's spider and used by search engines to describe your entire Web site and individual web pages. Meta Tags are either in the form of keywords or a descriptive phrase.

MGCP: Multimedia Gateway Control Protocol.

Microcontroller: A one chip computer. A microcontroller is a highly integrated chip which includes, on one chip, all or most of the parts needed for a controller (a a device that controls some process or aspect of an environment). A microcontroller typically includes a CPU, Memory (RAM, EPROM/PROM/ROM), an interrupt controller, and I/O (Input/Output) - serial and parallel timers.

Microfilter: Low-pass filters that block all signals above a certain frequency. It is a customer-installable low-pass filter with a RJ-11 jack on either end: the customer plugs the phone into one end and the plugs the other end into the wall jack. Microfilters prevent unwanted signals from interfering with other services, such as fax, answering machine, etc. When ordering DSL service that uses microfilter technology, there is no need for a splitter installation by the telco.

Micropayments: A business transaction type, which specializes in the sub-dollar range. Although each transaction is a small amount, it can add up to a sizable market because of global access of the Internet.

Microsoft: Obviously the creator of the predominant PC operating system, Windows. If you thought that AT&T was an aggressive monopolist, just wait until you see what Microsoft has in mind!

Mirroring: Keeping an exact or "mirror" copy of one disk on another,or making an exact duplicate of a system's data. In the NetWare environment, this is the process of keeping a second copy of the data on a second drive attached to the same "channel." Mirroring is often achieved via RAID setups or server-mirroring software, as in the case of NetWare.

Modulation: A process that varies some characteristics of a electrical carrier wave transporting information. See also 2B1Q, CAP, DMT, QAM, PAM.

Most Significant Bit (MSB): The IP address is composed of 32 bits, which consist of two parts: The Most Significant Bits (MSB), also referred to as High Order Bits, identity a particular network and the remaining bits specify a host on that network. IP address are typically written in format known as dotted decimal – four numbers separated by periods (i.e., 140.210.31.9). See also IP Address and TCP/IP.

Motherboard: A circuit board that everything (adapter cards, CPU, RAM, etc.) plugs into and therefore provides the actual physical connection between the different components that make up the computer.

MP: Multipoint Processor. Operates under the control of the MC and controls the actual data streams.

MP3: The most popular audio compression format on the Internet. A file extension for MPEG Audio Layer 3. Layer 3 is one of three coding schemes for the compression of audio signals, i.e., perceptual audio coding and psychoacoustic compression to the redundant and irrelevant parts of a sound signal and adds a MDCT (Modified Discrete Cosine Transform) that implements a filter bank, greatly increasing the frequency resolution.

MSB: See Most Significant Bit (MSB).

MTBF: Mean Time Between Failures, usually expressed in hours.

MTTR: Mean Time To Repair, usually expressed in hours.

Multimedia conferencing: Same as Conferencing. See Conferencing.

Multiplexer: Also spelled as "multiplexor." A communications device that combines several signals for transmission over a single medium. A multiplexer and a demultiplexer are frequently combined into a single device, thus enabling the device to process both outgoing and incoming signals.

Multipoint Protocol: A rule describing how several devices sharing a single communication line send data on a communication network. Multipoint protocol guards against packets dropped because of collisions on a channel and allows specific nodes to relay packets to other nodes. TCP/IP is a multipoint protocol.

Multiprocessing: The existence of two or more processors in one computer and an operating system's capability to take advantage of these processors. Symmetrical multiprocessing refers to the operating system's capability to assign tasks on the fly to the next available CPU. Asymmetrical multiprocessing requires the software designer to designate which CPU should be used for a given task.

Multitasking: The simultaneous execution of two or more programs in a single computer.

Multithreading: The concurrent processing of transactions or "threads" within one program. Because transactions can be processed in parallel, one transaction doesn't have to be completed before another one is started.

MVIP: Multi-Vendor Integration Protocol. MVIP is a family of standards to let telephony products from different vendors interoperate within a computer or group of computers. MVIP started in 1990 with a telephony bus for use inside a single computer. Picture a printed circuit card that fits into an empty slot in a personal computer. The slot carries information to and from the computer. This is called the data bus. Printed circuit cards that do voice processing typically have a second "bus" – the voice bus. That "bus" is actually a ribbon cable which connects one voice processing card to another. The ribbon cable is typically connected to the top of the printed circuit card, while the data bus is at the bottom. Has been replaced by H.100 and H.110.

N

NANP: North American Numbering Plan. The planning and coordination of North American area codes, now administered by Lockheed Martin, formerly a part of AT&Ts Bellcore.

NAP (Network Address Point): A public network exchange facility where ISPs can connect with one another in peering arrangements. NAPs are key components of the Internet backbone since connections within a NAP determine how traffic is routed. NAPs are also where most Internet congestion occurs.

NAT: (1) Network Address Translation. An Internet standard that enables a local-area network (LAN) to use one set of IP addresses for internal traffic and a second set of addresses for external traffic. Network Address Translation reduces the need for globally unique IP addresses and allows a business to use a single IP address when communicating with the outside world. (2) Network Address Table. It allows many computers (or other future digital devices or appliances) on a LAN to share the public IP address of the computer that's being used as a gateway. That public IP address can be static or it can be dynamically assigned to the gateway computer by a Dynamic Host Configuration Protocol (DHCP) server either at the ISP or somewhere on the Internet. See also DHCP, Gateway, LAN.

Natural Monopoly: An excuse that AT&T created to justify its death grip on telecommunications prior to Carterphone and the resultant opening of the telecommunications market. Since AT&T had never been granted a legal monopoly, it claimed that telecommunications should "naturally" be a monopoly, and protected from competition.

Navigation: Traveling from place to place on a Web page, a Web site or the Web, from information to information. It can also mean to search for information from a menu hierarchy or hypertext. Navigation in a hypertext environment (the Web) is a physical experience of scrolling, scanning and clicking, moving over the text with your mouse pointer and actively clicking on hyperlinks.

Netstat: A utility that provides statistics on the network components, i.e., it shows the network status. It displays the contents of various network-related data structures in various formats. For example, it displays: all connections and listening ports (server connections are normally not shown), ethernet statistics, addresses and port numbers in numerical form (rather than attempting name lookups), per-protocol statistics (shows connections for the protocol specified by protocol), and the contents of the routing table.

Network Address Tables: See NAT.

Network Address Translation: See NAT.

Network Interface Card (NIC): See NIC.

Network Interface Device (NID): A device that terminates a twisted pair copper wire from the local Central Office (CO); but the NID itself is typically located elsewhere. For instance, in a DSL installation, a telco-installed splitter is also sometimes referred to as a network interface device although a NID is actually the telephone line entering the premise.

Network Service Provider (NSP): A company that provides Internet access to ISPs. A NSP offers direct access to the Internet backbone and the Network Access Points (NAPs). Sometimes called access networks or backbone providers. See also Access Network, Internet Backbone, ISP, and NAP.

NEXT (Near End CrossTalk): Leakage of undesired local signals into the local retriever; could be from the companion transmitter or other nearby sources. In the majority of DSL deployments NEXT isn't a problem since there usually isn't as large a bundle of twisted pair copper wires at the far end (near the end user's premises). Therefore bandwidth traveling in that direction (from the CO to the end user - downstream) can be made greater.

Next Gen Telephony: A loose description of all the non-legacy telecommunications techniques, including VoIP, voice over DSL, and all forms of packet switching.

NIC (Network Interface Card): A printed circuit board comprising circuitry for the purpose of connecting a device to a network. It's usually inserted into a computer before a computer can be connected to a network, although there are USB NICs and PCIMIA NICs that don't require opening up the computer box. Although most NICs are designed for a particular type of network, protocol, and media, it's possible to find NICs that can serve multiple networks. See also PCIMIA Cards.

NID (Network Interface Device): See Network Interface Device.

Node: An endpoint of a network connection, or a junction common to two or more lines in a network. Nodes can be processors, controllers, or workstations.

Noise: An unwanted, undesirable electrical signal that is irregular and is riding on top of the desired signal. Such electrical noise is introduced into telephone lines by circuit components or natural disturbances, frequently degrading the line's performance.

O

OC: A term used to specify the speed of a fiber optic network conforming to the SONET standard. See Fiber Optic and SONET.

OCC: Other Common Carrier. An old term, originally created to differentiate between the activities of a common carrier, like AT&T and a specialized carrier, supposedly offering non-traditional telecommunications services.

ODBC (Open DataBase Connectivity): A standard database access method developed by Microsoft Corporation that allows databases such as dBASE, Microsoft Access, FoxPro and Oracle to be accessed by a common interface independent of the database file format.

ODBMS: See OODBMS.

OFDM (Orthogonal Frequency Division Multiplexing): A signal processing scheme from Lucent/Flarion that supports high data rates at very low packet and delay losses or latencies, over a distributed all-IP wireless network. OFDM was selected for Europe's Digital Audio Broadcast (DAB) to broadcast CD quality sound and multimedia over the airwaves for in-car and mobile applications.

OODBMS (Object-Oriented DataBase Management System aka ODBMS): A database management system (a program that lets one or more users simultaneously create and access data in a database) that supports the modeling and creation of data as an object.

OSI (Open System Interconnection): An ISO standard to facilitate worldwide communications. OSI defines a networking framework for implementing protocols in seven layers. Also referred to as the "OSI Reference Model" or "OSI Model."

Overvoltage: An abnormally high voltage, like a surge but lasting for a longer period of time.

P

p*64: (pronounced "p star 64") Same as H.261. This term is sometimes used for ITU-T recommendation H.261 because H.261 defines video signal compression / decompression in multiples of 64 Kbps (digital transmissions over modern landlines are allocated bandwidth in increments of 64 Kbps, the size of one digitized voice channel) where p = 1 to 30 or bandwidth equals 64 Kbps to 2.048 Mbps.

Packet-Switched: The new technique of routing and connecting telephone calls by sending packets of digitized voice in chunks (datagrams) back and forth.

Packet: A sub-unit of a data stream – a grouping of information that includes a header (containing information, such as the destination address) and, in most cases, user data.

Packet-switch PBX: An in-house telephone switching system that interconnects telephone extensions to each other, as well as to the outside telephone network.

Paddle Board: See IDE.

PAM (Pulse Amplitude Modulation): A modulation technique that offers low transmission latency when so-called "trellis coding" is used (which eliminates approximately half the latency found in an equivalent QAM system). See also Modulation and QAM.

Parity: Generally refers to any kind of error detection and/or correction scheme, even if it is not actually parity based. In the case of drive arrays, one drive is often referred to as the parity drive which contains data which can be used to re-create data on any one of the other drives assuming that the other drives remain operational. For example, you might have a five drive array where one drive is designated as the parity drive. If a data drive fails and is then replaced, the drive array controller will rebuild the data on that drive using the parity drive and the other functioning drives. Usually, some kind of exclusive OR scheme is used with the combined binary value of the original striped bytes or sectors used to recover data from a failed drive. Also, there is a loss of performance during the rebuild process. Unfortunately, if you have already lost one drive and yet another fails before the rebuild is complete, then your data is gone, unlike mirroring or duplexing.

Passive Backplane: Passive backplane is a technology where all of the active circuitry that is normally found on an "active" PC motherboard (such as the CPU) is moved onto a plug-in card. The motherboard itself is replaced with a passive backplane that has nothing on it other than connectors (this is why this technology is sometimes referred to as "slot cards"). The chance of a passive backplane failing is quite low.

PBX: Private Branch eXchange. A local telephone system that provides central office type services within a single location of an enterprise.

PC (Personal Computer): A computer designed for use by a single person versus simultaneous use by more than one person.

PCI: (Peripheral Component Interconnect) A local bus for PCs by Intel, providing a high-speed data path between the CPU and peripheral devices such as a video card, disk drive, etc. In a PC, the PCI bus usually coexists with an ISA or EISA bus – there are still lots of ISA "legacy" cards out there. There are usually four PCI slots on the motherboard and the rest are ISA or EISA slots. To increase the number of PCI slots requires special bridging circuitry. PCI runs at 33 MHz and can transfer data between the CPU and peripherals at up to 133 MBps, making it a much better performer than the ISA bus (8 MBps), and supports 32 and 64-bit data paths and bus mastering.

PCMCIA (Personal Computer Memory Card International Association) Card: Small, credit card-sized devices that meet the standard of the PCMCIA and used mainly with small computing devices, such as laptop (notebook) computers.

PDA (Portal Digital Assistant): A small device that can easily be held in one hand that combines computing, telephony and networking features. Also referred to as "palmtops," or "hand-held computers."

Peer-to-Peer: A type of network where each networked device has equivalent capabilities and responsibilities, which differs from a client/server network where some devices are dedicated to serving the other networked devices. Although peer-to-peer networks are simpler to set up, they don't offer the same performance capabilities.

PERL (Practical Extraction and Report Language): An interpreted scripting programming language which is used in writing CGI scripts.

PICMG: The PCI Industrial Computers Manufacturers Group (PICMG) is an organization originally formed by four companies to standardize the way the PCI bus would work on a passive backplane. They defined a specification that's mostly concerned with the interface between the CPU and the backplane and the electrical and logical interface.

PIM (Personal Information Manager): Software designed to enable users to organize random information. A PIM enables a user to enter various types of data, such as data from an address book, to-do lists, appointments and then link these bits of information together in useful ways.

Ping (Packet InterNet Groper)- A program used to test whether a particular network destination on the Internet is online by repeatedly bouncing a "signal" off a specified address and seeing how long that signal takes to complete the round trip. A common Unix network function that reports on whether another computer is currently up and running on the Internet as well as how long the ping takes to reach the computer.

Plug-ins: Software modules that run on the viewers' local machine and add to the functionality of an application, such as a browser.

Point-of-presence (POP): The point at which a line from a long distance carrier (IXC) connects to the line of the local telephone company or to the user if the local company is not involved. For online services and Internet providers, the POP is the local exchange users dial into via modem.

Point-to-Point Protocol (PPP): See PPP.

POP: (1) Post Office Protocol. A client e-mail application that's used to retrieve mail from a mail server. (2) Point of Presence. The physical place where a carrier installs his switch and terminates any network connections. In the old days, this was a toll office or a central office, today, it could be as modest as a router in a rack at a collocation facility.

POP: See Point-of-presence.

Pop-up Window: A second browser window that "pops up" when called by a link, a button or an action.

Port: A place of access to a device or network, used for input/output of digital and analog signals.

Portability: The capability to migrate an operating system from one hardware platform to another with a minimum of recoding. Windows NT portability refers to its capability to run on both RISC and Intel PC architectures.

POTS (Plain Old Telephone Service): A term commonly used to refer to standard telephony, as in placing and receiving telephone calls. POTS is more or less interchangeable with PSTN (Public Switched Telephone Network). See also PSTN.

Power conditioner: A combination voltage regulating transformer and isolation transformer. It provides smooth, regulated, noise-free AC-voltage. It takes care of voltage that is too high or too low. Power conditioners can kick out conditioned power at a constant 120 VAC, or let output voltage vary, so be sure you know what your equipment requires, before you buy.

Power Factor: The relationship of actual power to apparent power. In reference to a UPS, the relationship between watts and VA (volt-amperes). It is expressed as watts divided by volt-amperes (W/VA) and is usually in the range of 0.6-0.71.

PPP (Point-to-Point Protocol: A method for transmitting datagrams over serial point-to-point links. It is an encapsulation protocol for transporting IP traffic over point-to-point links. End users most commonly use PPP to dial out and connect their computing device to either the Internet or a corporate private network. PPP is a standard for the assignment and management of IP addresses, asynchronous (start/stop) and bit-oriented synchronous encapsulation, network protocol multiplexing, link configuration, link quality testing, error detection, and option negotiation for such capabilities as network-layer address and data-compression negotiation. PPP is one of the dominant methods for connecting a computer to the Internet.

PPPoE (Point-to-Point Protocol of Ethernet): A technology that specifies how a host computer interacts with a broadband modem (i.e. DSL model) to achieve access to a high-speed data network. PPPoE relies on two widely accepted standards: Ethernet and the point-to-point protocol (PPP). The popularity of PPPoE within the DSL service provider community is due to the ease of implementation. For the end user, PPPoE is as easy to set up as a standard Dialup Internet access account using an analog modem, and it requires no major changes in the operational model for ISPs and other DSL service providers.

Processor Architecture: The over all organizational structure of the processor. The main elements of any processor architecture are the selection and behavior of the structural elements and the selected collaborations that form larger subsystems that guide the workings of the entire processor.

Protocol: 1. A formal set of conventions governing the formatting and relative timing of message exchange between two communicating systems. 2. An agreed-upon format for transmitting data between two devices. An end user's computer or device must support the right protocol(s) for the end user to communicate with or access other computers.

Proxy Server: A server that rests between the client and the server to monitor and filter the traffic traveling between them. It can boost Web browser response time by storing copies of frequently accessed Web pages.

PSTN (Public Switched Telephone Network): The worldwide voice telephone network. Once only an analog system, the heart of most telephone networks today is all digital. In the US, most of the remaining analog lines are the ones from your house or office to the telephone company's central office (CO).

PTT: Post, Telephone and Telegraph. The name was originally given by the British to their telephone company, and became the name for the national telephone monopoly anywhere.

Public Exchange: See Central Office.

Pulse Amplitude Modulation: See PAM.

Pulse Width Modulation (PWM): Process of varying the width of a train of pulses by tying it to the characteristics of another signal.

Punch down: The act of pushing or punching wire down into metal teeth in a punch down block that strip the insulation from the wire to make a tight connection.

PVC (Permanent Virtual Circuit): A point-to-point connection that is established ahead of time.

Q

QAM (Quadrature Amplitude Modulation): A bandwidth-efficient line coding technique that works with single carrier signals where the data rate is divided in two and modulated onto two orthogonal carriers (I and Q) before being combined and transmitted. QAM generates two signals with a sine/cosine mixer and combines them in the analog domain. Some companies have found QAM to be the most viable line code for use in the highest bandwidth DSL of them all, VDSL (Very high bit rate Digital Subscriber Line).

QoS: Quality of Service. The level of technical quality provided by a service provider to their customer.

R

Rack Unit: A vertical shelving system to mount servers.

RADSL (Rate-Adaptive Digital Subscriber Line): A member of the ADSL family, RADSL uses CAP modulation techniques and offers automatic rate adaptation, a feature that allows actual transmission rates to adjust to line conditions and distance. RADSL can operate with symmetrical or asymmetrical send/receive channels, but asymmetrical configurations yield higher bandwidths because they reduce instances of NEXT (Near End Cross Talk). The term "RADSL" can also specifically refer to a proprietary modulation standard designed by Globespan Semiconductor. See also ADSL, CAP and NEXT.

RAID: (Redundant Array of Inexpensive/Independent Disks) A disk subsystem architecture that writes data across multiple hard disks to achieve redundancy and enhance fault resilience. The theory is that instead of buying gigantic, monolithic drives (SLEDs), you combine multiple smaller drives to save money, increase performance, and reliability. In practice, many RAID systems cost more and are not as fast as the SLEDs they replace. There are several types of RAID, each is identified by number.

RAID-0: Uses disk striping without parity information. RAID-0 is the fastest and most efficient array type but offers no data protection and is actually subject to the <f”i Century Expanded Italic”>opposite of fault tolerance, since the failure of <f”i Century Expanded Italic”>any disk will bring down the system. It's the one RAID level that's not really "RAID" at all!

RAID-1: A disk subsystem architecture that uses disk mirroring for 100% duplication of data. Uses disk mirroring or duplexing. This is the array of choice for performance-critical, fault-tolerant environments, especially if you don't want to buy more than two drives.

RAID-10: A combination of RAID-1 and RAID-5. Data is striped and mirrored across multiple drives.

RAID-2: A disk subsystem architecture that uses disk striping across multiple disks at the bit level with parity. In RAID-2, which includes error detection and correction, an array of four disks requires three parity disks of equal size. This is seldom used since error correction codes are embedded in sectors of almost all disk drives anyway. Still, some implementations exist for supercomputer storage.

RAID-3: Uses disk striping at the byte level with only one disk per array dedicated to parity information. This can be used in single-user environments and performs best when accessing long sequential records. However, RAID-3 does not allow multiple I/O operations to be overlapped and needs synchronized drives in order to prevent performance degradation involving short records.

RAID-4: Same as RAID 3 but stripes data in larger chunks (whole sectors or records). This allows multiple reads to be overlapped but not multiple writes. Like RAID-3, a dedicated disk stores parity information.

RAID-5: Same as RAID 4 but data is stripped in sector-sized blocks and parity data is also striped across the disks interleaved with the data. Its supports both overlapping reads and writes, but write performance is slightly degraded because of the need to update parity data.

RAID-6: Same as RAID 5, plus additional striping so two disks can fail simultaneously, redundant controllers, fans, power supplies, etc. An array providing striping of data across multiple drives and two parity sets for increased fault tolerance. Highly reliable but suffers from slow performance.

RAID-7: RAID 7 is not yet an industry-standard term, but rather a product name for an RAID-like approach for multiple-host, UNIX-based environments running on various hardware platforms, including those from DEC, Silicon Graphics, Sun Microsystems, Hewlett-Packard, IBM, and Sequent.

RAM: Acronym for (1) Random Access Memory, (2) Remote Access Multiplexer. See Remote Access Multiplexer.

RBOC (Regional Bell Operating Company): In compliance with a Consent Decree signed by AT&T on January 8, 1982, AT&T was required to divest itself of 22 of its telephone operating companies by December 30, 1983. Those 22 companies were melded into seven regional holding companies (RHC) with the presiding judge determining which Bell operating companies would join each RHC. As of January 1, 2001 the original seven RBOCs were compressed into only five RBOCs - Ameritech, Bell Atlantic (now Verizon bought out NYNEX), BellSouth, SBC Communications (formerly Southwestern Bell bought out the holding company for Pacific Bell) and Qwest (purchased USWest and thus Qwest is now a RBOC).

RDBMS (Relational DataBase Management System): A program that enables one or more people to simultaneously create, update and administer a relational database.

Real-Time: Occurring immediately (as opposed to simultaneously as in real time). The data is processed the moment it enters a computer, as opposed to BATCH processing, where the information enters the system, is stored and is operated on at a later time.

Rebuilding: Re-creation of data lost when a drive fails. Also called Reconstruction.

Rectifier/Charger: That part of a UPS that converts the incoming AC utility power to DC power for driving the inverter and charging the batteries.

Redundancy: Peripherals, computer systems and network devices that take on the processing or transmission load when other units fail. See fault tolerant, mirroring, RAID and backup.

Redundant: Duplicated, as in a redundant cabling system that provides a duplicate cabling route in case the first one goes down.

Refile: The redirecting of traffic from one country to another through a third country, for the purpose of getting the lower rates offered.

Regional Bell Operating Company (RBOC): See RBOC.

Remote Access Multiplexer (RAM): A system that terminates ADSL lines in a remote site. Also sometimes referred to as a "Remote DSLAM" although a RAM can generally multiplex fewer ADSL lines (less than 100) than a remote DSLAM, which can handle several hundred lines. See also DSLAM.

Resale: The business of acquiring facilities or services of primary carriers, and selling them to customers.

Retro-virus: A retro-virus is a virus that waits until all possible backup media are infected too, sot that it is not possible to restore the system to an uninfected state.

RISC (Reduced Instruction Set Computer): A microprocessor that is designed to favor the speed at which individual instructions execute over the robustness of the instructions; i.e., it performs a smaller simpler set of operating commands so that the computer can operate at a higher speed.

Risk analysis: The process of systematically identifying security risks, determining their magnitude, and identifying areas needing safeguards. Risk analysis is a part of risk management.

Rotational Latency: The delay time from when a disk drive's read/write head is on-track and when the requested data rotates under it.

Router: A highly intelligent device that provides an interface between two networks. As a hardware device. Routers are the central switching offices of the Internet, intranets and WANs; but, routers are also used to connect LANs in the SOHO and home environment.

RTCP (RTP Control Protocol): A protocol that carries status and control information via TCP so that the quality of data delivery can be monitored involving large multicast networks.

RTP (Real-time Transport Protocol): An IP protocol that supports real-time transmission of voice and video.

S

Scattering: When Spanning is used under NetWare entire blocks of data are scattered among the drives which have been spanned into a single volume.

Script: A type of computer code that can be directly executed by a program that understands the language in which the script is written.

SCSA: Signal Computing System Architecture is an open architecture introduced by Dialogic Corporation in 1993 for transmitting signals, voice and video from card to card and from chassis to chassis. It is the common set of standards that telecommunication system manufacturers and computing system manufacturers can use to create computer telephony systems. Has been replaced by H.100 and H.110.

SCSI: Small Computer Systems Interface. Pronounced "scuzzy," SCSI is a bus that allows computers to communicate with any peripheral device that carries embedded intelligence. It evolved form the old Shugart Associates Standard Interface (SASI) bus. The SCSI standard is covered by the American National Standards Institute (ANSI) and has developed from standard SCSI-1 into the immensely popular SCSI-2, then "Wide" SCSI (which expands the 8-bit data path to 16 or 32 bits), then SCSI-3 (which has better error control), a "Fast and Wide" SCSI and now SCSI160.

SDMT (Synchronized DMT): A modulation technique currently in favor with the VDSL community. SDMT layers time-domain ("ping-pong") processing on DMT, while loop-timing all lines to the same network clock (so that lines in the same binder "ping" and "pong" at the same time to eliminate crosstalk). All lines in the same cable are synchronized, thus avoiding NEXT. See also Crosstalk, NEXT and VDSL.

SDSL (Symmetric or Single-line Digital Subscriber Line): An umbrella term for a number of supplier-specific implementations over a single copper pair providing variable rates of symmetric service. SDSL can deliver high-speed bandwidth over a single-pair of copper phone lines, at the same speed in both the upstream and downstream directions including support for voice.

Sealed Lead-Acid Battery : A battery containing a liquid electrolyte that has no opening for water replenishment

Search engine: A database system designed to index Internet addresses via a schema that allows submission of a URL and through a defined process the search engine includes the submitted URL into its index.

Security Policy: A set of rules and procedures regulating the use of information, including its processing, storage, distribution, and presentation. Also, the set of laws, rules, and practices that regulate how an organization manages, protects, and distributes sensitive information.

Segment: The smallest read/write unit supported by some levels of RAID. Similar to a NetWare "block." Can also mean a set of slots, as in a SEGMENTED BACKPLANE.

Segmented Backplane: Segmented or "split" backplanes allows for multiple systems in a single chassis. This not only distributes processing but also saves on cost and rack space.

Self-NEXT: The common term used for when near-end crosstalk (NEXT) becomes more significant due to higher data rates and limitations of echo canceling equipment, thus causing NEXT to become more severe when many subscriber loops within one cable carry these high-speed signals. See also Crosstalk, NEXT and FEXT.

Series H Recommendations for Line Transmission of Non-telephone Signals: This is an ITU-T recommendation that encompasses what we call video conferencing. "H" recommendations include H.320 (see glossary definition), H.261 (see glossary definition), H.221 framing, H.233 encryption, H.323 for real-time packet based multimedia communications over LANs and the Internet, H.324 for crude multimedia communication over ordinary analog telephone connections, and H.242 for setting up and disconnecting audiovisual calls using digital channels up to 2 Mbps in bandwidth. Although called "recommendations" these technical specifications can for all intents and purposes be treated as standards.

Server cabinet: A metal cabinet designed to house rack-mounted servers (some also house tower configured systems). The cabinet will usually have a slotted front door, perforated steel rear door and top panel, with room for fans or blowers, and lift off side panels.

Server: A computer that manages network resources. For example, a Web server has a high-speed permanent connection to the Internet and subsystems to protect against power outages, hackers and system crashes. A database server manages and processes the database and database queries.

SET (Secured Electronic Transactions): A standard that enables secure credit card transactions on the Internet thereby making the theft of credit card numbers via the Internet much more difficult.

Settlement Rate: The actual amount that one international carrier has to pay the other correspondent for the excess of traffic. The settlement rate is equal to one half the Accounting Rate. If the number of minutes is EXACTLY equal in both directions, neither party has to pay.

SFT: System Fault Tolerant, meaning that a complete system has been established with fault-tolerant features such as redundancy.

SHDSL (Single-pair High-bit-rate DSL): A multirate DSL technology that allows data rates from 192 Kbps to 2.3 Mbps and can transport T-1/E1, ISDN, ATM and IP signals' version of DSL. SHDSL enables DSL service providers to provide symmetric, high-speed data transmission over existing copper pairs. The SHDSL standard was adopted in early 2001 by ITU (ITU reference code G.991.2) and eventually may spur Voice-over-DSL (VoDSL) service.

Signaling: Sending control signals that start and stop a transmission or other operation. The signals are the commands that request an operation to be performed. In telephony, a control signal is sent to establish a call, and later one is sent to tear down (disconnect) the call.

Sinewave: A fundamental waveform produced by periodic, regular oscillation that expresses the sine or cosine of a linear function of time or space or both.

Single Phase: The portion of a power source that represents only a single phase of the three phases that are available.

SIP: Session Initiation Protocol. A telecommunications protocol originally designed to allow two IP devices to link up over a transmission circuit. It has been focused on supporting Internet Telephony.

SLA: Service Level Agreement. The primary agreement between a telecommunications service provider and a user or reseller, which defines the service provided, and a way of measuring the quality of service provided.

SLED: Single Large Expensive Disk – the opposite of RAID.

SLIP (Serial Line Internet Protocol): A simple protocol for connecting to the Internet although PPP has recently become the more common protocol used when connecting to the Internet.

SMTP (Simple Mail Transfer Protocol): A TCP/IP protocol for sending e-mail between servers. The majority of the e-mail systems in use today send mail use SMTP to send mail via the Internet.

SNMP (Simple Network Management Protocol): A network monitoring and control protocol. Data is passed from SNMP agents that are hardware and/or software processes reporting activity in a network device such as a hub, router, bridge, etc. to the computer administering the network. The SNMP agents return information contained in a MIB (Management Information Base), which is a data structure that defines what is obtainable from the device and what can be controlled, i.e., turned off, on, etc.

SNR (Signal to Noise Ratio): Also written as "S/N." SNR is a measure of signal strength relative to background noise. The ratio is usually measured in decibels (dB).

Software Array: An array that uses control programs (such as Windows NT or NetWare's NLMs) and the server's CPU to perform array functions, such as rebuilding data. An inexpensive technique, but performance usually not as good as with dedicated hardware controllers.

Software: Computer instructions or data: anything that can be stored electronically is software.

Spanning: How early NetWare combines multiple physical drives into a single volume, a process vaguely resembling RAID-1. It is software driven rather than hardware driven. This differs from regular RAID-0 striping in that complete blocks are written to each drive, rather than dividing one block up among several drives.

SPARC: Sun Microsystems' open RISC-based architecture for microprocessors. SPARC is the basis for Sun's own computer platforms and it's licensed to third parties.

Spareset: See Hot Set.

Specialized Common Carrier: An early form of carrier that was intended to offer telecommunications services not provided by incumbent carriers like AT&T. In reality, they often just provided competitive services.

SPECmark: The Systems Performance Evaluation Cooperative MARK, is a suite of benchmarks that test integer and floating point performance of a computer. Results are called SPECint and SPECfp numbers. A VAX-11\780 is considered to be a "one-SPECmark machine."

Spectral Compatibility: A situation where there's little crosstalk between one DSL line and another as they sit together in a cable binder. See also Cable Binder and Crosstalk.

Speed Dialing: A method for dialing complete sequences of telephone numbers and control codes by dialing a abbreviated number of digits.

Spider: A special program (also referred to as "bot" or "crawler") utilized by search engines to index Web sites. There are two spider classes – deep and shallow. The deep spider drills through the entire Web site and then either finds the URL and copies the file or finds a single directory within the URL and copies the file. A shallow spider can do one of two things, it can either spider the URL given and stop, or only spider those URLs it finds within a single level of directories.

Spike: An in-phase impulse causing spontaneous increases in voltage. Spikes are very fast impulses, less than around 100 microseconds, of high-voltage electricity ranging from 400 volts to 5,600 volts superimposed on the normal 120V AC electrical sine wave.

Spindle Synchronization: A process that coordinates all the disks in an array to use a single drive's spindle synchronization pulse so that rotation of disk drive platters will occur in unison (sector zero goes by the read/write heads on all the drives at the same time). This is done to improve short I/O for RAID-2 and RAID-3

Splitter: A device attached to the phone line near where it enters the customer premise. The splitter forks the phone line: one branch hooks up to the original premise telephone wiring and the other branch heads to the DSL modem. Besides splitting the POTS line, a splitter acts as a low pass filter, allowing only 0-4kHz frequencies to pass to/from the telephone and thus eliminating the 4kHz interference between phones and DSL modems. See also Network Interface Device(NID)..

Splitterless DSL: See ADSL Lite, G.lite, and Universal ADSL.

SPS: A term referring to a stand-by or offline type UPS.

SQL (Structured Query Language) pronounced "sequel": A database language used for creating, maintaining and viewing database data.

SS7: Acronym for Signaling System 7. SS7 is the signaling protocol that has become the worldwide standard. SS7 provides many services in addition to signaling, enabling enhanced features like Caller ID. SS7is an out of band signaling service, allowing maximum efficiency of telecommunication channels.

SSA: Serial Storage Architecture is a powerful high-speed serial interface from IBM designed to connect high-capacity data storage devices, subsystems, servers and workstations.

SSL (Secure Socket Layer): A transport level technology developed by Netscape for authentication and data encryption between a Web server and a Web browser.

SSL Server Certificate: Also known as a Digital Certificate, it is a small piece of unique data used by encryption and authentication software which enables SSL encryption on a Web Server. This allows a Web site to accept credit card orders securely.

Standard Bodies: Committees working under various trade and international organizations to work out agreed principals of specifications and/or protocols for use within an industry or on a global basis.

Static IP Address: A 32-bit numeric identifier written as four numbers separated by periods (commonly referred to as "dotted decimal"), with each 4-number set being within the range of zero to 255 that's assigned to a computer by an Internet Service Provider (ISP) to be its permanent address on the Internet. See also Dynamic IP Address and IP Address.

Storyboard: The pictorial representation of the screen elements and their operations for every Web page, which taken as a whole, constitutes a Web site.

Streaming media: Strictly speaking "streaming media" is streaming video with sound. However, common usage has morphed this term to mean both streaming video and streaming audio whether accessed together or separately, i.e., a movie clip or the latest top 100 tune. With streaming media, a Web user does not have to wait to download a large file before seeing the video or hearing the sound. Instead, the media is sent in a continuous stream and is played as it arrives. However, the end user does need a "media player" - software that uncompresses and sends the streaming data to the monitor and/or speakers.

Striping: (pronounced with a long I) This is the process of equally dividing a single logical block (which can be a byte, sector or record) into multiple physical blocks on multiple disk drives. The goal behind this is increased throughput by dividing the data up among multiple parallel data paths, the implication being that each drive has its own controller which can operate in parallel with the other drives controllers. Often this is done as part of a parity scheme. Spindle synchronization is required for true striping to work most efficiently.

Subnet Mask: A 32-bit number in IP form that's used by your company's routing equipment and computers that helps your equipment route IP data packets to the proper subnetwork. See IP Address and Subnetwork.

Subnet: A commonly used term for subnetwork. See Subnetwork.

Subnetwork: An identifiable, separate part of a network. Typically, a subnet may represent all the networked devices at one geographic location, in one building, or on the same LAN. Also referred to as "Subnet."

Surge: A sudden change (usually an increase) in power line voltage. A surge is similar to a spike, but is of longer duration and is generally more immediately destructive.

Switch: A telecommunications term for a PBX, central office, or any other device that facilitates ad hoc communications between parties. In the old days, a Central Office was comprised of physical switches, giving rise to the name "switch". Even though these devices are almost entirely solid state today, the term has stuck, and has been extended to the function provided by routers and gateways.

Switched 56: A dial-up digital service provided by local and long distance telephone companies, which uses a 64 Kbps channel, but one bit per byte is used for in band signaling, leaving 56 Kbps for data.

Switching Time: The amount of time it takes a stand-by or offline type UPS to switch from utility output to inverter output after the UPS senses a power interruption. Normally expressed in milliseconds. See also Transfer Time.

Symmetric Multiprocessing (SMP): SMPs consist of multiple processors in a single computer sharing a common memory pool, bus architecture and all I/O devices, They are in a sense the opposite of CLUSTERS.

Symmetric: A system in which data speed or quantity is the same in both directions, averaged over time.

Synchronization: RAID Level 3 requires spindle synchronization for optimum operation. This is impossible with drives from different manufacturers and difficult with two drives from a single company. The new SCSI-3 standard will support "masterless synchronization" where drives will coordinate this process automatically.

Synchronous Data Transfer: In telecommunications, transmission in which data bits are sent at a fixed rate, with the transmitter and receiver synchronized. Synchronized transmission eliminates the need for start and stop bits with every character. When used in the context of disk subsystems it refers to a data transfer method in which each byte or word is "clocked" from the sender to the receiver at a fixed rate. This is faster than asynchronous data transfer because the time used to acknowledge each byte or word is eliminated.

Systat: A program owned by SPSS that provides powerful statistical techniques through its convenience in handling data, selecting and defining procedures to use, and formatting output.

System Fault Tolerance (SFT): The ability of a system to remain fully operational regardless of component failures.

T

T.120: A data conferencing protocol. T.120 is the most important transmission protocol standard for document conferencing over transmission media ranging from analog phone lines to the Internet.

T-1: Technically, this service is actually called T1.5, meaning that it provides a 1.544 Megabit per second transmission speed. T-1 travels over a single pair of copper wires, and is the standard for carrier communications services in the US, Canada, Japan and Hong Kong. T-1s can be connected directly to most PBXs, routers and data communications equipment.

T3: A 44.736 Mbps point-to-point dedicated line provided by the telephone companies, which provides 672 64-Kbps voice or data channels. T3 channels are widely used on the Internet.

Tables: A collection of data arranged in rows and columns and in which each item is arranged in relation to the other items.

Targeted text link: Text you can click on that will transport you to a specific section of a Web site.

Tariff: The public filing by a regulated telecommunications company, detailing its service offerings, conditions and pricing. These tariffs, filed with the Federal Communications Commission in the US, and the PTT or regulatory authority elsewhere, represent the formal offering by a common carrier to the public in general. Prior to the onset of competition, tariffs guaranteed that every customer would be treated equally.

Tcl Scripting Language (pronounced "tickle"): An open-source tool command scripting language that can be embedded within existing C++ code. As such, Tcl is used in the development of many Web sites.

TCP (Transmission Control Protocol): A transport layer, connection-oriented, end-to-end protocol that provides reliable, sequenced and unduplicated delivery of bytes to a remote or local user.

TCP/IP (Transmission Control Protocol/ Internet Protocol): A networking protocol (the Internet's protocol) that provides communication across interconnected networks, between computers with diverse hardware architectures and various operating systems.

TDM: Time Division Multiplexing. A method for carrying multiple communications paths over one physical connection by sequentially allocating the available bandwidth in short chunks of time.

Telecommute: See Telework.

Telephone Line Surge Suppressor: Guards against surges and spikes in your phone lines. Telephone line surge suppressors have higher clamping levels than AC power suppressors to accommodate ring voltage. Around 240V is the standard.

Telephony: The science of converting sound into electrical signals, transmitting it within cables or via radio and reconverting it back into sound. It refers to the telephone industry in general.

Telnet: A terminal-remote host protocol. A program that lets one computer connect to another computer on the Internet.

Text-only Browser: A browser that cannot handle hypermedia files. For example, Lynx is a text-only browser that lets you travel from one link on the Web to the next, in sequential order. Lynx gives access to all of the information that the graphical browsers can, just without the pictures or sounds. Netscape and Internet Explorer are graphical browsers that let you see pictures and hear sound.

Threat Event: A specific type of threat event as often specified in a risk analysis procedure.

Threat Monitoring: Analysis, assessment, and review of audit trails and other data collected for the purpose of searching out system events that may constitute violations or attempted violations of system security.

Three Phase: An electrical system with three different voltage lines with sinewave waveforms that are 120 degrees out of phase from one another.

TLD (Top Level Domain): Domains of which .com, .net and .org are the most common.

Toll Office: An old term for a long distance switch.

TPC-C: This is a benchmark developed by the industry-wide Transaction Processing Performance Council (TPC). The TPC-C benchmark defines a rigorous standard for calculating performance and price/performance measured by transactions per minute (tpmC) and $/tpmC, respectively.

Traceroute: Software to help you analyze what's happening on an Internet connection by showing the full connection path between one Web site and another Internet address. A common Unix network function that reports the number of hops, or intermediate routers, between a computer and a remote server.

Traffic Allowance: Refers to how many bytes can be transferred from a Web site per month, i.e., number of megabytes sent to a Web site's visitors' browsers.

Transfer Time: The amount of time it takes a stand-by or offline type UPS to sense a power interruption and switch from utility output to inverter output. Normally expressed in milliseconds. See also Switching Time.

Transformer: A device used to change the voltage of AC power or to isolate a circuit from its power source.

Transient: An abnormal and irregular electrical event, such as a surge or sag; any high-speed, short duration increase or decrease impairment superimposed on a circuit. Transients can interrupt or halt data exchange on a network.

Transport node: A network junction or connection point of a telecommunication provider.

Trunk: A communications channel between two points. It generally refers to a high-bandwidth, fiber-optic line between telephone switching centers (central offices).

T-span: A 24-channel group, which makes up one T1 line. See T1.

Tunneled VPN: A bi-directional virtual private network that encapsulates data and transmits relatively securely across an untrusted network.

Twisted pair: A thin-diameter wire (22 to 26 guage) commonly used for telephone and network cabling. The wires are twisted around each other to minimize interference from other twisted pairs in the cable.

U

Undervoltage: An abnormal low voltage lasting for a longer period of time than a sag.

Unified Messaging: A system for consolidating all the various message sources - voice, fax, email, paging, etc - into a common location where it can be accessed as one source. Easier said than done!

UNIX: The multi-user computer operating system created by AT&T in the early 1970s. UNIX or its derivatives are the basis for most of the AT&T and Lucent switching products, and runs most of their network. Today, UNIX is the predominant operating system running Internet servers.

UPS: An Uninterruptible Power Supply device provides a steady source of electric energy to a piece of equipment. UPS are typically used to provide continuous power in case you lose commercial power. A UPS is typically a bank of wet cell batteries engineered to power a phone system up to eight hours without any re-charging. A UPS system can also include a gasoline-powered generator. They're designed to function in one of three ways: 1) A continuous on-line UPS is one in which the load is continually drawing power through the batteries, battery charger and inverter and not directly from the AC supply. No switchover or transfer time. 2) A steady off-line/standby UPS normally has the load connected to the AC supply. When the voltage drops of shuts off, the UPS transfers the load to battery power without any user intervention. This is one of the most affordable types, generally. Their slower transfer times don't favor sensitive, high-speed computing equipment. 3) Line interaction UPS is one that fixes sags and spikes in line voltage with transformers that don't rely on the battery. The switch to battery power happens at a lower threshold, about 90V (less than the 100V typical of standby systems).

Up-selling: Offering customers additional recommendations when they are placing an order.

URL (Uniform Resource Locator): The global address of resources on the Internet. The first part of the address indicates what protocol to use, and the second part specifies the IP address or the domain name where the resource is located.

V

VA: See Volt-Ampere

Virtual Private Network: (VPN) A restricted-use, logical (i.e., artificial or simulated) computer network that is constructed from the system resources of relatively public, physical (i.e., real) network (such as the Internet), often by using encryption (located at hosts or gateways), and often by tunneling links of the virtual network across the real network.

Virus Scanner: A software program which can search out, locate, and possibly remove a virus. [AFSEC] (see also virus-detection tool, risk, virus)

Virus: A hidden, self-replicating section of computer software, usually malicious logic, that propagates by infecting – i.e., inserting a copy of itself into and becoming part of – another program. A virus cannot run by itself; it requires that its host program be run to make the virus active.

Virus-detection tool: Software that detects and possibly removes computer viruses, alerting the user appropriately. (see also virus scanner, risk, virus)

VoIP: Voice Over IP (Internet Protocol). The routing of calls over internets (lower case "i") or THE Internet using data communications connected devices. This is a really horrible definition, and you should read the entire book to see what it is all about.

Volt: The quantitative measure describing electrical force or potential.

Voltage Regulator: A device that provides constant or near-constant output voltage even when input voltage fluctuates.

Volt-Ampere: The unit of measure of apparent power that is the traditional unit of measure for rating UPSs. Compare this to watts, which is the unit of measure of "actual" power.

VPN (Virtual Private Network): A secure, encrypted connection between two points across the Internet. It can act as an intranet or extranet, but uses the Internet as the networking connection. Most VPNs are built and run by Internet service providers.

Vulnerability Security weakness in an information system or components (e.g. system security procedures, hardware design, or internal controls) that could be exploited to produce an information-related misfortune, such as a violation in system security policy.

Vulnerability Analysis: Systematic examination of

an information system or product to determine the adequacy of security measures, identify security deficiencies, provide data from which to predict the effectiveness of proposed security measures, and confirm the adequacy of such measures after implementation.

W

WAN (Wide Area Network): A network that is geographically scattered with a broader structure than a LAN. It can be privately owned or leased, but the term usually implies public networks.

War Dialer: A computer program that automatically dials a series of telephone numbers to find lines connected to computer systems, and catalogs those numbers so that a cracker can try to break into the systems. A cracking tool that's also called a Demon dialer.

Watts: The unit of measure of actual power. Compare to volt-amperes (VA), which is the unit of measure of apparent power.

Waveform: The graphic form of an electrical parameter.

Web address: See Domain Name.

Web hosting service: A third party that leases space on its Web servers and use of its other hardware such as UPS, backup, its technical staff, etc., so the lessee's Web site can be accessed over the Internet.

Web page editor: A plain text editor, such as Notepad for Windows offers a place to type in your HTML code so that you can post the file on your Web site. A more complicated editor can do just about everything for you (so there is no need to know HTML) just drag and drop text, images, etc. onto your page, and the editor writes the code.

Web Server: A computer with data and specific software to operate a Web site.

Web Site: Data residing on a computer which has software running on it that allows the download and presentation of the data to another computer that is permanently connected to the Internet.

Web: A subset of the Internet that in today's world is accessed via a Web browser.

Whois: A common Unix network function that queries databases for information about domain names, IP address assignments, and individual names.

Worm: A computer program which replicates itself and is self-propagating. Worms, as opposed to viruses, are meant to spawn in network environments. Network worms were first defined by Shoch and Hupp of Xerox in a March 1982 issue of the journal ACM Communications. The Internet worm of November 1988 is perhaps the most famous; it successfully propagated itself on over 6,000 systems across the Internet.

WTO: World Trade Organization. The United Nations body charged with the mission of facilitating trade among the nations of the world. The successor to GATT, which in 1997, reached an accord that is designed to open telecommunications markets around the world.

WYSIWYG (What You See Is What You Get) pronounced wiz-e-wig : An application that enables you to see on the computer monitor exactly what will appear when the document is printed.

X

XML (eXtensible Markup Language): A system for organizing and tagging elements of a document specifically designed for Web documents. It enables designers to create their own customized tags to provide functionality not available with HTML. XML also has the ability to enable the structured exchange of data between computers attached to the Web, thus allowing one Web server to talk to another Web server. This means manufacturers and merchants can begin to quickly swap data, such as pricing, stock-keeping numbers, transaction terms and product descriptions.

XOR. Short for eXclusive OR, this binary logic function is sometimes used to generate parity or error correction data. Arithmetically, it is an add without carry between bits.

Index of Sidebars

Index of Graphics